Reconnaissance for Ethical Hackers

Focus on the starting point of data breaches and explore essential steps for successful pentesting

Glen D. Singh

BIRMINGHAM—MUMBAI

Reconnaissance for Ethical Hackers

Group Product Manager: Pavan Ramchandani

Publishing Product Manager: Prachi Sawant

Senior Editor: Divya Vijayan

Technical Editor: Rajat Sharma

Copy Editor: Safis Editing

Project Coordinator: Sean Lobo

Proofreader: Safis Editing

Indexer: Sejal Dsilva

Production Designer: Nilesh Mohite

Marketing Coordinator: Marylou De Mello

First published: August 2023

Production reference: 1070623

Published by Packt Publishing Ltd.

Grosvenor House

11 St Paul's Square

Birmingham

B3 1RB, UK.

ISBN 978-1-83763-063-9

www.packtpub.com

I would like to dedicate this book to the people in our society who have always worked hard in their field of expertise and who have not been recognized for their hard work, commitment, sacrifices, and ideas, but who, most importantly, believed in themselves when no one else did. This book is for you. Always have faith in yourself. With commitment, hard work, and focus, anything is possible. Never give up because great things take time.

– Glen D. Singh

Contributors

About the author

Glen D. Singh is an information security author and cybersecurity lecturer. His areas of expertise are cybersecurity operations, offensive security tactics and techniques, and enterprise networking. He holds a **Master of Science** (**MSc**) in cybersecurity and many industry certifications from top awarding bodies such as EC-Council, Cisco, and Check Point.

Glen loves teaching and mentoring others while sharing his wealth of knowledge and experience as an author. He has written many books, which focus on vulnerability discovery and exploitation, threat detection, intrusion analysis, incident response, network security, and enterprise networking. As an aspiring game changer, Glen is passionate about increasing cybersecurity awareness in his homeland, Trinidad and Tobago.

I would like to thank God, the preserver of the universe, for all His divine grace and guidance. I would also like to thank Prachi Sawant, Divya Vijayan, Aryaa Joshi, and the wonderful team at Packt Publishing, who have provided amazing support throughout this journey. To the technical reviewers, Rishalin Pillay and Lendl Smith, thank you for your outstanding contribution to making this an amazing book.

About the reviewers

Lendl Smith is a cybersecurity professional with over 11 years of experience in IT and brings a unique perspective to the field with a solid background in development and programming. Lendl currently serves as the Group IT Security Analyst at ANSA McAL Limited, where he is responsible for implementing and maintaining the group's security strategy by proactively identifying and mitigating threats. He is a Certified Ethical Hacker with a Master's in cybersecurity, specializing in penetration testing, Active Directory hardening, and threat intelligence. Lendl extends his gratitude to Glen Singh for the opportunity to serve as a technical reviewer for this book, and thanks him for his ongoing contributions to the field of cybersecurity.

Rishalin Pillay is a seasoned professional with years of experience in various cybersecurity fields such as Offensive security, Cloud security, Threat hunting, and Incident response. He is also an active author on Pluralsight and has authored several courses including *Red Team Tools* and *Threat Protection*, as well as two books titled *Learn Penetration Testing* and *Offensive Shellcode from Scratch*. Rishalin has contributed as a technical contributor to multiple books on topics such as Dark web analysis, Kali Linux, SecOps, and study guides covering networking and Microsoft.

I'd like to thank my wife and son for their continued support in all my efforts to make the Cybersecurity industry great!

Table of Contents

3

Understanding Passive Reconnaissance 65

4

Domain and DNS Intelligence 99

5

Organizational Infrastructure Intelligence 149

6

Imagery, People, and Signals Intelligence 191

Part 2: Scanning and Enumeration

7

Working with Active Reconnaissance 247

8

Performing Vulnerability Assessments 283

9

Delving into Website Reconnaissance 317

10

Implementing Recon Monitoring and Detection Systems 349

Index 397

Other Books You May Enjoy 408

Preface

Cybersecurity is one of the most interesting topics and demanding fields in the world. As the world continues to evolve, the same can be said for our technological advances to help humans improve the way they perform tasks. However, there are many systems and networks around us that contain hidden security weaknesses that are taken advantage of by adversaries such as hackers. As a cybersecurity author and lecturer, I've heard from many professionals, enthusiasts, and students about the importance of finding a book that guides the reader to thoroughly understand how to efficiently perform reconnaissance techniques and procedures to identify and reduce the attack surface of their organizations.

Reconnaissance is the first phase of any cyber-attack performed by an adversary. The attacker needs to understand the infrastructure of the target, identify whether any security vulnerabilities exist and how to exploit them, and what attack vectors can be used to carry out the attack on the target. Without such intelligence about the target, the hacker will experience difficulties in compromising the potential victim. As an aspiring ethical hacker, it's essential to understand the **Tactics, Techniques, and Procedures** (**TTPs**) that are commonly used by real hackers to discover hidden security vulnerabilities and apply those TTPs to help improve the cyber defenses of your organization.

Organizations commonly leak too much sensitive data about themselves on the internet without realizing how such data can be leveraged by a threat actor in planning a future attack on their target. Learning reconnaissance-based techniques and procedures helps ethical hackers to identify how organizations are leaking data, determine the potential impact and cyber-risk to an organization if an attacker were to leverage the leaked data to execute a cyber-attack, and how to mitigate and implement countermeasures to improve the cyber defenses of the company.

Over the years, I've researched and developed a lot of cybersecurity-related content, and one of the most important elements of being an ethical hacker and penetration tester is the need to keep up to date with the ever-changing cybersecurity landscape. There are new tools, techniques, and procedures that are being developed and used by cybersecurity professionals in the industry to ensure they are at least one step ahead of cyber-criminals and help secure their organizations' assets. As a result, ethical hackers and penetration testers need to be well equipped with the latest knowledge, techniques, skills, and tools to efficiently perform reconnaissance-based attacks and **Open Source Intelligence** (**OSINT**) penetration testing to determine the attack surface of their organization.

During the writing process of this book, I've used a student-centric and learner-friendly approach to ensure all readers are able to easily understand the most complex topics, terminologies, and why there is a need to identify security vulnerabilities in organizations, systems, and networks.

This book begins by introducing you to the importance of reconnaissance and how both cybersecurity professionals and adversaries use it to identify vulnerable points of entry in a company. Then, you'll be taken through an exciting journey learning how to apply reconnaissance-based TTPs that are commonly used by adversaries to efficiently collect and analyze publicly available data to create a profile about their targets' systems and network infrastructure. You will learn how to set up a sock puppet and anonymize your internet-based traffic to conceal your identity as an ethical hacker to reduce your threat level during reconnaissance assessments.

Furthermore, you'll discover how people and organizations are leaking data about themselves and how adversaries can leverage it to improve their cyber-attacks and threats. You'll also learn how to leverage OSINT and common tools to identify exposed systems and networks within organizations, gather leaked employees' credentials, and perform wireless signals intelligence to better understand how a potential hacker can compromise their targets.

In addition, you will gain hands-on skills in performing active reconnaissance to identify live systems, open ports, running services, and operating systems, and perform vulnerability assessments to identify how an attacker can identify security vulnerabilities on a system and what organizations can do to mitigate the threat. Furthermore, you'll learn how to identify the attack surface of a target's website and infrastructure and discover additional assets owned by the same target. Lastly, you'll discover how to leverage Wireshark and popular open source tools to identify reconnaissance-based attacks and threats on a network as a cybersecurity professional.

Upon completing this book, you'll have been taken on an amazing journey from beginner to expert by learning, understanding, and developing your reconnaissance-based skills in ethical hacking and penetration testing as an aspiring cybersecurity professional in the industry.

Who this book is for

This book is designed for ethical hackers, penetration testers, law enforcement, and cybersecurity professionals who want to build a solid foundation and gain a better understanding of how reconnaissance-based attacks threaten organizations and their assets. Ethical hackers and penetration testers will find this book very useful when gathering and analyzing intelligence to gain insights into how a real threat actor will be able to compromise their targets.

In addition, law enforcement and ethical hackers can use the knowledge found within this book to find persons of interest and understand how organizations are leaking data that led to a cyber-attack. Furthermore, cybersecurity professionals will find this book useful in identifying the attack surface of their organizations and discovering exposed and vulnerable assets on their network, while understanding the behavior of adversaries.

What this book covers

Chapter 1, Fundamentals of Reconnaissance, introduces the reconnaissance phase in offensive security and how it helps organizations improve their cyber defenses.

Chapter 2, Setting Up a Reconnaissance Lab, focuses on setting up systems within a virtualized environment for practicing active reconnaissance techniques.

Chapter 3, Understanding Passive Reconnaissance, helps you to understand how adversaries can leverage OSINT to improve their attacks while anonymizing their identity on the internet.

Chapter 4, Domain and DNS Intelligence, teaches you how to efficiently collect and analyze domain-related information about a targeted organization to identify security vulnerabilities.

Chapter 5, Organizational Infrastructure Intelligence, focuses on collecting and analyzing publicly available data to profile an organization's network infrastructure and its employees.

Chapter 6, Imagery, People, and Signals Intelligence, teaches you how to use reconnaissance techniques to find and locate people, organizations, and wireless network infrastructure.

Chapter 7, Working with Active Reconnaissance, focuses on discovering hosts on a network, enumerating vulnerable systems, and performing wireless reconnaissance on a targeted network.

Chapter 8, Performing Vulnerability Assessments, teaches you how to set up and perform vulnerability assessments using common tools in the industry.

Chapter 9, Delving into Website Reconnaissance, explores various tools and techniques used by adversaries to identify the attack surface on websites and domains.

Chapter 10, Implementing Recon Monitoring and Detection Systems, focuses on identifying suspicious network traffic using Wireshark and Security Onion.

To get the most out of this book

To get the most out of this book, it's recommended to have a solid foundation in networking, such as understanding common network and application protocols of the TCP/IP, IP addressing, routing and switching concepts, and the roles and functions of networking devices and security appliances. Knowledge of virtualization technologies such as hypervisors and their components will be beneficial as most labs are built within a virtualized environment to reduce the need to purchase additional systems.

Software/hardware covered in the book	
Kali Linux 2022.4	Oracle VM VirtualBox
Kali Linux ARM 2023.1	Oracle VirtualBox Extension Pack
Trace Labs OSINT VM 2022.1	Vagrant 2.3.3
OWASP JuiceShop	7-Zip

Software/hardware covered in the book	
Metasploitable 3 v0.1.0	VMware Workstation 17 Pro
Security Onion 2.3	
TOR and TOR Browser	
Recon-ng	
Nessus Essentials	
SpiderFoot	
Sherlock	
Sn1per	
Amass	
Raspberry Pi 3 B+	
Alfa AWUS036NHA - Wireless B/G/N USB Adapter	
VK-162 G-Mouse USB GPS Dongle Navigation Module	

All labs and exercises were built on a system running Windows 11 Home as the host operating system, a multicore processor with virtualization enabled, 16 GB of RAM, and 400 GB of free storage for the virtual machines. Oracle VM VirtualBox was the preferred choice when choosing a hypervisor as it provides great virtual networking capabilities and it's free, however, VMware Workstation Pro was also used to set up the threat detection system at the end of the book.

If you are using the digital version of this book, we advise you to type the code yourself or access the code from the book's GitHub repository (a link is available in the next section). Doing so will help you avoid any potential errors related to the copying and pasting of code.

After completing this book, equipped with your imagination and newfound skills, attempt to create additional lab scenarios and even extend your lab environment with additional virtual machines to further improve your skillset. This will help you with continuous learning while developing your skills as an aspiring ethical hacker.

Download the color images

We also provide a PDF file that has color images of the screenshots and diagrams used in this book. You can download it here: `https://packt.link/E4kdf`.

Conventions used

There are a number of text conventions used throughout this book.

`Code in text`: Indicates code words in text, database table names, folder names, filenames, file extensions, pathnames, dummy URLs, user input, and Twitter handles. Here is an example: "For instance, `cybersecurity filetype:pdf` will provide results that contain the word *cybersecurity* and a PDF file."

A block of code is set as follows:

```
interface wlan0
    static ip_address=192.168.4.1/24
    nohook wpa_supplicant
```

When we wish to draw your attention to a particular part of a code block, the relevant lines or items are set in bold:

```
kali@kali:~$ sudo apt update
kali@kali:~$ git clone https://github.com/sherlock-project/sherlock
```

Any command-line input or output is written as follows:

```
kali@kali:~$ sudo apt update
```

Bold: Indicates a new term, an important word, or words that you see onscreen. For instance, words in menus or dialog boxes appear in **bold**. Here is an example: "Next, the **Editing Wired connection 1** window will appear; select the **IPv6 Settings** tab, change **Method** to **Disabled**, and click on **OK**."

> **Tips or important notes**
> Appear like this.

Disclaimer

The information within this book is intended to be used only in an ethical manner. Do not use any information from the book if you do not have written permission from the owner of the equipment. If you perform illegal actions, you are likely to be arrested and prosecuted to the full extent of the law. Neither Packt Publishing nor the author of this book takes any responsibility if you misuse any of the information contained within the book. The information herein must only be used while testing environments with proper written authorization from the appropriate persons responsible.

Get in touch

Feedback from our readers is always welcome.

General feedback: If you have questions about any aspect of this book, email us at `customercare@packtpub.com` and mention the book title in the subject of your message.

Errata: Although we have taken every care to ensure the accuracy of our content, mistakes do happen. If you have found a mistake in this book, we would be grateful if you would report this to us. Please visit `www.packtpub.com/support/errata` and fill in the form.

Piracy: If you come across any illegal copies of our works in any form on the internet, we would be grateful if you would provide us with the location address or website name. Please contact us at `copyright@packt.com` with a link to the material.

If you are interested in becoming an author: If there is a topic that you have expertise in and you are interested in either writing or contributing to a book, please visit `authors.packtpub.com`.

Share your thoughts

Once you've read Reconnaissance For Ethical Hackers, we'd love to hear your thoughts! Scan the QR code below to go straight to the Amazon review page for this book and share your feedback.

`https://packt.link/r/1837630631`

Your review is important to us and the tech community and will help us make sure we're delivering excellent quality content.

Download a free PDF copy of this book

Thanks for purchasing this book!

Do you like to read on the go but are unable to carry your print books everywhere?

Is your eBook purchase not compatible with the device of your choice?

Don't worry, now with every Packt book you get a DRM-free PDF version of that book at no cost.

Read anywhere, any place, on any device. Search, copy, and paste code from your favorite technical books directly into your application.

The perks don't stop there, you can get exclusive access to discounts, newsletters, and great free content in your inbox daily

Follow these simple steps to get the benefits:

1. Scan the QR code or visit the link below

`https://packt.link/free-ebook/9781837630639`

2. Submit your proof of purchase
3. That's it! We'll send your free PDF and other benefits to your email directly

Part 1: Reconnaissance and Footprinting

In this section, you will learn about the importance of reconnaissance and how ethical hackers use it to identify the attack surface of organizations, locate persons of interest, and perform wireless signals intelligence.

This part has the following chapters:

- *Chapter 1, Fundamentals of Reconnaissance*
- *Chapter 2, Setting Up a Reconnaissance Lab*
- *Chapter 3, Understanding Passive Reconnaissance*
- *Chapter 4, Domain and DNS Intelligence*
- *Chapter 5, Organizational Infrastructure Intelligence*
- *Chapter 6, Imagery, People, and Signals Intelligence*

1

Fundamentals of Reconnaissance

As an aspiring ethical hacker, penetration tester, or red teamer, reconnaissance plays an important role in helping cybersecurity professionals reduce organizations' digital footprint on the internet. These digital footprints enable adversaries such as hackers to leverage publicly available information about a target to plan future operations and cyber-attacks. As more organizations and users are connecting their systems and networks to the largest network infrastructure in the world, the internet, access to information and the sharing of resources are readily available to everyone. The internet has provided the platform for many organizations to extend their products and services beyond traditional borders to potential and new customers around the world. Furthermore, people are using the internet to enroll and attend online classes, perform e-commerce transactions, operate online businesses, and communicate and share ideas with others.

Nowadays, using the internet is very common for many people. For instance, if an organization is looking to hire an employee to fill a new or existing role, the recruiter simply posts the job vacancy with all the necessary details that are needed for an interested candidate. This enables anyone with internet access to visit various job forums and recruiting websites to seek new career opportunities and easily submit an application via the online platform. Information that's posted and available online enables adversaries to collect and leverage specific details about the targeted organization. Such details help hackers to determine the type of network infrastructure, systems, and services that are running on the internal network of a company without breaking in. This book will teach you all about how threat actors and ethical hackers are able to leverage publicly available information in planning future operations that lead to a cyber-attack.

During the course of this chapter, you will gain a solid understanding of the importance of reconnaissance from both an adversary and cybersecurity professional's perspective, and why organizations need to be mindful when connecting their systems and network to the internet. Furthermore, you will learn the fundamentals of attack surface management, why it's important to organizations, and how cybersecurity professionals use it to reduce the risk of a possible cyber-attack on their networks. Lastly, you will discover the tactics, techniques, and procedures that are commonly used by threat actors, adversaries, ethical hackers, and penetration testers during the reconnaissance phase of an attack.

In this chapter, we will cover the following topics:

- What is ethical hacking?
- Importance of reconnaissance
- Understanding attack surface management
- Reconnaissance tactics, techniques, and procedures

Let's dive in!

What is ethical hacking?

The term *hacking* is commonly used to describe the techniques and activities that are performed by a person with malicious intentions, such as a *hacker*, to gain unauthorized access to a system or network. Since the early days of telephone systems, computers, and the internet, many people have developed a high level of interest in determining how various devices and technologies operate and work together. It's quite fascinating that a person can use a traditional landline telephone to dial the telephone number of another person and establish a connection for a verbal conversation. Or even using a computer to send an email message to someone else, where the email message can be delivered to the intended recipient's mailbox almost instantaneously compared to traditional postal operations.

Due to the curiosity of people around the world, the idea of disassembling a system to further understand its functions created the foundation of hacking. Early generations of hackers sought to understand how systems and devices work, and whether there was any flaw in the design that could be taken advantage of to alter the original function of the system. For instance, during the 1950s and 1960s in the United States, a security vulnerability was found in a telephone system that enabled users to manipulate/alter telephone signals to allow free long-distance calls. This technique was known as **phreaking** in the telecommunication industry. Specifically, a person could use whistles that operated at 2600 MHz to recreate signals that were used as the telephone routing signals, thus enabling free long-distance calling to anyone who exploited this flaw. However, telecommunication providers had implemented a solution known as **Common Channel Interoffice Signaling** (**CCIS**) that separated the signals from the voice channel. In this scenario, people discovered a security vulnerability in a system and exploited it to alter the operation of the system. However, the intention varied from one person to another, whether for fun, experimental, or even to gain free long-distance calling.

Important note

A **vulnerability** is commonly used to describe a security flaw or weakness in a system. An **exploit** is anything that can be used to take advantage of a security vulnerability. A **threat** is anything that has the potential to cause damage to a system. A **threat actor** or **adversary** is the person(s) who's responsible for the cyber-attack or creating a threat.

A very common question that is usually asked is why someone would want to hack into another system or network. There are various motives behind each hacker, for instance, many hackers will break into systems for fun, to prove a point to others, to steal data from organizations, for financial gain by selling stolen data on the dark web, or even as a personal challenge. Whatever the reason is, hacking is illegal around the world as it involves using a computing system to cause harm or damage to another system.

While hacking seems all bad on mainstream media, it's not all bad because cybersecurity professionals such as ethical hackers and penetration testers use similar techniques and tools to simulate real-world cyber-attacks on organizations' networks with legal permission and intent to discover and resolve hidden security vulnerabilities before real cyber-attacks occur in the future. Ethical hackers are simply good people and are commonly referred to as *white-hat hackers* in the cybersecurity industry, who use their knowledge and skills to help organizations find and resolve their hidden security weaknesses and flaws prior to a real cyber-attack. Although threat actors and ethical hackers have similar skill sets, they have different moral compasses, with threat actors using their skills and abilities for malicious and illegal purposes and ethical hackers using their skills to help organizations defend themselves and safeguard their assets from malicious hackers.

The following are common types of threat actors and their motives:

- **Advanced Persistent Threat (APT) groups** – The members of an APT group design their attacks to be very stealthy and undetectable by most threat detection systems on a targeted network or system. The intention is to compromise the targeted organization and remain on its network while exploiting additional systems and exfiltrating data.
- **Insider threats** – This is an attacker who is inside the targeted organization's network infrastructure. This can be a hacker who is employed within the company and is behind the organization's security defense systems and has direct access to vulnerable machines. In addition, an insider threat can be a disgruntled employee who intends to cause harm to the network infrastructure of the company.
- **State actors** – These are cybersecurity professionals who are employed by a nation's government to focus on national security and perform reconnaissance on other nations around the world.
- **Hacktivists** – These are persons who use their hacking skills to support a social or political agenda such as defacing websites and disrupting the availability of or access to web servers.

- **Script kiddie** – This type of hacker is a novice and lacks the technical expertise in the industry but follows the tutorials or instructions of experts to perform cyber-attacks on targeted systems. However, since this person does not fully understand the technicalities behind the attack, they can cause more damage than a real hacker.
- **Criminal syndicates** – This is an organized crime group that focuses on financial gain and each person has a specialized skill to improve the attack and increase the likelihood of success. Furthermore, this group is usually well funded to ensure they have access to the best tools that money can buy.
- **White hat** – These are cybersecurity professionals such as ethical hackers, penetration testers, and red teamers who use their skills to help organizations prevent cyber-attacks and threats.
- **Gray hat** – These are people who use their hacking skills for both good and bad. For instance, a gray hat threat actor could be a cybersecurity professional who uses their skills in their day job to help organizations and at night for malicious reasons.
- **Black hat** – These are typical threat actors who use their skills for malicious reasons.

Ethical hackers, penetration testers, and red team operators always need to obtain legal permission from authorities before engaging in simulating any type of real-world cyber-attacks and threats on their customers' systems and network infrastructure, while ensuring they remain within scope. For instance, the following agreements need to be signed between the cybersecurity service provider and the customer:

- **Non-Disclosure Agreement (NDA)**
- **Statement of Work (SOW)**
- **Master Service Agreement (MSA)**
- **Permission of Attack**

The NDA is commonly referred to as a *confidentiality agreement*, which specifies that the ethical hacker, penetration tester, or red teamer will not disclose, share, or hold on to any private, confidential, sensitive, or proprietary information that was discovered during the security assessment of the customer's systems and network infrastructure.

However, the SOW documentation usually contains all the details about the type of security testing that will be performed by the ethical hacker/service provider for the customer and the scope of the security testing, such as the specific IP addresses and ranges. It's extremely important that ethical hackers do not go beyond the scope of security testing for legal reasons. Furthermore, the SOW will contain the billing details, duration of the security testing, disclaimer and liability details, and deliverables to the customer.

The MSA is a general agreement that contains the payment details and terms, confidentiality and work standards of the provider, limitations and constraints, and delivery requirements. This type of agreement helps the cybersecurity service provider to reduce the time taken for any similar work that needs to be provided to either new or existing customers. In addition, the MSA document can be customized to fit the needs of each customer as they may require unique or specialized services.

Permission of attack is a very important agreement for ethical hackers, penetration testers, and red teamers as it contains the legal authorization that is needed to perform the security testing on the customer's systems and network infrastructure. Consider this agreement, in the form of a document, as the *get-out-of-jail card* that is signed by the legal authorities, which indicates the granting of permission to the service provider and its employee(s) who are performing ethical hacking and penetration testing services on the customer's systems and network.

Mindset and skills of ethical hackers

Threat actors are always seeking new and advanced techniques to compromise their target's systems and networks for legal purposes. For instance, there are different types of hackers and groups around the world, and each of these has its own motive and rationale for their cyber-attacks:

- Personal accomplishment/challenge, such as proving they have the skills and capabilities to break into an organization and its systems
- Financial gain, such as stealing confidential data from organizations and selling it on various dark web marketplaces
- Supporting a social or political agenda such as defacing and compromising websites that are associated with a social/political movement
- Cyber warfare, such as compromising the **Industrial Control Systems** (**ICS**) that manage the critical infrastructure of a country

While there are many cybersecurity companies around the world who are developing and improving solutions to help organizations defend and safeguard their assets from cyber criminals, attacks, and threats, there's also a huge demand for cybersecurity professionals in the industry. It's already noticeable through mainstream media platforms that it's only a matter of time before another organization is the target of threat actors. In an online article published by the World Economic Forum on January 21, 2015, *What does the Internet of Everything mean for security?*, the former executive chairman and CEO of Cisco Systems, John Chambers, said, "*There are two types of companies: those who have been hacked, and those who don't yet know they have been hacked.*" Each day, this statement is becoming more evident, and more of a reality, as many companies are reporting data breaches, and some reports indicate attackers were *living off the land* for many days or even months before the security incident was detected and contained.

The need for ethical hacking skills and knowledge is ever growing around the world, as leadership teams within small to large enterprises are realizing their assets need to be protected and ethical hackers and penetration testers can help discover and remediate hidden security vulnerabilities, reduce the attack surface, and improve the cyber defenses of their company against cyber criminals and threats. Ethical hackers have the same skill set and expertise as malicious attackers such as threat actors, however, the difference is their intention. Ethical hackers have a good moral compass and choose to use their skills for good reasons, whereas threat actors use their skills and knowledge for bad reasons, such as causing harm and damage to systems for illegal purposes.

The following are common technical skills of ethical hackers in the cybersecurity industry:

- Administrative-level skills with various operating systems such as Windows and Linux
- Solid foundational knowledge of networking, such as routing and switching
- Understanding the fundamentals of common security principles and best practices
- Familiar with programming languages such as Go and Python, and scripting languages such as Bash and PowerShell
- Familiarity with virtualization, containerization, and the cloud

While the preceding list of foundational skills seems a bit intimidating, always remember the field of cybersecurity and learning is like a marathon and not a sprint. It's not how quickly you can learn something, but ensuring you're taking the time you need to fully understand and master a topic.

The following are non-technical skills of ethical hackers:

- Being proficient in oral and written communication between technical and non-technical persons
- Being an *out-of-the-box* thinker
- Being self-motivated and driven to learn about new topics and expand knowledge
- Ensuring you understand the difference between using knowledge for good and bad intentions

Ethical hackers use the same techniques, tools, and procedures as real threat actors to meet their objectives and discover hidden security vulnerabilities in systems. There's a proverb that says if you want to catch a thief, you need to think like one. This proverb applies to ethical hacking – if you want to find the security vulnerabilities that real hackers are able to discover and exploit, then you need to adapt your mindset while using the same techniques, tools, and procedures to help you do the same, with legal permission and good intentions.

The following diagram shows the EC-Council's five stages of ethical hacking:

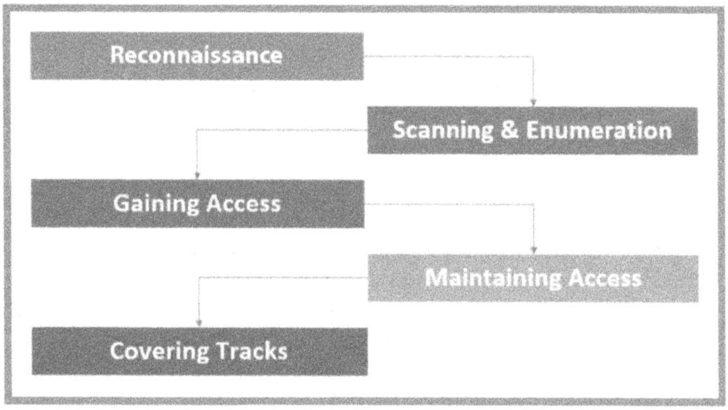

Figure 1.1 – Stages of hacking

As shown in the preceding diagram, ethical hackers and threat actors start with reconnaissance on their target, then move on to scanning and enumeration, then onward to gaining access and establishing a foothold in the system by maintaining access, and then covering tracks to remove any evidence of an attack. Since this book is based on the concept of *Reconnaissance for Ethical Hackers*, we'll focus on reconnaissance, scanning, and enumeration during the course of it.

The importance of reconnaissance

The first phase of ethical hacking is **reconnaissance** – the techniques and procedures that are used by the ethical hacker to collect as much information as possible about the target to determine their network infrastructure, cyber defenses, and security vulnerabilities that can be compromised to gain unauthorized access and improve attack operations accordingly. From a military perspective, reconnaissance plays an important role in planning and launching an attack on a target. Collecting information about the target helps the attacker to determine the points of entry, type of infrastructure, assets owned, and the target's strengths and weaknesses.

To put it simply, reconnaissance helps ethical hackers to gain a deeper understanding of an organization's systems and network infrastructure before launching an attack. The collected information can be leveraged to identify any security vulnerabilities that can be exploited, thus enabling the ethical hacker to compromise and gain a foothold in the targeted systems. For instance, using reconnaissance techniques enables the ethical hacker to identify any running services and open ports and the service and software versions on a system, all of which can be used to identify and determine potential attack vectors on the target.

In addition, using reconnaissance techniques such as **Open Source Intelligence** (**OSINT**) enables the ethical hacker to passively collect information about their target that's publicly available on the internet. Such information may contain usernames, email addresses, and job titles of employees of the targeted organization. This information can be leveraged to create various *social engineering* attacks and *phishing* email campaigns that are sent to specific employees within the targeted company.

The following screenshot shows an example of employees' information that's publicly available on the internet:

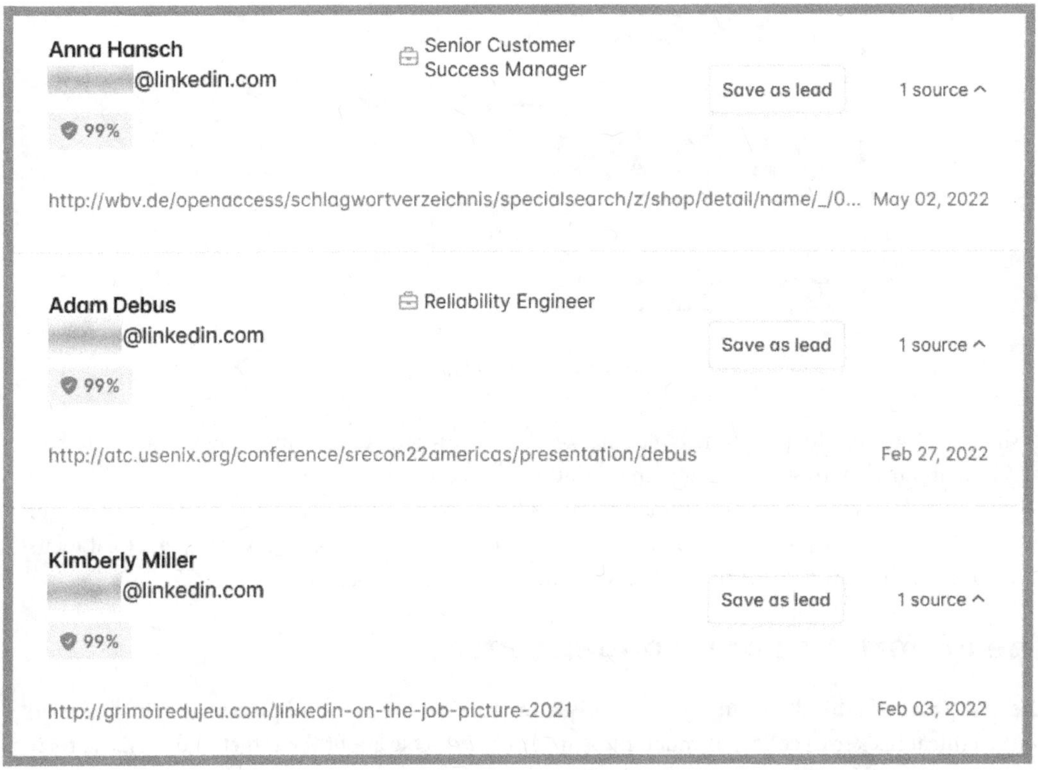

Anna Hansch
@linkedin.com
99%
Senior Customer Success Manager
Save as lead
1 source ^
http://wbv.de/openaccess/schlagwortverzeichnis/specialsearch/z/shop/detail/name/_/0... May 02, 2022

Adam Debus
@linkedin.com
99%
Reliability Engineer
Save as lead
1 source ^
http://atc.usenix.org/conference/srecon22americas/presentation/debus Feb 27, 2022

Kimberly Miller
@linkedin.com
99%
Save as lead
1 source ^
http://grimoiredujeu.com/linkedin-on-the-job-picture-2021 Feb 03, 2022

Figure 1.2 – Employees' data

As shown in the preceding screenshot, these are various employees of a specific organization. Their names, email addresses, and job titles are publicly known on the internet. A threat actor could look for patterns in their email addresses to determine the format that's used for all employees of the company. For instance, let's imagine there's an employee whose name is John Doe and his email address is `jdoe@companyname.com` and another employee is Jane Foster with an email address of `jfoster@companyname.com`. This information shows a pattern and format for employees within the same organization: `{f}{lastname}@domain-name.com`, where `f` is the initial letter of the person's first name followed by their last name and the company's domain name. Such information can help an ethical hacker to send phishing email campaigns to specific email addresses of high-profile employees of the targeted organization.

Reconnaissance helps organizations to reduce the risk of being compromised by a threat actor and improve their cyber defenses. By enabling an ethical hacker to perform reconnaissance techniques and procedures on an organization's systems and network infrastructure, the organization can efficiently identify security vulnerabilities and take the necessary measures to remediate and resolve them before they are discovered and exploited by adversaries. Furthermore, reconnaissance helps organizations to both identify and keep track of potential threat actors, enabling the company to gain a better understanding of the cybersecurity threat landscape while implementing and improving proactive countermeasures to safeguard their assets, systems, and networks. Hence, reconnaissance is not only important to adversaries but cybersecurity professionals use the gathered information to help organizations.

Reconnaissance is divided into the following types:

- **Passive reconnaissance**
- **Active reconnaissance**

Passive reconnaissance enables the ethical hacker to leverage OSINT techniques to gather information that's publicly available from various sources on the internet without making direct contact with the target.

The following are some examples of OSINT data sources:

- Job websites
- Online forums
- Social media platforms
- Company registry websites
- Public **Domain Name System** (**DNS**) servers

It's important for ethical hackers to use similar techniques and procedures as adversaries during their security assessments to provide real-world experience to their customers. In addition, it also helps the organization to determine whether its security team and solutions are able to detect any security intrusions that are created by the ethical hacker. If the security team were unable to detect any actions that were performed during the ethical hacking and penetration testing assessment, it's a good sign for the ethical hacker as their techniques were stealthy enough to bypass and evade any threat detection systems on the network. However, this means the organization's security team needs to improve their threat monitoring and detection strategies and tune their sensors to catch any security-related anomalies.

Active reconnaissance involves a more direct approach by the threat actor and ethical hacker to gather information about the target. In active reconnaissance, the ethical hacker uses scanning and enumeration techniques and tools to obtain specific details about the targeted systems and networks. For instance, to determine running services and open ports on a server, the ethical hacker can use a network and port scanning tool such as **Nmap** to perform host discovery on a network. However, active reconnaissance increases the risk of triggering security sensors and alerting the security team about a possible reconnaissance-based attack being performed.

In the next section, you will learn how cybersecurity professionals, including ethical hackers, leverage the information that is collected during reconnaissance to help organizations improve their security posture and manage their attack surfaces.

Understanding attack surface management

The **attack surface** is simply the number of potential security vulnerabilities that can be exploited to gain access to a system, network, and organization using attack vectors. If organizations are unable to identify their security vulnerabilities and implement countermeasures, they are simply leaving themselves susceptible and exposed to cyber-attacks and threats. **Attack Surface Management** (**ASM**) is not a new study in the cybersecurity industry, rather it's a new focus for cybersecurity professionals and organizations around the world. ASM is a strategy that's used by cybersecurity professionals that enables them to focus on identifying, analyzing, and reducing the attack surface of an organization. As a result, by reducing the attack surface of an organization, it reduces the risk of being compromised by cyber-attacks and threats while safeguarding its assets, resources, and sensitive information.

Adopting ASM within an organization enables the security team to identify and prioritize security vulnerabilities based on their vulnerability score and potential impact. The **Common Vulnerability Scoring System** (**CVSS**) is commonly referenced within many vulnerability scanning tools to provide vulnerability of between `0` and `10`, where `0` is the least impact and `10` is critical. These scores help cybersecurity professionals to apply high priority and resources to remediate security vulnerabilities with higher severity.

For instance, the following screenshot shows the base metrics of the CVSS calculator:

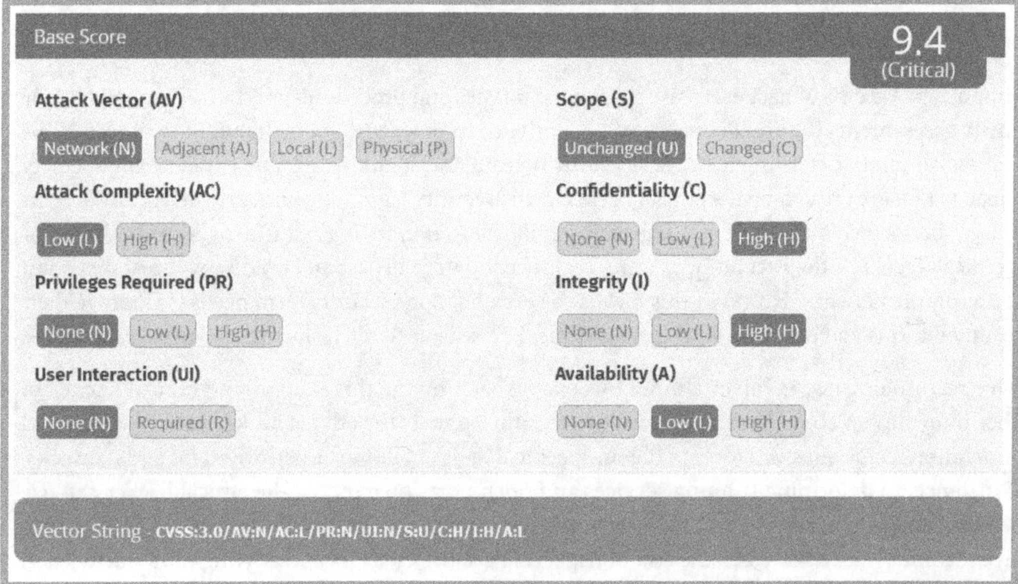

Figure 1.3 – CVSS calculator

As shown in the preceding snippet, the metrics within the base score influence the vulnerability score. For instance, if an attacker can compromise a security vulnerability on a targeted system over a network, where the attack complexity is low and does not require any user interaction or escalated privileges, where the impact will greatly affect the confidentiality and integrity of the system, the CVSS calculator provides a vulnerability score of 9.4. Keep in mind, these scores are assigned to a vulnerability based on the criticality and impact on the system.

> **Tip**
>
> To learn more about the CVSS calculator, please see `https://www.first.org/cvss/calculator/3.1`.

The following snippet shows the results of a Nessus vulnerability scan, displaying the number of security flaws and their scores:

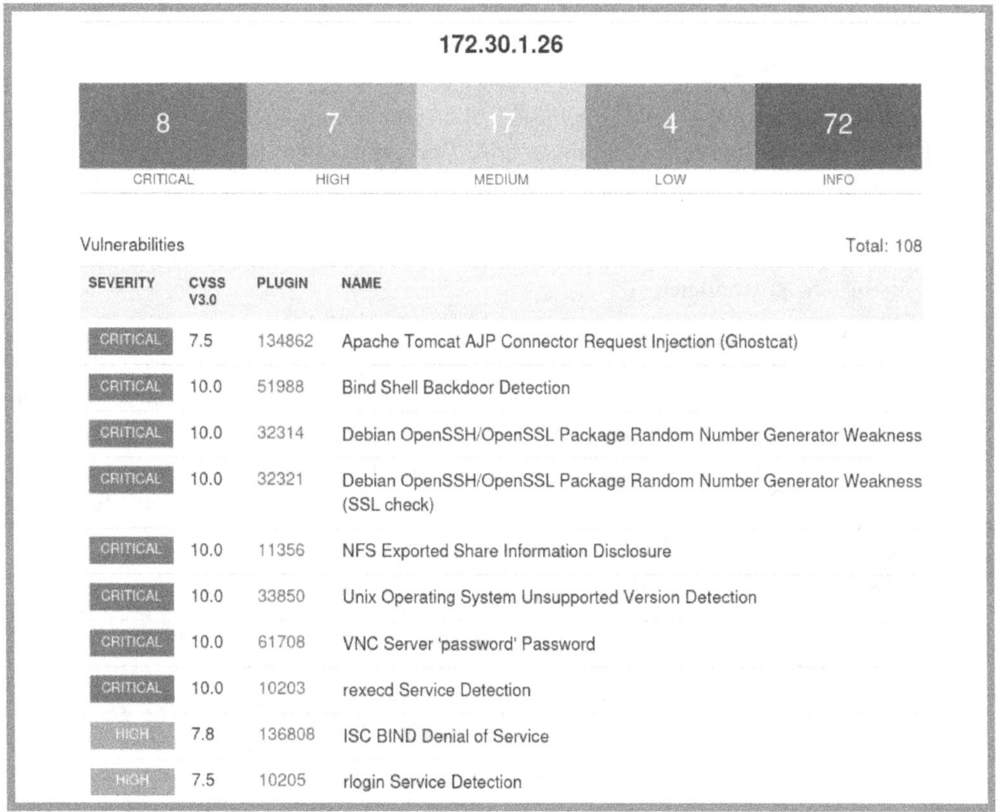

SEVERITY	CVSS V3.0	PLUGIN	NAME
CRITICAL	7.5	134862	Apache Tomcat AJP Connector Request Injection (Ghostcat)
CRITICAL	10.0	51988	Bind Shell Backdoor Detection
CRITICAL	10.0	32314	Debian OpenSSH/OpenSSL Package Random Number Generator Weakness
CRITICAL	10.0	32321	Debian OpenSSH/OpenSSL Package Random Number Generator Weakness (SSL check)
CRITICAL	10.0	11356	NFS Exported Share Information Disclosure
CRITICAL	10.0	33850	Unix Operating System Unsupported Version Detection
CRITICAL	10.0	61708	VNC Server 'password' Password
CRITICAL	10.0	10203	rexecd Service Detection
HIGH	7.8	136808	ISC BIND Denial of Service
HIGH	7.5	10205	rlogin Service Detection

Figure 1.4 – Nessus scan results

As shown in the preceding snippet, the CVSS scores were referenced from the CVSS calculator.

It's important to recognize that cybersecurity professionals may identify a security vulnerability that is critical to the operation of the organization and its business processes but has a low potential impact. There can be security vulnerabilities that are less critical to the operation of the business but have a greater potential impact if they're exploited by a threat actor. Therefore, ASM helps organizations in prioritizing security vulnerabilities based on their impact levels while allocating their resources to remediating the most critical security vulnerabilities first.

Additionally, organizations that implement ASM are able to better identify and track changes to their attack surfaces. For instance, if an organization installs a new update to an existing system, this new update could introduce new security vulnerabilities and potentially change the attack surface, enabling a threat actor to use new techniques to compromise the system. Similarly, if an organization implements a new system or application on its network infrastructure, it has the potential of bringing new security flaws to the attack surface. However, ASM enables cybersecurity professionals to track changes that are being made to the attack surface of the organization while ensuring the security team is aware of any new security vulnerabilities that are introduced during this process. Furthermore, the organization can take the necessary actions to remediate these security vulnerabilities before they can be exploited by a threat actor.

Another benefit of ASM is its capability of helping organizations efficiently monitor their attack surface and identify any suspicious activities. This improves real-time threat detection and response within the company, enabling the security team to take immediate action to prevent, contain, or remediate the threat from systems and networks. Lastly, when ASM is implemented properly, it helps security teams to identify whether any malicious activities or threats that evaded security solutions have gone undetected on their systems and networks.

The following are the major benefits of ASM within the cybersecurity industry:

- **Reducing risk** – Organizations that adopted ASM are able to identify and reduce their own attack surfaces, thereby reducing the risk of potential cyber-attacks and threats, and protecting their assets from threat actors. Hence, by identifying and remediating security vulnerabilities, it becomes more difficult for threat actors to compromise systems and gain a foothold.
- **Prioritization** – ASM helps companies to prioritize their resources to remediate security vulnerabilities that are more critical than others.
- **Continuous monitoring** – For organizations to ensure their attack surface is small, continuous monitoring and maintenance are needed. This helps both cybersecurity professionals and organizations to always be aware of any new security vulnerability that may exist, either due to a new implementation or an upgrade to a system, therefore, taking the necessary actions needed to mitigate any security vulnerabilities before they can be exploited.
- **Improving incident response** – ASM helps security teams to efficiently identify and respond to security incidents on their network in real time, as a result, reducing the impact and spread of a threat.

- **Compliance** – There are regulatory standards and frameworks that are needed within organizations that operate in certain industries. For instance, organizations that operate in the payment card industry need to ensure their systems and networks are compliant with the **Payment Card Industry Data Security Standard** (**PCI DSS**). Being compliant means the organization's systems and networks have the specific security controls in place to ensure data is protected.
- **Cost-effectiveness** – Since ASM helps organizations to improve the identification and remediation of security vulnerabilities, it reduces the risk of data breaches and increases the availability of systems that are critical to the organization.

The following are key steps that organizations and cybersecurity professionals can use to get started with ASM:

1. **Asset management** – Ensure all assets within your organization are properly tracked and entered into your inventory. These may include computers, servers, applications, and mobile devices. This helps organizations to better understand which assets are to be protected and identify security flaws in them.
2. **Identifying and mapping the attack surface** – At this stage, the cybersecurity professionals are to identify security vulnerabilities and map the attack surface of the organization. This stage includes potential attack vectors that could be used to deliver an exploit and points of entry such as open ports and vulnerable running services on systems and networks.
3. **Assessing risk** – This stage focuses on assessing the risk of each security vulnerability and its impact on the organization. This phase helps with prioritizing and focusing on the most critical security vulnerabilities, then on less critical vulnerabilities.
4. **Implementing security controls** – This phase focuses on implementing security controls and solutions to remediate and mitigate security vulnerabilities that were identified in the previous stages. Here, the security team will implement network security devices, threat monitoring and prevention solutions, network segmentation, and so on.
5. **Monitoring and maintenance** – For ASM to be effective, continuous monitoring of all assets, systems, and devices is required. It's important to continuously monitor and maintain security controls that are responsible for mitigating cyber-attacks and threats from exploiting security vulnerabilities. In addition, continuous monitoring and maintenance help ensure security controls are effective in safeguarding the assets of the organization.
6. **Continuously perform reconnaissance** – To identify new security vulnerabilities on the attack surface, organizations need to continuously perform reconnaissance on their assets, systems, and network infrastructure. Once new security vulnerabilities are identified, the lifecycle of ASM is repeated, taking the necessary steps to mitigate the security risk.

In addition to using the preceding key steps, there are several tools that will help both cybersecurity professionals and organizations with ASM:

- **Vulnerability scanners** – These are specialized, automated tools that help cybersecurity professionals identify security vulnerabilities in a system and provide recommendations on how to remediate the issue. Furthermore, these tools provide severity ratings, vulnerability scores, and potential impact.
- **Network scanners and mappers** – This type of tool helps cybersecurity and networking professionals to determine live hosts, open service ports, and running applications on host devices. In addition, they help organizations to map their entire network infrastructure and identify unauthorized devices that are connected to the company's network.
- **Configuration management tools** – This type of tool helps organizations track and manage their configurations on systems and networks. It also helps cybersecurity professionals to identify new security vulnerabilities such as misconfigurations that are introduced onto a device after a new change.
- **Application security testing tools** – These are specialized tools that are commonly used by cybersecurity professionals to perform security testing on applications and software to identify any unknown security flaw.
- **Attack Surface Reduction** (**ASR**) tools – These tools are designed to help organizations reduce their attack surfaces. It works by identifying and denying any malicious network traffic and disabling unnecessary services on systems and protocols.
- **Risk management tools** – Risk management tools enable organizations to both track and manage the risk as it's associated with their attack surface. Furthermore, this tool helps cybersecurity professionals to monitor the effectiveness of the security controls that are in place to prevent cyber-attacks and threats.
- **Security Information and Event Management** (**SIEM**) – This is a security solution that collects, aggregates, and analyzes security-related log messages generated from systems and devices within an organization to identify any potential cyber-attack and threat in real time.

While these tools are simply recommendations, it's important to remember no single tool has the capability of providing complete coverage of the attack surface of an organization. Therefore, a combination of different tools, techniques, and procedures is required to ensure the organization can effectively manage its attack surface. Furthermore, as many tools are software-based, it's important they are regularly updated to ensure they have the capability of detecting the latest security vulnerabilities and threats in the industry.

In the next section, you will learn about the tactics, techniques, and procedures that are used by adversaries during the reconnaissance phase of a cyber-attack.

Reconnaissance tactics, techniques, and procedures

As you have learned thus far, before an adversary launches an attack against an organization, they need to perform reconnaissance to gather as much information as possible on the target to determine its attack surface (points of entry). While there are many techniques that are used by both threat actors and ethical hackers, MITRE has created its well-known **MITRE ATT&CK** framework, which outlines the **Tactics, Techniques, and Procedures (TTPs)** of adversaries that are based on real-world events. These TTPs are commonly used by cybersecurity professionals, researchers, and organizations to both develop and improve their threat modeling and cyber defenses.

MITRE ATT&CK includes reconnaissance TTPs that help us to better understand the methods that are used by adversaries to collect information about their targets prior to launching an attack. These TTPs are also used by ethical hackers to efficiently identify security vulnerabilities and how a threat actor could compromise the attack surface of their client's network infrastructure.

The following are common reconnaissance TTPs that are used by adversaries:

- **Active scanning** – During active scanning, adversaries use various scanning tools to collect information about the target that can be leveraged in future operations. These scanning tools send special probes to targeted systems and networks to determine live hosts, operating systems, open ports, and running services on the host machine. Active scanning is an active reconnaissance technique that involves scanning IP network blocks and public IP addresses of the target, vulnerability scanning to identify security weaknesses that can be exploited, and wordlist scanning to retrieve possible passwords for future password-based attacks against the target.
- **Gathering victim host information** – This technique enables the attacker to collect specific details about the target's devices such as their hostnames, IP addresses, device types/roles, configurations, and operating systems. Additionally, the adversary is able to collect hardware, software, and client configuration details that can be used to improve the plan of attack. This technique involves using a combination of both active and passive reconnaissance as a threat actor can gain a lot of intelligence from OSINT alone and can perform active reconnaissance to identify specific details that are not easily available on the internet.
- **Gathering victim identity information** – This technique focuses on collecting details about the target's identity – personal data such as employees' names, email addresses, job titles, and users' credentials. This type of information can be collected using passive reconnaissance and leveraged for future social engineering attacks and gaining access to the target's systems.
- **Gathering victim network information** – Adversaries can use passive reconnaissance techniques to collect information on the target's network infrastructure such as IP ranges, domain names, domain registrar details (physical addresses, email addresses, and telephone numbers), and DNS records. However, active reconnaissance techniques will help the attacker to better identify the target's network topology, networking devices, and security appliances. Such information helps the adversary to better understand the target's network infrastructure.

- **Gathering victim organization information** – This technique enables adversaries to collect specific information about the target's organization such as names of departments, business operations and processes, and employees' roles and responsibilities. Such information can be collected using passive reconnaissance. Furthermore, adversaries use this technique to determine physical locations, business relations, and operating hours.
- **Phishing for information** – Adversaries send phishing email messages to employees of the target organization with the intention of tricking a victim into performing an action such as downloading and installing malware on their system or even revealing sensitive information such as their user credentials. Adversaries can use spear phishing services from online service providers, insert malicious attachments in email messages, and insert obfuscated links within the body of the email message. Since the attacker is using a direct approach, this is an active reconnaissance technique.
- **Searching closed sources** – The adversary may attempt to collect information about the target from closed sources, where the information is available as a paid subscription (passive reconnaissance). Such information includes threat intel vendors such as private details from threat intelligence sources that can be used to compromise the target. Furthermore, adversaries can purchase information about the target from *Dark Web* marketplaces/black markets.
- **Searching open technical databases** – There are many public online sources that enable anyone to collect information about a target. This technique focuses on leveraging public information that can be used to improve the plan of attack against an organization. For instance, the adversary can leverage public DNS records, WHOIS data (domain registration details), digital certificates (help identify sub-domains), and public databases that contain IP addresses, open ports, and server banner details about the target. This is another passive reconnaissance technique to collect information about the target.
- **Searching open websites and domains** – Adversaries use this technique to search various online websites and platforms such as social media, internet search engines, and code repositories (such as GitHub) to collect information that can be used to compromise the target. Searching open websites and domains is another passive reconnaissance technique for collecting public information.
- **Searching victim-owned websites** – This technique is used by the adversary to search the target's websites for any details that can be leveraged, such as organizational details, physical locations, email addresses of employees, high-profile employees, and even employees' names and contact details. This is an active reconnaissance technique since the attacker establishes a direct connection to the target's asset.

These are common strategies used by threat actors, and it helps ethical hackers to efficiently identify security vulnerabilities within organizations. Additionally, keep in mind that reconnaissance TTPs are continuously expanding as adversaries are developing new techniques and tools to compromise organizations. However, cybersecurity professionals and organizations can leverage reconnaissance TTPs to improve cyber defenses, identify and remediate security vulnerabilities, and reduce their attack surface and risk of a cyber-attack.

Summary

In this chapter, you have learned the importance of ethical hacking and how it helps organizations to improve their security posture. You have also discovered why threat actors spend a lot of time collecting information about their targets and how it can be leveraged to identify security vulnerabilities. Furthermore, you have learned why ethical hackers use similar techniques and strategies to help organizations identify and remediate their security vulnerabilities before a real cyber-attack occurs.

In addition, you have explored the need for attack surface management within the cybersecurity industry and how it helps organizations improve their defenses against cyber-attacks and threats. Lastly, you have gained an insight into reconnaissance TTPs that are commonly observed around the world as it helps security professionals and organizations to improve their threat modeling and strategies in safeguarding their assets from cyber criminals.

I hope this chapter has been informative for you and helpful on your journey in the cybersecurity industry. In the next chapter, *Setting Up a Reconnaissance Lab*, you will learn how to construct a security lab environment that will be safe for performing active reconnaissance and vulnerability assessments on your personal computer.

Further reading

- Basics of footprinting and reconnaissance: `https://www.eccouncil.org/cybersecurity-exchange/ethical-hacking/basics-footprinting-reconnaissance/`
- Attack surface management: `https://www.crowdstrike.com/cybersecurity-101/attack-surface-management/`
- MITRE Reconnaissance: `https://attack.mitre.org/tactics/TA0043/`
- What does the Internet of Everything mean for security?: `https://www.weforum.org/agenda/2015/01/companies-fighting-cyber-crime/`

2

Setting Up a Reconnaissance Lab

Learning about various cybersecurity topics is always exciting as there are so many new technologies and content that are quite interesting, especially using various tools and techniques to discover sensitive details on vulnerable systems and applications. However, obtaining sensitive details about systems such as their operating systems, known vulnerabilities, open service ports, and running services requires the ethical hacker to send specially crafted probes to the target. To put it simply, the ethical hacker can create a script or use an automated software-based scanning tool to send probes and analyze the responses to determine whether security vulnerabilities exist on the target. However, scanning is illegal and should not be performed on any system without obtaining legal permission from the necessary authorities.

As an aspiring or seasoned ethical hacker, it's important to set up your own lab environment as it enables you to perform intrusive security testing on your own devices, without worrying about causing harm or damage to someone else's systems.

During this chapter, you will learn how to set up and build your personal lab environment to learn reconnaissance techniques and hone your skills as a cybersecurity professional. You will learn how to deploy a hypervisor to create virtual machines and implement virtual networking. Additionally, you'll learn how to get both Kali Linux and Trace Labs OSINT virtual machines up and running. Lastly, you will learn how to set up both a vulnerable web application and server to learn active reconnaissance safely on your personal lab. Active reconnaissance requires the ethical hacker to perform various types of scans on a target that are considered to be intrusive and illegal due to the type of information that can be collected about the target.

In this chapter, we will cover the following topics:

- Lab overview and technologies
- Setting up a hypervisor and virtual networking
- Deploying Kali Linux
- Deploying an OSINT virtual machine
- Implementing vulnerable systems

Let's dive in!

Technical requirements

To follow along with the exercises in this chapter, please ensure that you have met the following hardware and software requirements:

- Oracle VM VirtualBox: `https://www.virtualbox.org/`
- Oracle VM VirtualBox Extension Pack: `https://www.virtualbox.org/wiki/Downloads`
- 7-zip: `https://www.7-zip.org`
- Kali Linux 2022.4: `https://www.kali.org/get-kali/`
- Trace Labs OSINT VM 2022.1: `https://www.tracelabs.org/initiatives/osint-vm`
- Vagrant: `https://www.vagrantup.com/`
- OWASP Juice Shop: `https://owasp.org/www-project-juice-shop/`
- Metasploitable 3: `https://app.vagrantup.com/rapid7/boxes/metasploitable3-win2k8`

Lab overview and technologies

Building a personal lab environment enables you to explore and improve your technical skills without the need to acquire expensive hardware or pay for a subscription service. A personal lab environment provides a lot of scalability and flexibility options such as enabling you to add and remove systems as needed based on your learning objectives and goals. For instance, by leveraging the power of virtualization technologies, you can create an entire virtual network with multiple virtual machines running on your personal computer without any hidden expenses or need for physical hardware components.

To better understand the components that are needed for our lab environment, let's take a look at the role and function of each major component:

- **Hypervisor** – The hypervisor will enable you to create virtualized environments for installing and running multiple operating systems simultaneously on top of your existing hardware on your personal computer. In the lab environment, **Oracle VM VirtualBox** will be the preferred hypervisor as it's free and works with many host operating systems and hardware.
- **Virtual networking** – The hypervisor contains virtual networking components such as a **virtual Switch (vSwitch)** and **virtual Network Interface Card (vNIC)**, which enables a virtual machine to intercommunicate with other virtual machines on the same hypervisor and systems on the physical network.
- **Attacker machine** – The attacker machine(s) will be used to perform passive and active reconnaissance and scanning techniques. We'll use **Kali Linux 2022.4** (not Kali Linux Purple) and **Trace Labs OSINT VM** as these are free and contain all the necessary tools you'll need to learn and perform reconnaissance and scanning.
- **Vulnerable server** – To perform active reconnaissance and scanning on a target, a vulnerable server is needed within the lab environment. The preferred choice will be **Metasploitable 3 Windows VM** as it's designed for learning about ethical hacking and penetration testing.
- **Vulnerable web application** – To safely perform active reconnaissance on a web application, we'll set up **OWASP Juice Shop**, a vulnerable web application that's designed to help cybersecurity professionals better understand the **OWASP Top 10: 2021 Web Application Security Risks** in the industry.

The following diagram shows the lab topology that will be built and run on your personal computer:

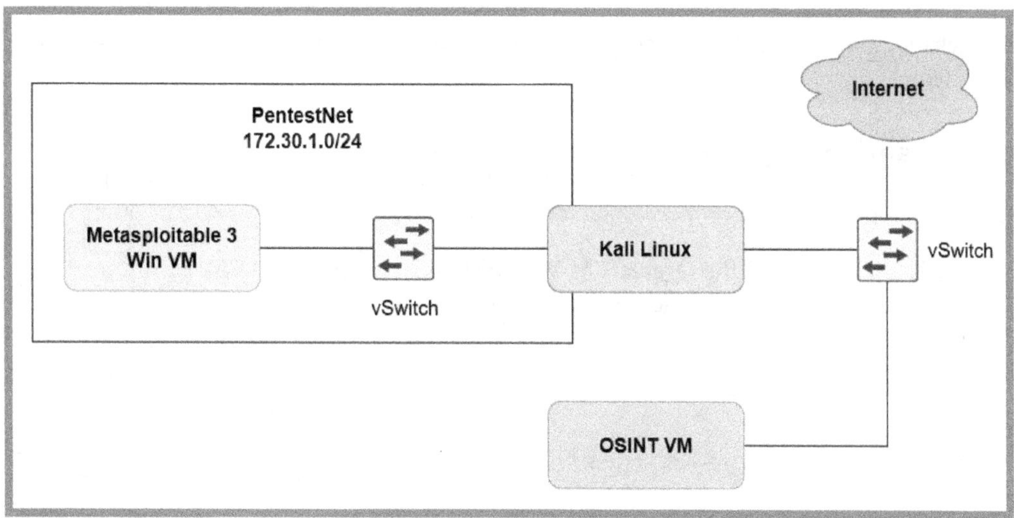

Figure 2.1 – Lab infrastructure

The hypervisor will be installed on your host operating system, which enables you to create virtual machines. Kali Linux, Trace Labs OSINT VM, and Metasploitable 3 Windows VM will run as virtual machines on the hypervisor when needed. Additionally, the hypervisor will assign vNICs to each virtual machine to interface with the virtual and physical networks via vSwitches. The **PentestNet** (`172.30.1.0/24`) is a virtually isolated network between Kali Linux and the Metasploitable 3 Windows VM machines that will be strictly used for active reconnaissance. However, Kali Linux and the Trace Labs OSINT VM will be connected to the internet.

In the next section, you will learn how to set up the hypervisor and virtual network environments on your computer.

Setting up a hypervisor and virtual networking

There are many hypervisors that are available in the industry; however, we'll be using Oracle VM VirtualBox as it's free and simple to use while containing all the important features and components needed for our lab environment. In this section, you will learn how to set up a hypervisor and virtual networking.

Before getting started, please ensure your system meets the following requirements:

- Ensure the **VT-x/AMD-V** feature is supported on your processor
- Ensure the virtualization feature is enabled on the processor via the BIOS/UEFI

Most modern processors support VT-x (Intel) or AMD-V (AMD), a technology that enables the host operating system and applications to leverage the virtualization features of the **Central Processing Unit** (**CPU**). However, some computer vendors do not enable the operating system to leverage the virtualization feature of the processor. Hence, it's important to manually access the BIOS/UEFI and ensure virtualization is enabled under the processor menu, or else the hypervisor may not work as expected.

Part 1 – setting up the hypervisor

To get started with setting up the hypervisor on your host operating system, please use the following steps:

1. Firstly, to download the **Oracle VM VirtualBox** application, go to `https://www.virtualbox.org/` and click on **Download VirtualBox**, as shown here:

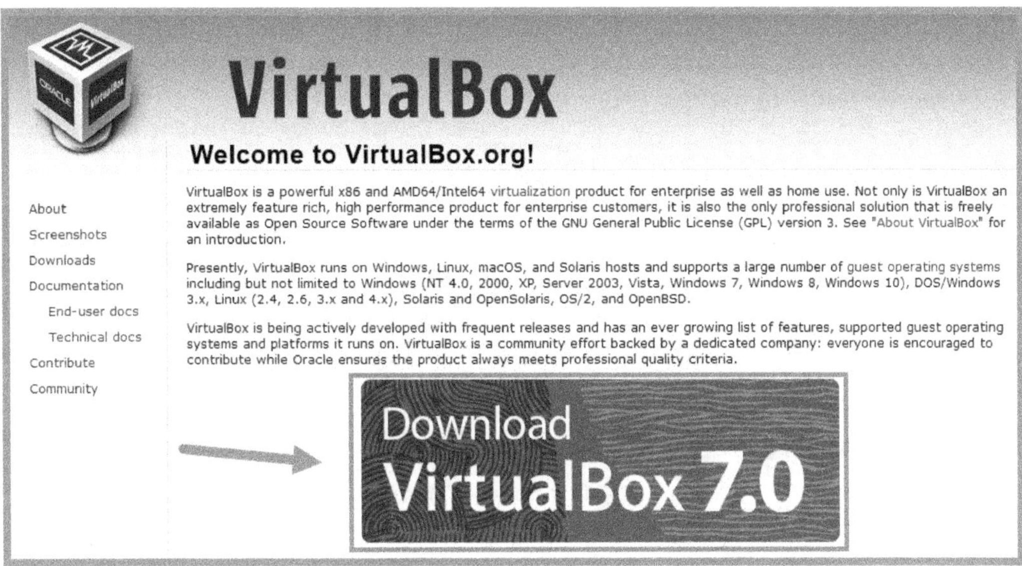

Figure 2.2 – VirtualBox website

2. Next, you'll be redirected to the **Downloads** page at `https://www.virtualbox.org/wiki/Downloads`. Click on **Windows hosts** to download the VirtualBox 7.0.4 platform package on your computer:

VirtualBox binaries

By downloading, you agree to the terms and conditions of the respective license.

If you're looking for the latest VirtualBox 6.1 packages, see VirtualBox 6.1 builds.

VirtualBox 7.0.4 platform packages

- Windows hosts
- macOS / Intel hosts
- Developer preview for macOS / Arm64 (M1/M2) hosts
- Linux distributions
- Solaris hosts
- Solaris 11 IPS hosts

The binaries are released under the terms of the GPL version 3.

Figure 2.3 – VirtualBox platform packages

3. Once the platform package has been downloaded, scroll down on the same **Downloads** page to the **VirtualBox 7.0.4 Oracle VM VirtualBox Extension Pack** section and click on **All supported platforms** to download the extension pack onto your system:

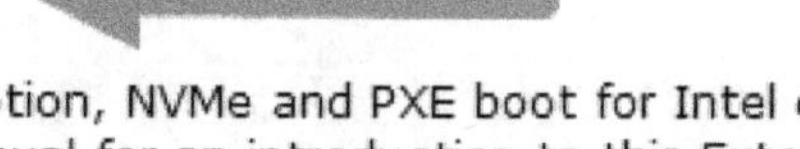

VirtualBox 7.0.4 Oracle VM VirtualBox Extension Pack
• All supported platforms
Support VirtualBox RDP, disk encryption, NVMe and PXE boot for Intel cards. See this chapter from the User Manual for an introduction to this Extension Pack. The Extension Pack binaries are released under the VirtualBox Personal Use and Evaluation License (PUEL). Please install the same version extension pack as your installed version of VirtualBox.

Figure 2.4 – VirtualBox extension pack

4. Next, install the Oracle VM VirtualBox platform package that was downloaded during *Step 2*. During the installation process, use the default setting by the installer to complete this step.
5. Next, to install the VirtualBox 7.0.4 Oracle VM VirtualBox Extension Pack, right-click on the package and select **Open With** > **VirtualBox Manager**. Accept the user agreement and proceed with the installation.

At this point, both VirtualBox Manager and VirtualBox Extension Pack are installed on the host operating system of your computer. Next, you will create the *PentestNet* virtual network using the networking capabilities of VirtualBox.

Part 2 – creating a virtual network

The *PentestNet* virtual network will enable Kali Linux to communicate with the Metasploitable 3 Windows VM within our lab environment. To get started with this exercise, please use the following steps:

1. On your Windows host operating system, open the Command Prompt.
2. Next, use the following commands to change the working directory to the installation location of VirtualBox and create a **Dynamic Host Configuration Protocol** (**DHCP**) server within VirtualBox Manager:

```
C:\Users\Glen> cd C:\Program Files\Oracle\VirtualBox
C:\Program Files\Oracle\VirtualBox> vboxmanage dhcpserver
add --network=PentestNet --server-ip=172.30.1.1
--lower-ip=172.30.1.20 --upper-ip=172.30.1.50
--netmask=255.255.255.0 --enable
```

The following snippet shows the result of executing the preceding commands on the Windows Command Prompt:

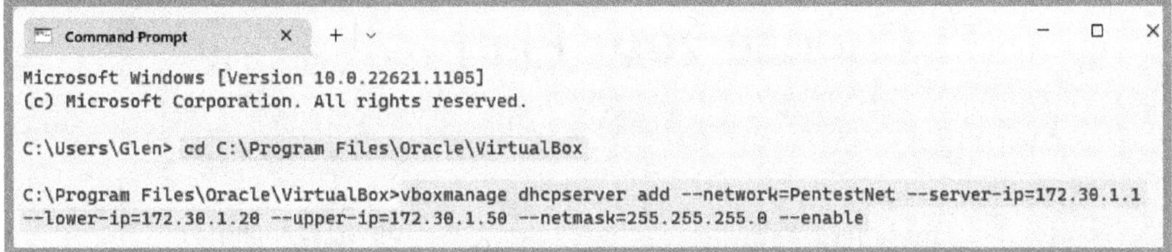

Figure 2.5 – Configuring virtual networking

Once the commands are executed correctly, the *PentestNet* virtual network will be automatically created and VirtualBox Manager will enable us to connect any virtual machine on it and receive an IP address from the virtual DHCP server.

> Tip
>
> To learn more about VirtualBox networking commands, please see `https://www.virtualbox.org/manual/ch06.html`.

Having completed this section, you have learned how to set up a hypervisor and create a virtual network using Oracle VM VirtualBox. Next, you will learn how to both deploy and get up and running with Kali Linux as a virtual machine.

Deploying Kali Linux

Kali Linux is the most popular Linux-based penetration testing operating system within the cybersecurity industry. It contains all the essential tools an ethical hacker and penetration tester would need to discover and exploit security vulnerabilities on systems, networks, and applications.

Part 1 – setting up Kali Linux as a virtual machine

To get started with deploying Kali Linux as a virtual machine, use the following steps:

1. Firstly, go to the official Kali Linux website at `https://www.kali.org/get-kali/` and click on **Virtual Machines**, as shown in the following screenshot:

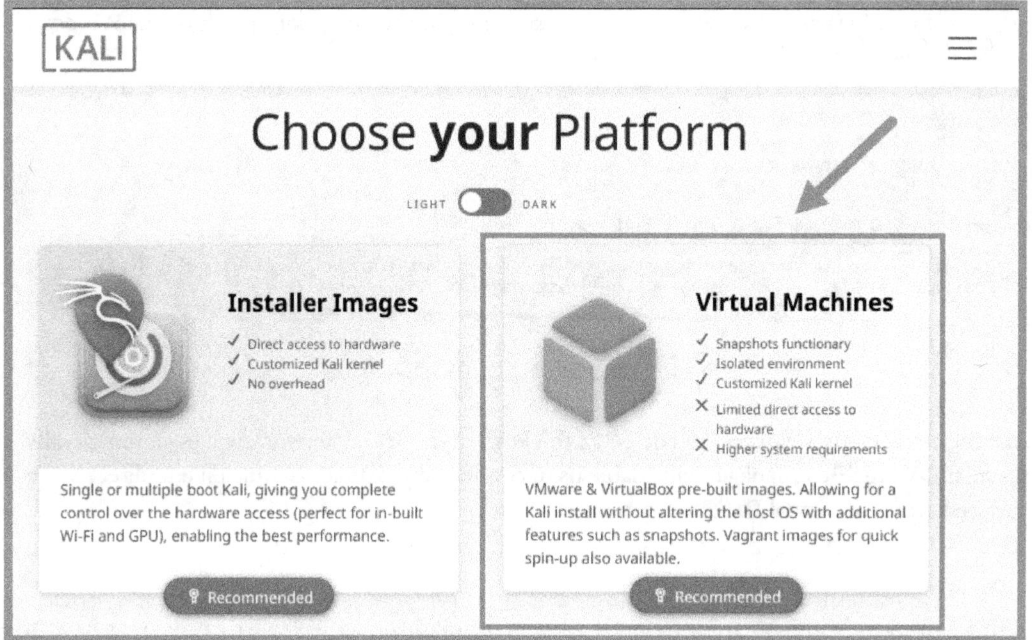

Figure 2.6 – Kali Linux download page

2. Next, click on **VirtualBox 64** to download the Kali Linux VirtualBox file, as shown in the following screenshot:

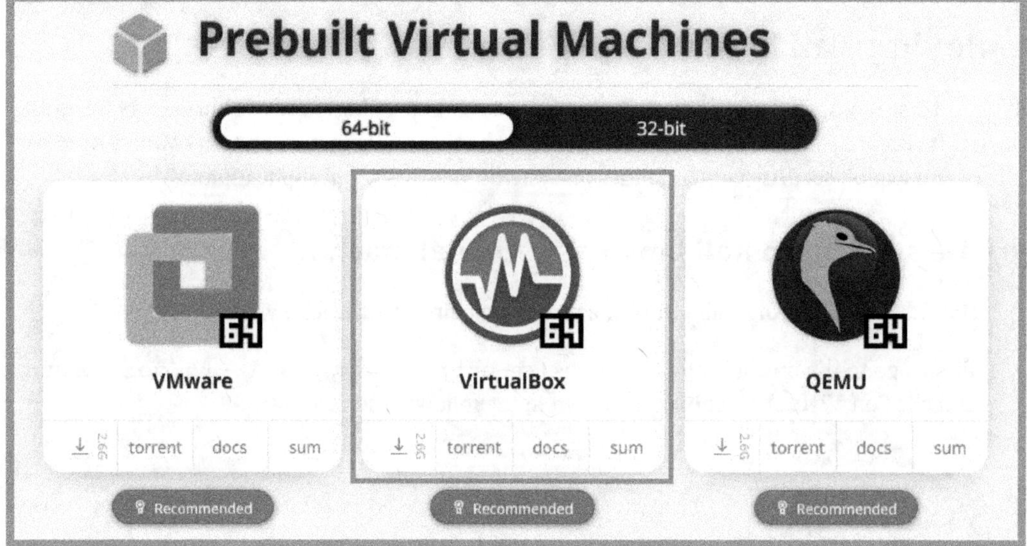

Figure 2.7 – Kali Linux virtual file

This will be a compressed (zipped) file that contains the Kali Linux virtual hard disk and setup configurations and enables us to easily import it into VirtualBox Manager.

3. After downloading the Kali Linux VirtualBox file, you will need an unzipping application such as **7-Zip** to extract the contents of the downloaded file. You can download 7-Zip from `https://www.7-zip.org/` and install it.
4. Next, right-click on the Kali Linux zipped file and select **Show more options**, as shown in the following screenshot:

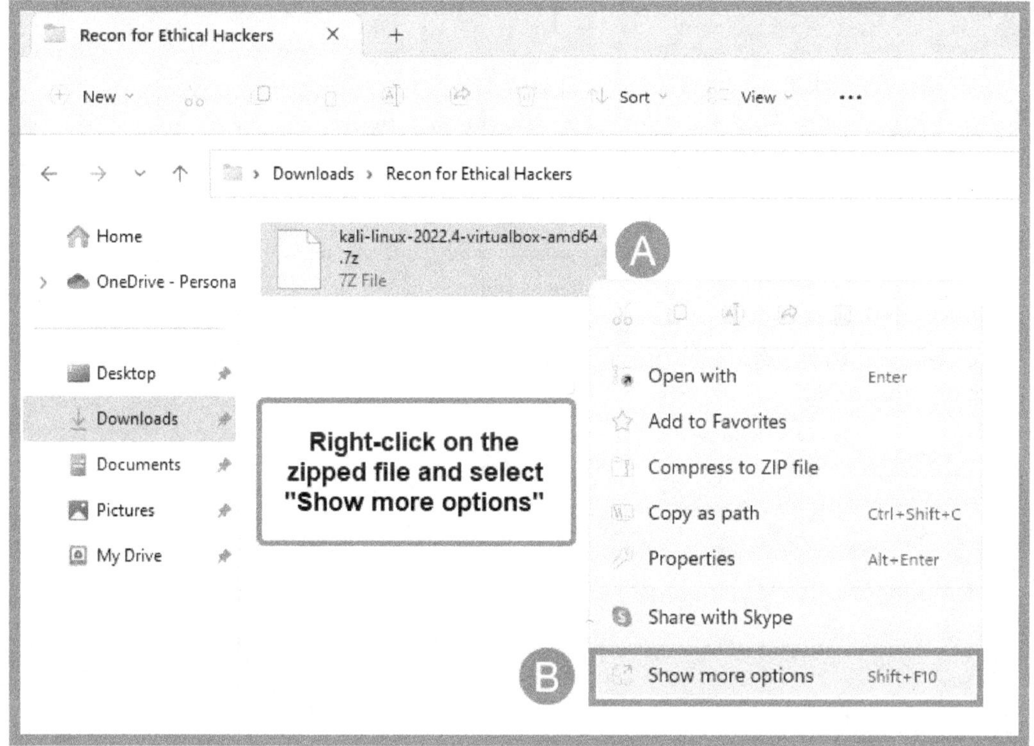

Figure 2.8 – Displaying more options

5. Next, select **7-Zip** > **Extract Here** to extract the contents within the present working directory:

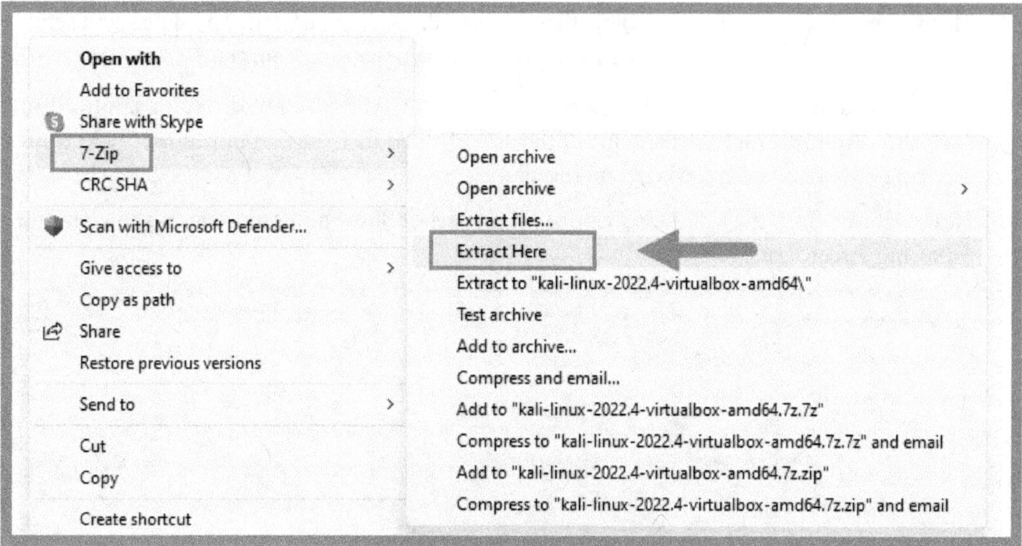

Figure 2.9 – 7-Zip extract options

Once the extract process begins, the following window will appear showing the status until completion:

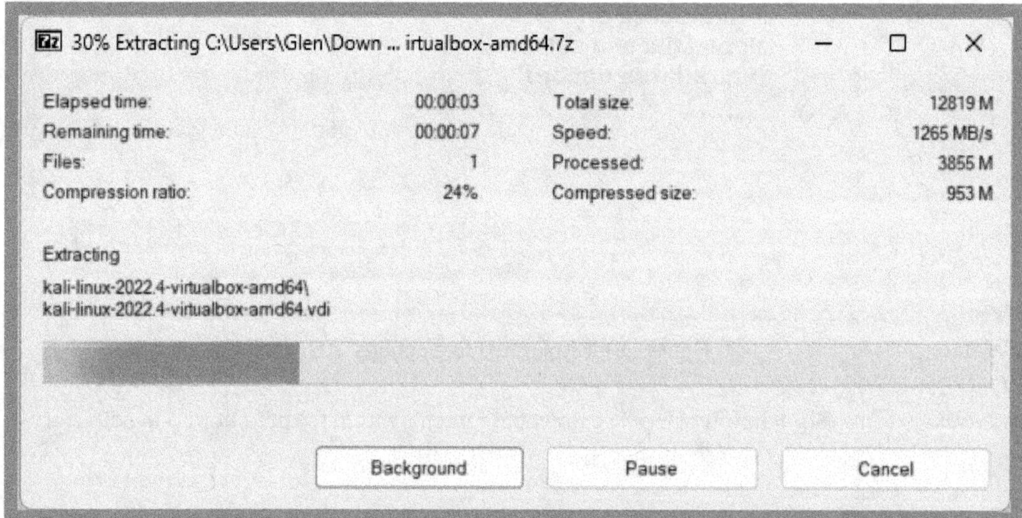

Figure 2.10 – Extraction status window

Once the extraction is completed, the contents will be available within the same directory.

6. Next, open the **VirtualBox Manager** application, and click on **Machine** > **Add...**, as shown in the following screenshot:

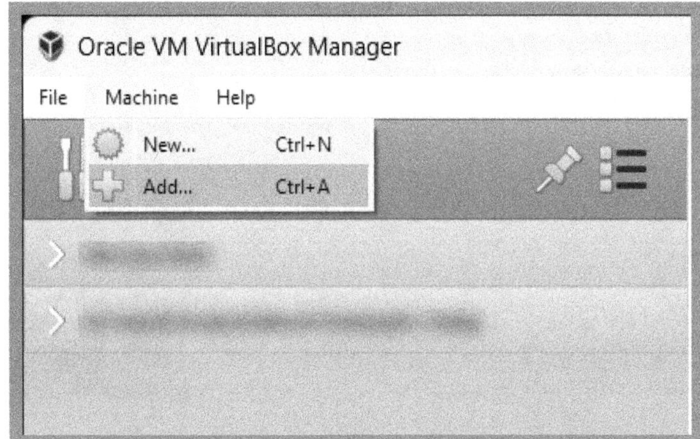

Figure 2.11 – Adding a new virtual machine

7. A new window will appear. Navigate to the location of the extracted folder, select the **kali-linux-2022.4-virtualbox-amd64** file, and click on **Open**, as shown in the following screenshot:

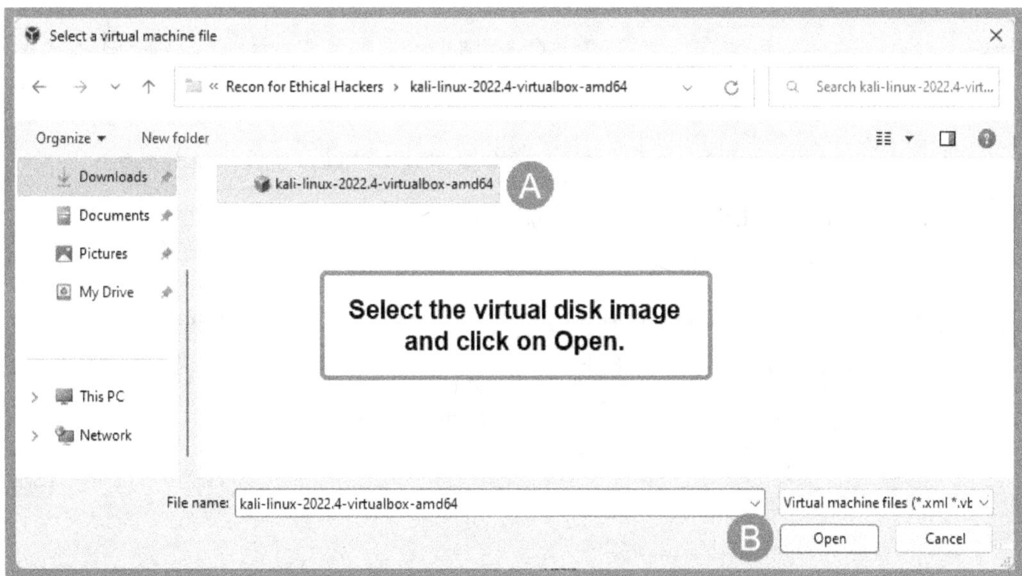

Figure 2.12 – Selecting the virtual machine file

Once you've clicked on **Open**, the Kali Linux virtual machine will automatically be imported into VirtualBox Manager.

8. Next, we need to assign one network adapter to our Kali Linux virtual machine for internet connectivity and another to access the *PentestNet* virtual network. Select the Kali Linux virtual machine and click **Settings**, as shown in the following screenshot:

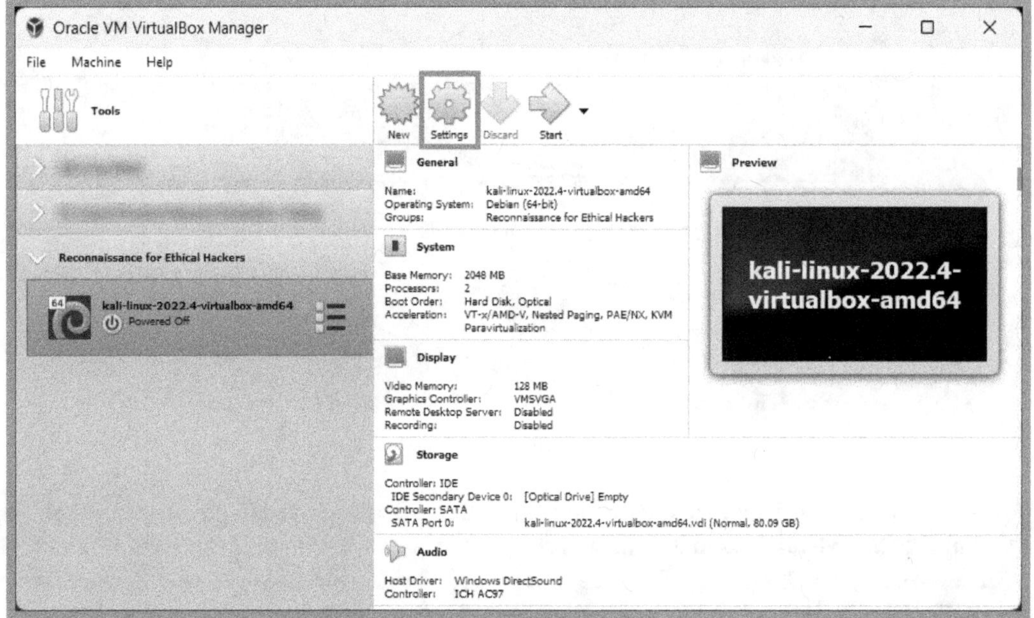

Figure 2.13 – Virtual machine settings

9. Next, select the **Network** category and use the following configurations for **Adapter 1**:

   - **Enable Network Adapter**: Check the box to enable
   - **Attached to**: **Bridged Adapter**
   - **Name**: Select the physical network adapter that is connected to the internet
   - **Cable Connected**: Yes – check the box

   The following snippet shows the configurations of **Adapter 1**:

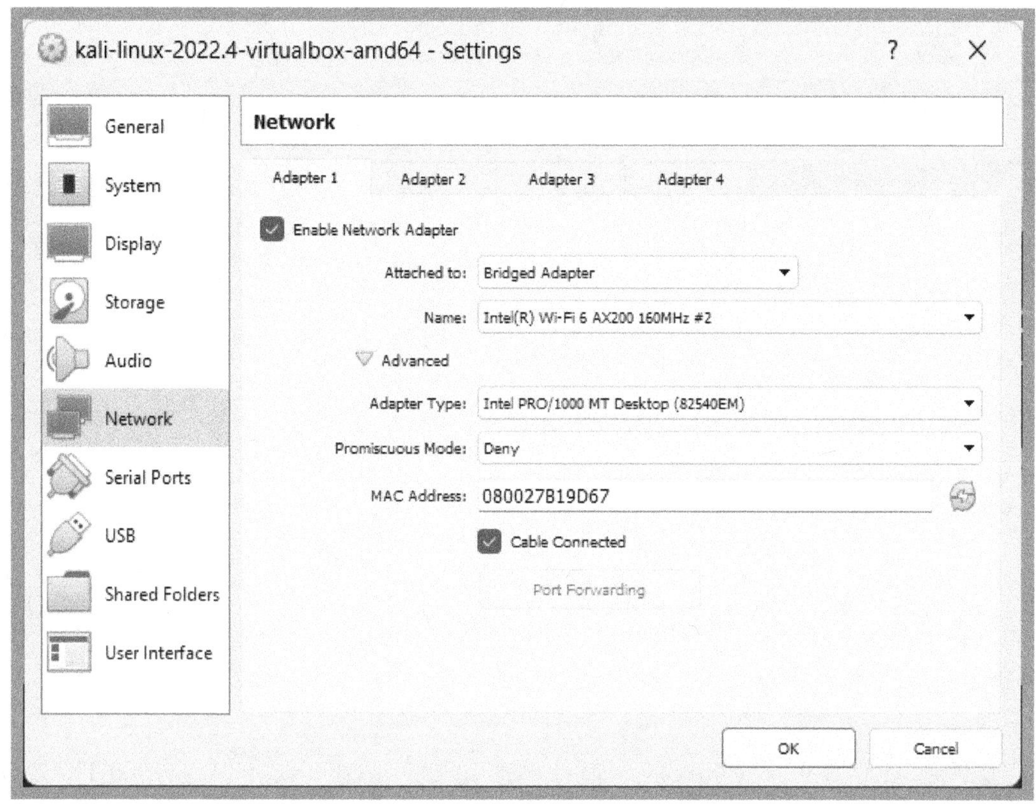

Figure 2.14 – Network Adapter 1 settings

10. Next, select **Adapter 2** and use the following configurations:

    - **Enable Network Adapter**: Check the box to enable
    - **Attached to**: **Internal Network**
    - **Name**: Manually type `PentestNet` into the field
    - **Cable Connected**: Yes – check the box
    - **Promiscuous Mode**: **Allow All**

The following snippet shows the configurations of **Adapter 2**:

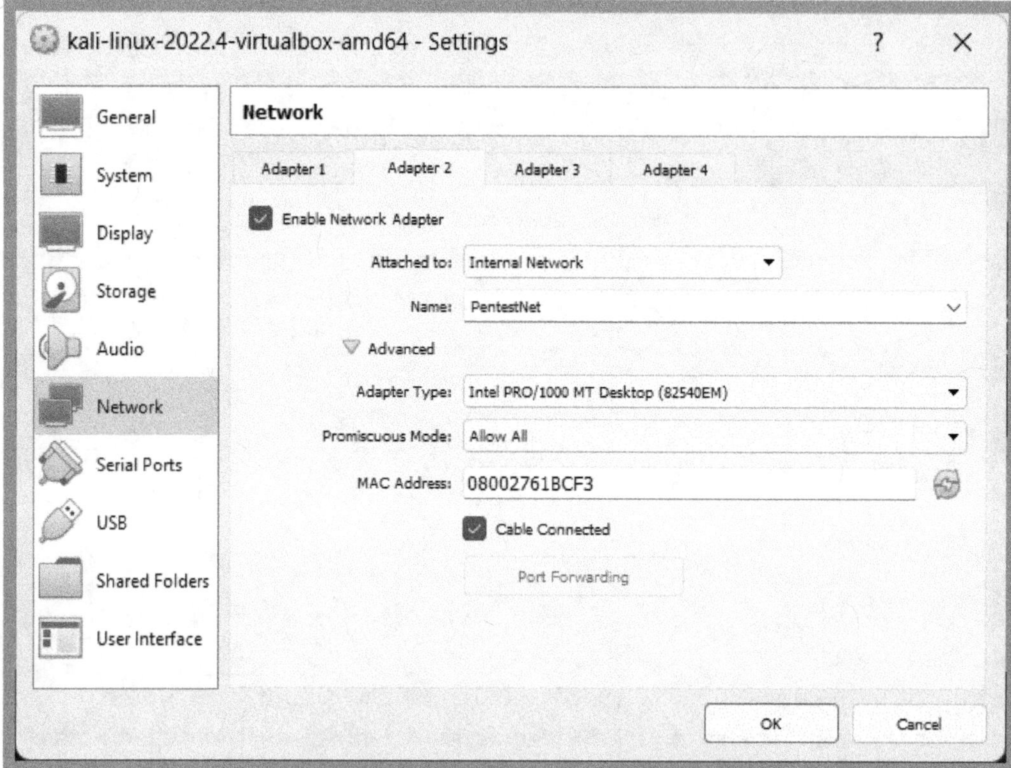

Figure 2.15 – Configuring Adapter 2

11. Lastly, click on **OK** to save the configurations of the virtual machine.

Part 2 – getting started with Kali Linux

In my years of experience in deploying Kali Linux as a virtual machine, sometimes there's no internet connectivity when it's connected to both an IPv4 and IPv6 network. Therefore, we'll need to disable IPv6 within Kali Linux to bypass this issue for our future exercises. Additionally, it's good practice to change the default user credentials, and verify whether **Domain Name System** (**DNS**) is working as expected and there's internet connectivity on the virtual machine.

To get started with this exercise, please use the following steps:

1. Let's power on the Kali Linux virtual machine; select **Kali Linux** and click on **Start**, as shown in the following screenshot:

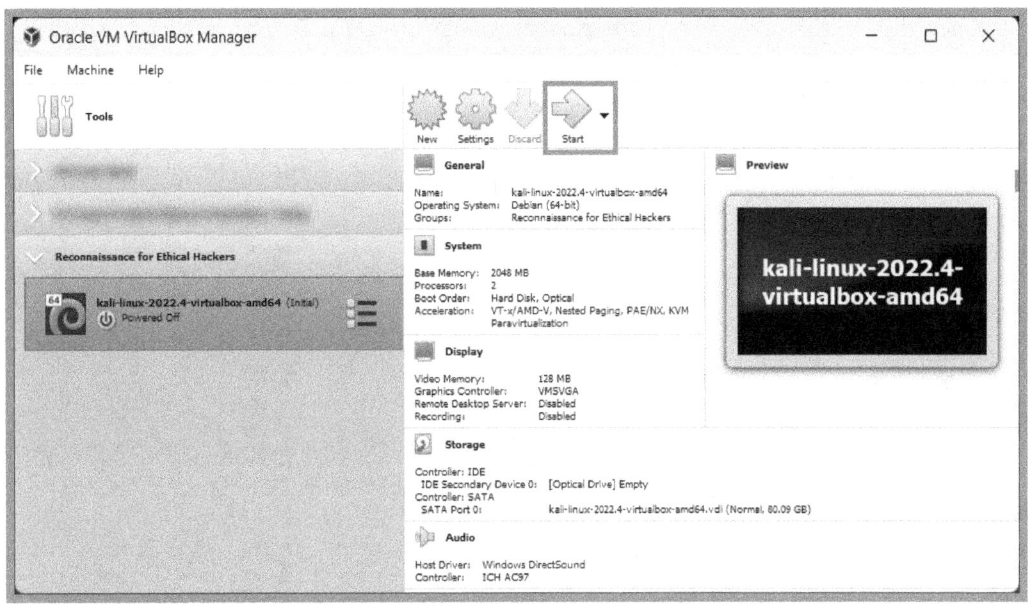

Figure 2.16 – Booting Kali Linux

2. Next, to log in to Kali Linux, use the default user credentials of `kali` for both **Username** and **Password**, as shown in the following screenshot:

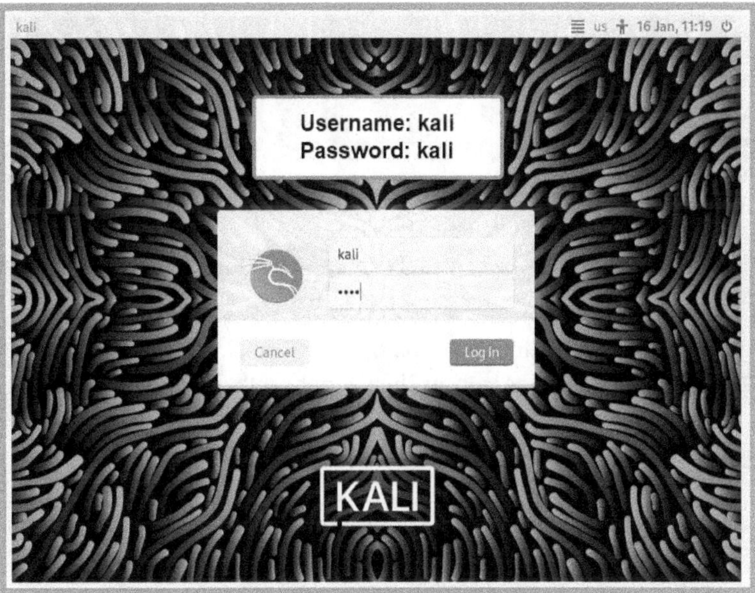

Figure 2.17 – Login interface

3. Once you're logged in, the display of the guest operating system (Kali Linux) may not automatically scale to fit your monitor's resolution. To automatically scale the display, select **View** and toggle with **Auto-resize Guest Display**, as shown in the following screenshot:

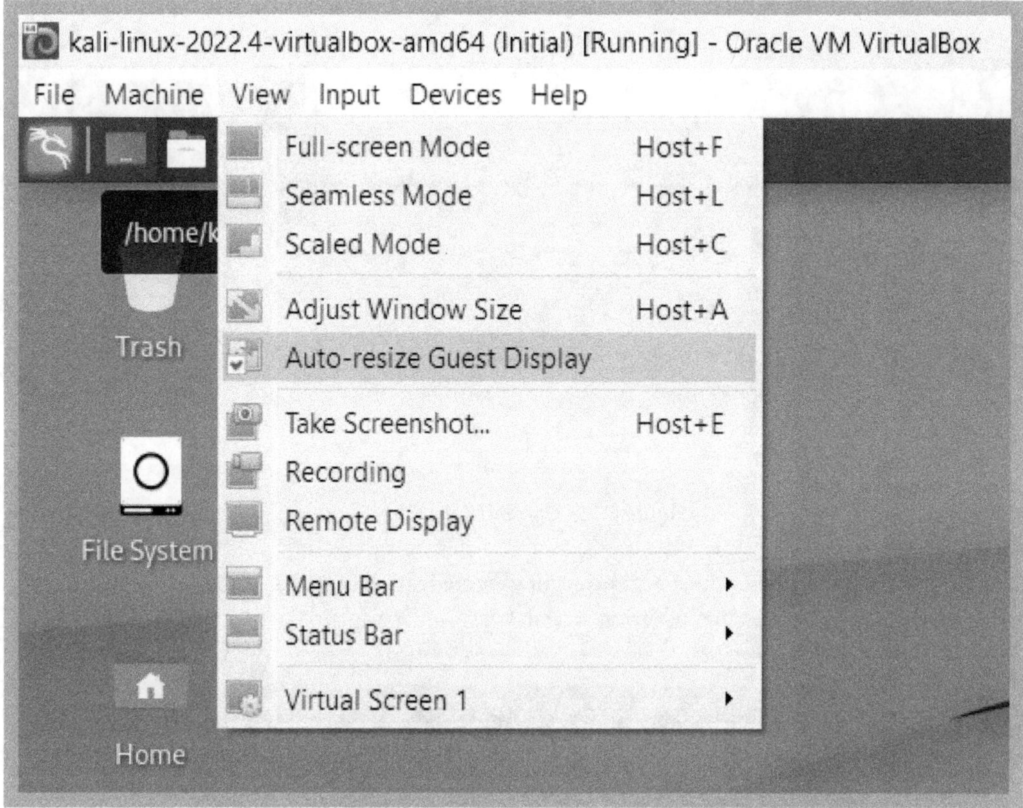

Figure 2.18 – Scaling the display

After toggling with the **Auto-resize Guest Display** option, the display will automatically scale to fit your monitor's resolution onward, even after reboots.

4. Next, to disable IPv6 on Kali Linux, click on the Kali Linux icon in the top-left corner to open the menu and select the **Settings** icon, as shown in the following screenshot:

Figure 2.19 – Kali Linux menu

As shown in the preceding snippet, the pre-installed tools are categorized based on the phases of ethical hacking and penetration testing.

5. When the **Settings** menu appears, click on **Advanced Network Configuration**, as shown in the following screenshot:

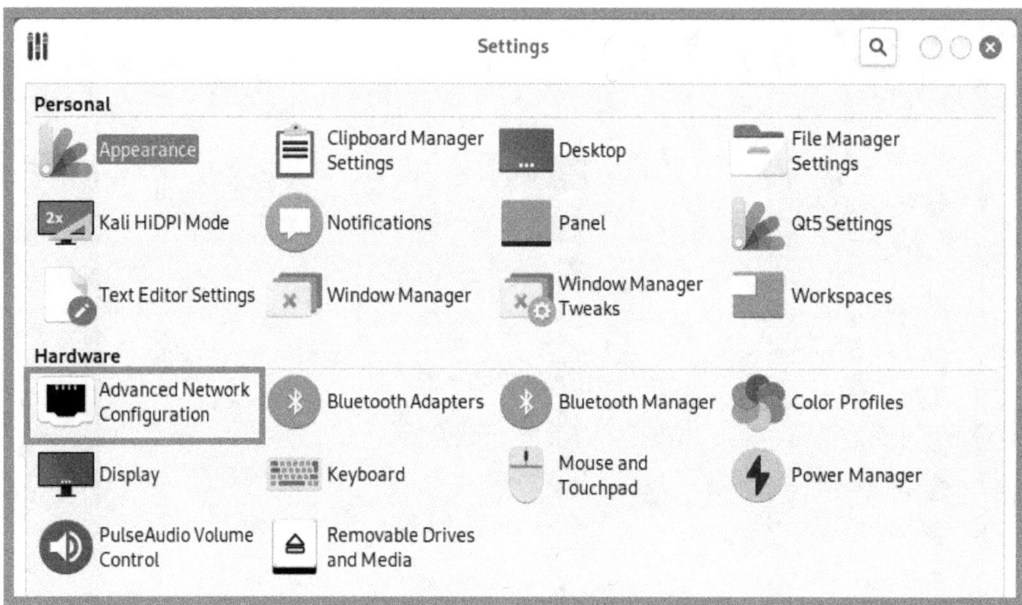

Figure 2.20 – Settings menu

6. Next, the **Network Connections** window will appear; select **Wired connection 1** and click the *gear* icon as shown in the following screenshot:

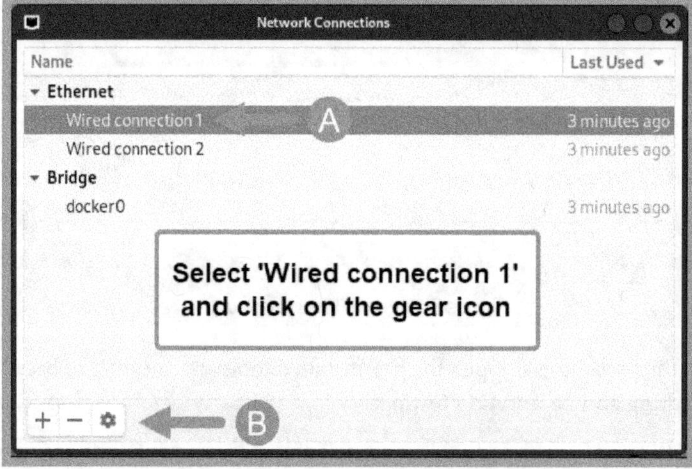

Figure 2.21 – Network connections window

If you recall on the virtual machine settings of Kali Linux, **Adapter 1** is connected to the physical network with the internet connection, hence we selected **Wired connection 1**, as these are the same network interfaces.

7. Next, the **Editing Wired connection 1** window will appear; select the **IPv6 Settings** tab, change **Method** to **Disabled**, and click on **Save**, as shown in the following screenshot:

Figure 2.22 – Disabling IPv6

8. Next, close both the **Network Connections** and **Settings** windows.
9. Lastly, restart Kali Linux for the changes to take effect. The power options are located in the top-right corner of the Kali Linux user interface.

Part 3 – changing the password and testing connectivity

To get started with changing the default password and testing network connectivity, please use the following steps:

1. Firstly, log in to Kali Linux by using the default user credentials (`kali/kali`).
2. Next, click on the Kali Linux icon in the top-left corner to expand the Kali Linux menu and select **Terminal Emulator**, as shown in the following screenshot:

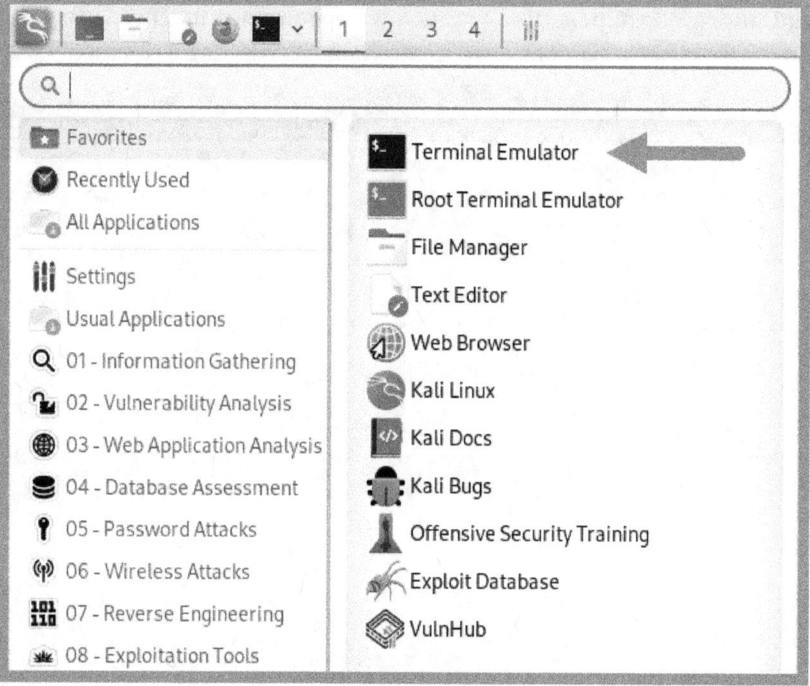

Figure 2.23 – Locating the Terminal

3. Next, use the following command to change the default password for the `kali` user account:

```
kali@kali:~$ passwd
```

When you're entering passwords on a Linux Terminal interface, it's invisible for security reasons, as shown in the following screenshot:

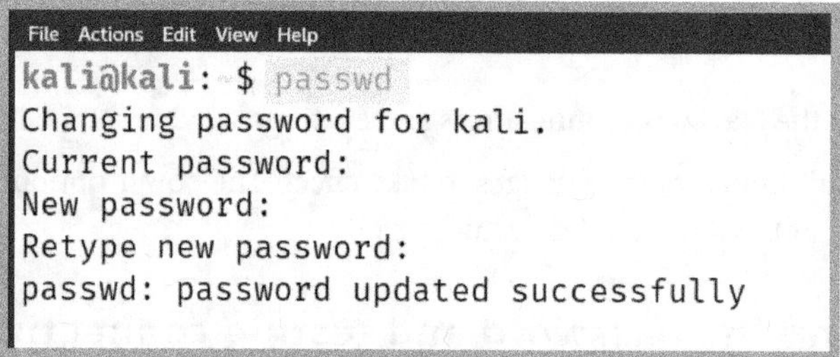

Figure 2.24 – Changing passwords

4. Next, use the following commands to display the IP addresses, **Media Access Control** (**MAC**) addresses, and status of the network adapters on Kali Linux:

```
kali@kali:~$ ip address
```

The following snippet shows the two network adapters on Kali Linux – `eth0`, which is connected to the internet, and `eth1`, which is connected to the *PentestNet* network on the `172.30.1.0/24` subnet:

```
File Actions Edit View Help
kali@kali: $ ip address
1:      <LOOPBACK,UP,LOWER_UP> mtu 65536 qdisc noqueue state UNKNOWN group default qlen 1000
    link/loopback                   brd
    inet 127.0.0.1/8 scope host lo
       valid_lft forever preferred_lft forever
    inet6 ::1/128 scope host
       valid_lft forever preferred_lft forever
2:        <BROADCAST,MULTICAST,UP,LOWER_UP> mtu 1500 qdisc fq_codel state    group default qlen 1000
    link/ether                   brd
    inet 172.16.17.35/24 brd 172.16.17.255 scope global dynamic noprefixroute eth0
       valid_lft 86378sec preferred_lft 86378sec
3:        <BROADCAST,MULTICAST,UP,LOWER_UP> mtu 1500 qdisc fq_codel state    group default qlen 1000
    link/ether                   brd
    inet 172.30.1.42/24 brd 172.30.1.255 scope global dynamic noprefixroute eth1
       valid_lft 579sec preferred_lft 579sec
```

Figure 2.25 – Checking network adapters

The `eth0` network adapter should have an IP address from your personal network, while `eth1` is assigned an IP address from the *PentestNet* virtual network.

5. Next, to test DNS resolution and internet connectivity on Kali Linux, use the following command to send four **Internet Control Message Protocol** (**ICMP**) **echo request** messages to Google's web server:

```
kali@kali:~$ ping www.google.com -c 4
```

The following snippet shows that Kali Linux was able to resolve the hostname to an IP address and successfully reach Google's web server on the internet:

```
File Actions Edit View Help

kali@kali: $ ping www.google.com -c 4
PING forcesafesearch.google.com (216.239.38.120) 56(84) bytes of data.
64 bytes from any-in-2678.1e100.net (216.239.38.120): icmp_seq=1 ttl=111 time=50.9 ms
64 bytes from any-in-2678.1e100.net (216.239.38.120): icmp_seq=2 ttl=111 time=50.5 ms
64 bytes from any-in-2678.1e100.net (216.239.38.120): icmp_seq=3 ttl=111 time=50.7 ms
64 bytes from any-in-2678.1e100.net (216.239.38.120): icmp_seq=4 ttl=111 time=50.5 ms

— forcesafesearch.google.com ping statistics —
4 packets transmitted, 4 received, 0% packet loss, time 3043ms
rtt min/avg/max/mdev = 50.463/50.654/50.922/0.183 ms
```

Figure 2.26 – Testing internet connectivity

6. Next, use the following commands to update the package repository list on Kali Linux:

```
kali@kali:~$ sudo apt update
```

The following snippet shows the execution of the preceding commands:

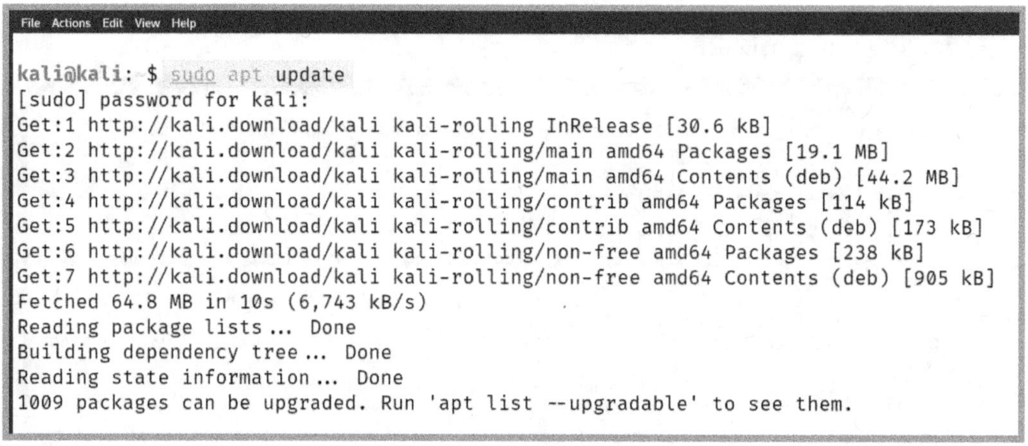

```
File  Actions  Edit  View  Help

kali@kali: $ sudo apt update
[sudo] password for kali:
Get:1 http://kali.download/kali kali-rolling InRelease [30.6 kB]
Get:2 http://kali.download/kali kali-rolling/main amd64 Packages [19.1 MB]
Get:3 http://kali.download/kali kali-rolling/main amd64 Contents (deb) [44.2 MB]
Get:4 http://kali.download/kali kali-rolling/contrib amd64 Packages [114 kB]
Get:5 http://kali.download/kali kali-rolling/contrib amd64 Contents (deb) [173 kB]
Get:6 http://kali.download/kali kali-rolling/non-free amd64 Packages [238 kB]
Get:7 http://kali.download/kali kali-rolling/non-free amd64 Contents (deb) [905 kB]
Fetched 64.8 MB in 10s (6,743 kB/s)
Reading package lists ... Done
Building dependency tree ... Done
Reading state information ... Done
1009 packages can be upgraded. Run 'apt list --upgradable' to see them.
```

Figure 2.27 – Updating package repository

Updating the package repository list on Kali Linux ensures it will download both the latest software packages and updates from the official sources.

Having completed this section, you have learned how to deploy and set up Kali Linux as a virtual machine. In the next section, you will learn how to set up a dedicated virtual machine for collecting **Open Source Intelligence** (OSINT).

Deploying an OSINT virtual machine

The Trace Labs OSINT VM is a customized version of Kali Linux that's designed for learning and performing reconnaissance. It has all the essential pre-installed software-based tools an ethical hacker will need to collect information about their targets. In addition, the Trace Labs OSINT VM is free for everyone to download and set up as a virtual machine on a hypervisor.

Part 1 – setting up OSINT VM

To get started setting up Trace Lab OSINT VM, please use the following steps:

1. Firstly, go to `https://www.tracelabs.org/initiatives/osint-vm` and click the **Download OVA** button, as shown in the following screenshot:

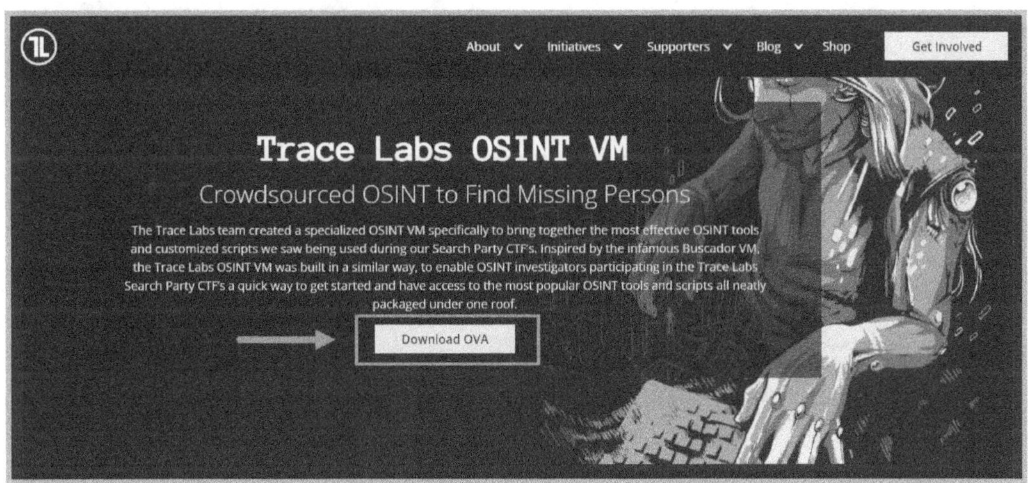

Figure 2.28 – Trace Labs download page

2. Next, select the **TL OSINT VM 2022.1 OVA** release file to download:

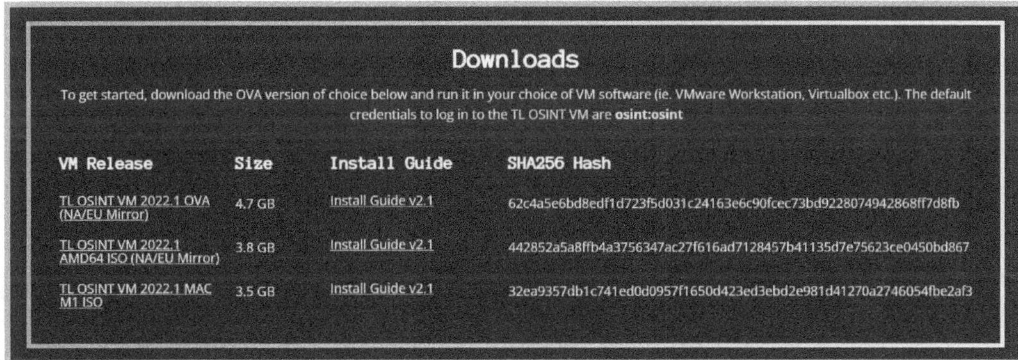

VM Release	Size	Install Guide	SHA256 Hash
TL OSINT VM 2022.1 OVA (NA/EU Mirror)	4.7 GB	Install Guide v2.1	62c4a5e6bd8edf1d723f5d031c24163e6c90fcec73bd9228074942868ff7d8fb
TL OSINT VM 2022.1 AMD64 ISO (NA/EU Mirror)	3.8 GB	Install Guide v2.1	442852a5a8ffb4a3756347ac27f616ad7128457b41135d7e75623ce0450bd867
TL OSINT VM 2022.1 MAC M1 ISO	3.5 GB	Install Guide v2.1	32ea9357db1c741ed0d0957f1650d423ed3ebd2e981d41270a2746054fbe2af3

Figure 2.29 – OSINT VM OVA file

3. After the OVA file has been downloaded onto your computer, open **VirtualBox Manager**, select **Tools**, and click on **Import**, as shown in the following screenshot:

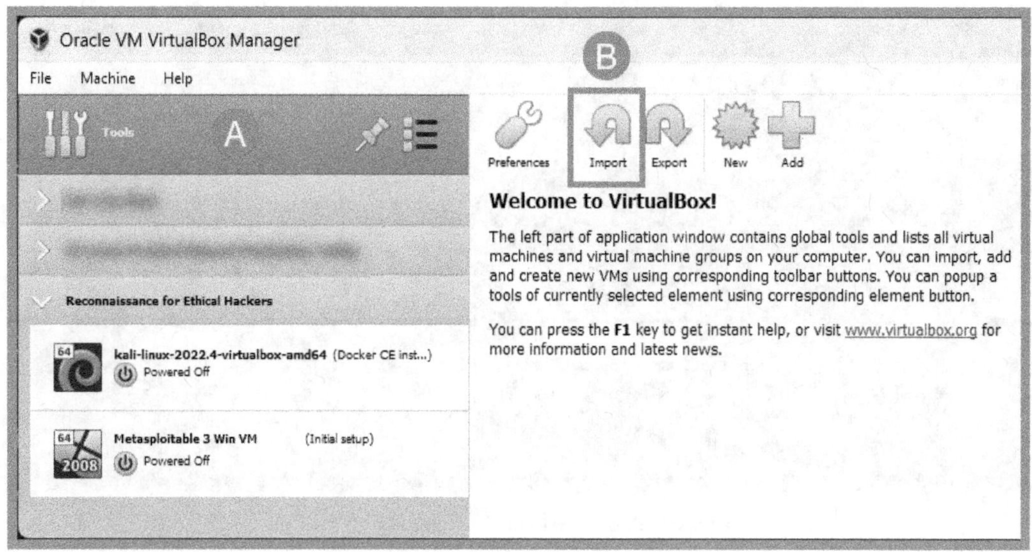

Figure 2.30 – VirtualBox Manager

4. Next, a new window will appear; simply navigate to the location where the `OSINT VM OVA` file is stored, select it, and click on **Open**, as shown in the following screenshot:

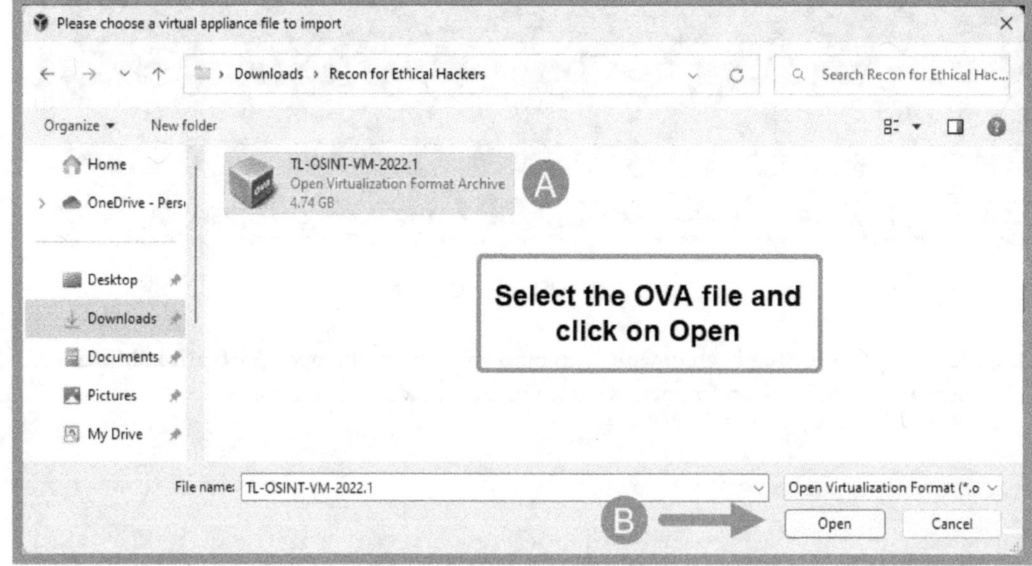

Figure 2.31 – OSINT VM OVA file

5. Next, the **Import Virtual Appliance** window will appear; click on **Next**:

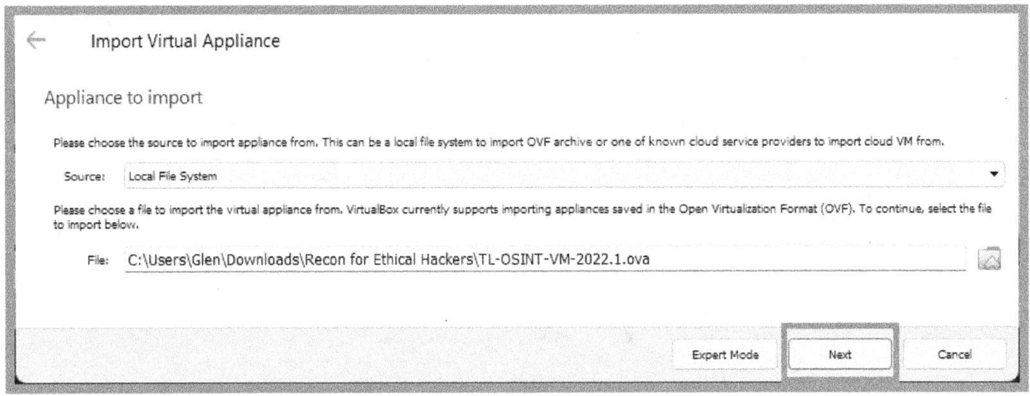

Figure 2.32 – Importing the virtual appliance

6. Next, the configurations window will appear; click on **Import**:

Import Virtual Appliance

Appliance settings

These are the virtual machines contained in the appliance and the suggested settings of the imported VirtualBox machines. You can change many of the properties shown by double-clicking on the items and disable others using the check boxes below.

Virtual System 1

Name	TL OSINT VM 2022.1
Guest OS Type	Oracle (64-bit)
CPU	2
RAM	2048 MB
DVD	☑
USB Controller	☑
Sound Card	☑ ICH AC97
Network Adapter	☑ Intel PRO/1000 MT Desktop (82540EM)
Storage Controller (IDE)	PIIX4
Storage Controller (IDE)	PIIX4
Storage Controller (SATA)	AHCI
Virtual Disk Image	TL OSINT VM 2022.1-disk001.vmdk
Base Folder	D:\VirtualBox VM
Primary Group	/

Machine Base Folder: D:\VirtualBox VM

MAC Address Policy: Generate new MAC addresses for all network adapters

Additional Options: ☑ Import hard drives as VDI

Appliance is not signed

Restore Defaults | Import | Cancel

Figure 2.33 – Appliance settings window

Once the import process begins, the following status window will appear:

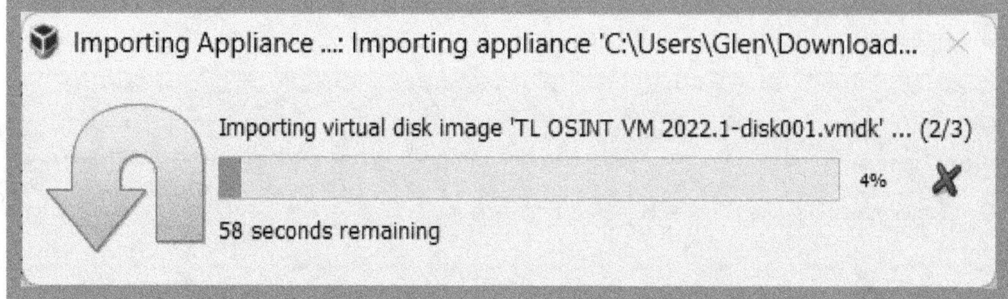

Figure 2.34 – Importing status window

After the import process is completed, the OSINT VM will appear within VirtualBox Manager.

7. Next, let's connect the OSINT VM to the physical network by bridging the vNIC of the virtual machine to the physical NIC of your computer. Select the OSINT VM and click on **Settings**:

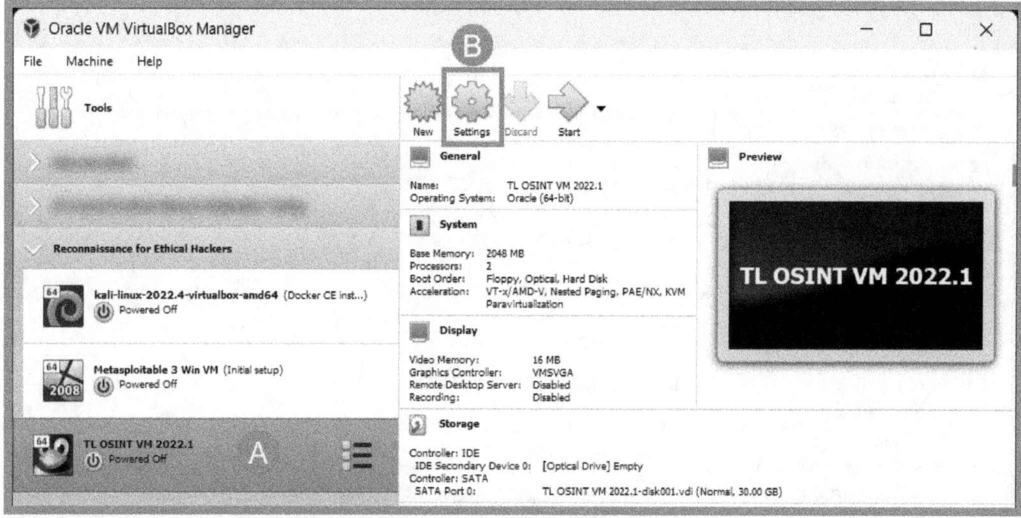

Figure 2.35 – VirtualBox Manager

> **Tip**
>
> Within the OSINT VM settings, you can increase the allocated memory to the virtual machine. Within the virtual machine, go to **Settings** > **System** > **Motherboard** > **Base Memory**. If you have additional memory on your host computer, consider increasing the memory allocation to 4,096 MB (4GB) of RAM.

8. Next, click on **Network** and use the following settings for **Adapter 1**:

   - **Enable Network Adapter**: Check the box to enable
   - **Attached to**: **Bridged Adapter**
   - **Name**: Select the physical network adapter that is connected to the internet
   - **Cable Connected**: Yes – check the box

   The following snippet shows the configurations:

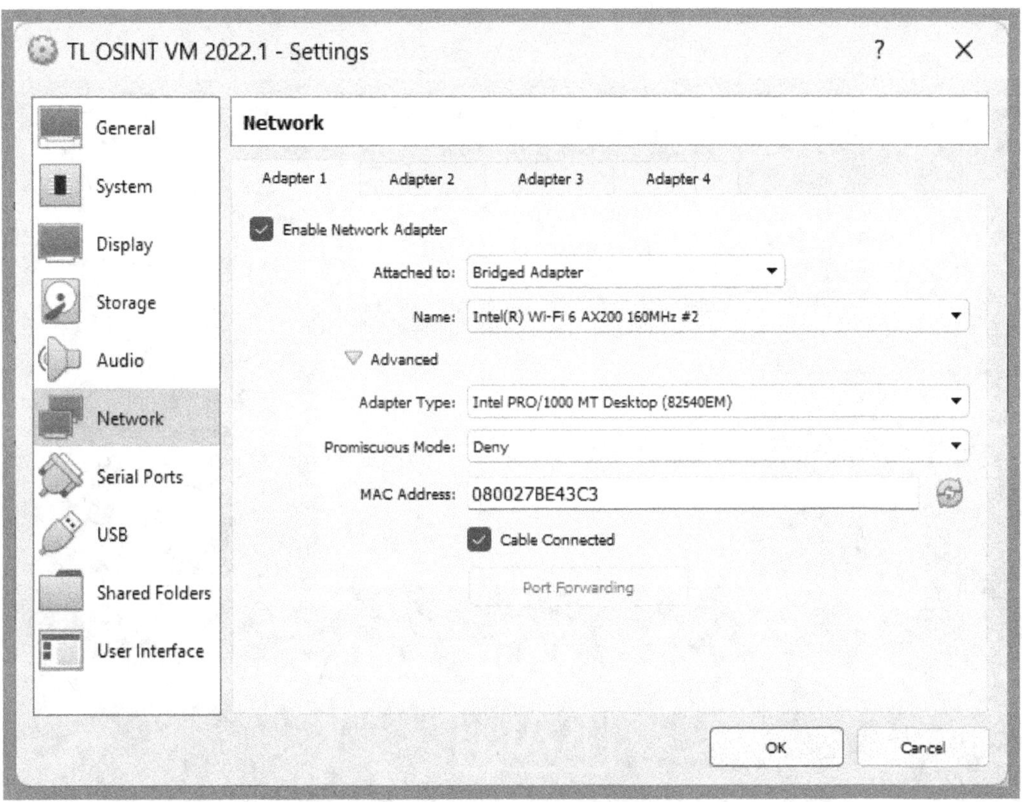

Figure 2.36 – Network adapter configurations

9. Next, click on **OK** to save the configurations for the OSINT VM.

Part 2 – getting started and testing connectivity

To get started with OSINT VM, please use the following steps:

1. Open **VirtualBox Manager**, select the TL OSINT VM, and click on **Start** to power on the virtual machine.
2. Once the OSINT VM is powered on, log in using `osint` for both **Username** and **Password**, as shown in the following screenshot:

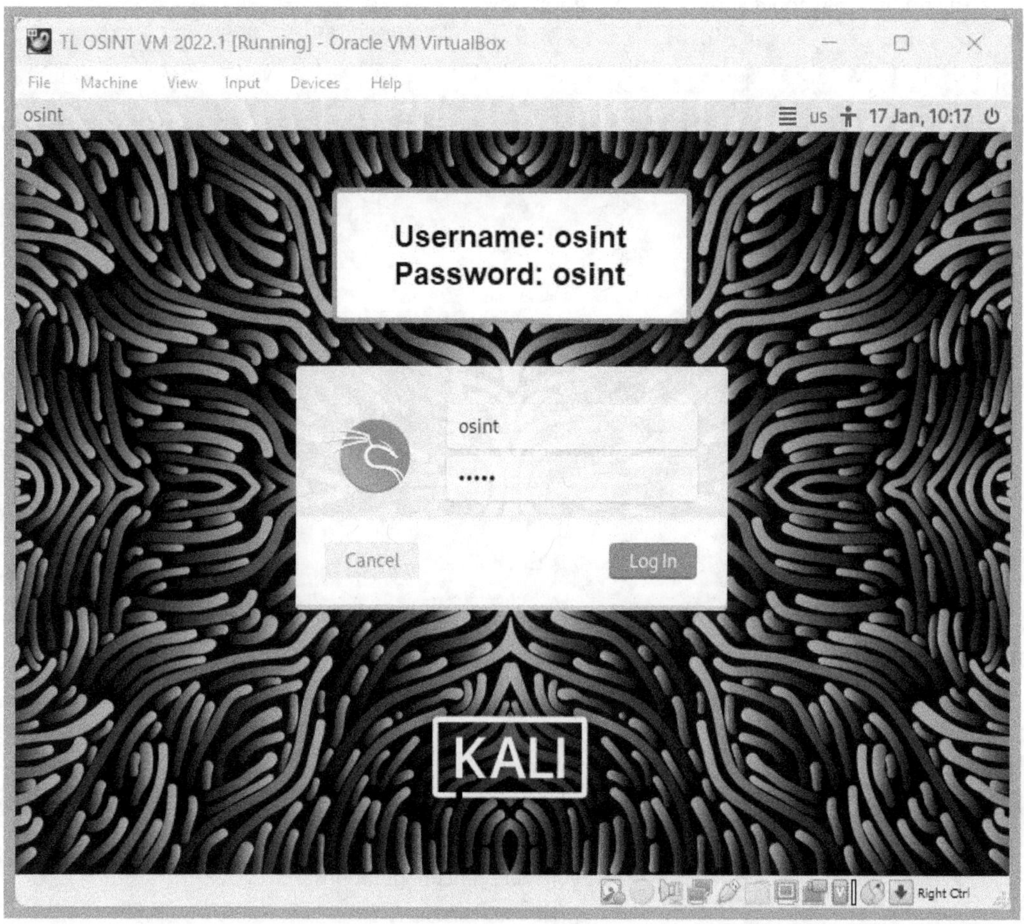

Figure 2.37 – OSINT VM

3. Once you're logged in, use the steps in the previous section to disable IPv6 on the OSINT VM and restart for the changes to take effect.
4. (Optional) To change the default password of the `osint` user account, open the Terminal and use the `passwd` command.

5. Next, open the Terminal and use the following commands to identify the network adapters and their IP addresses, and test connectivity to the internet:

```
osint@osint:~$ ip address
osint@osint:~$ ping www.google.com -c 4
```

The following snippet shows the expected results:

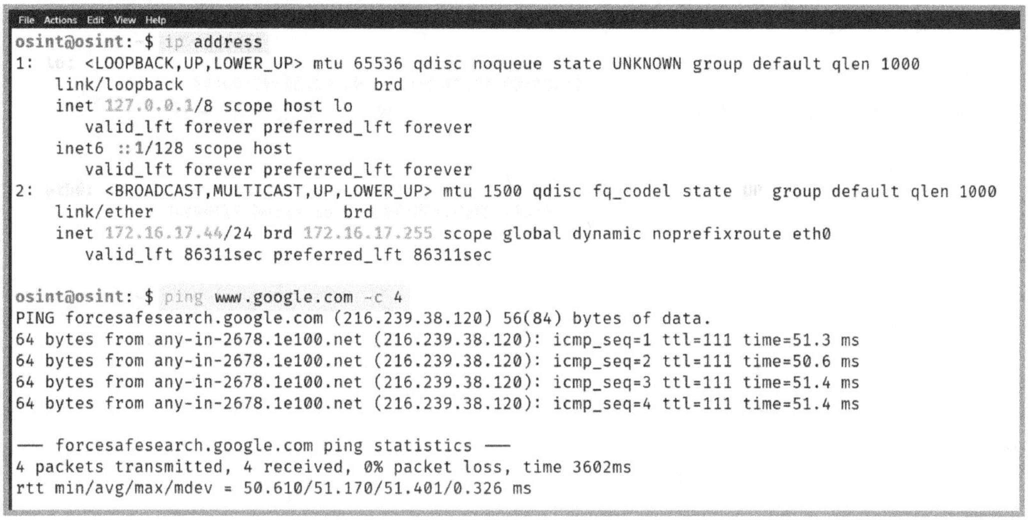

```
File Actions Edit View Help
osint@osint: $ ip address
1: lo: <LOOPBACK,UP,LOWER_UP> mtu 65536 qdisc noqueue state UNKNOWN group default qlen 1000
    link/loopback                     brd
    inet 127.0.0.1/8 scope host lo
       valid_lft forever preferred_lft forever
    inet6 ::1/128 scope host
       valid_lft forever preferred_lft forever
2: eth0: <BROADCAST,MULTICAST,UP,LOWER_UP> mtu 1500 qdisc fq_codel state UP group default qlen 1000
    link/ether                        brd
    inet 172.16.17.44/24 brd 172.16.17.255 scope global dynamic noprefixroute eth0
       valid_lft 86311sec preferred_lft 86311sec

osint@osint: $ ping www.google.com -c 4
PING forcesafesearch.google.com (216.239.38.120) 56(84) bytes of data.
64 bytes from any-in-2678.1e100.net (216.239.38.120): icmp_seq=1 ttl=111 time=51.3 ms
64 bytes from any-in-2678.1e100.net (216.239.38.120): icmp_seq=2 ttl=111 time=50.6 ms
64 bytes from any-in-2678.1e100.net (216.239.38.120): icmp_seq=3 ttl=111 time=51.4 ms
64 bytes from any-in-2678.1e100.net (216.239.38.120): icmp_seq=4 ttl=111 time=51.4 ms

— forcesafesearch.google.com ping statistics —
4 packets transmitted, 4 received, 0% packet loss, time 3602ms
rtt min/avg/max/mdev = 50.610/51.170/51.401/0.326 ms
```

Figure 2.38 – OSINT VM connectivity test

As shown in the preceding snippet, the `eth0` network adapter has an IP address from the DHCP server on the physical network. This IP address will be different based on your network settings. Additionally, the OSINT VM was able to successfully perform DNS resolution and was able to connect to Google's web server on the internet.

> **Tip**
>
> Executing some commands on Kali Linux and OSINT VM requires the usage of `sudo` followed by the commands to invoke root privileges, such as `sudo ip address`. To use the root account, use the `sudo su -` command on the Terminal.

Having completed this section, you have deployed the OSINT VM as a virtual machine. In the next section, you will learn how to set up a vulnerable web application and server within the lab environment.

Implementing vulnerable systems

In this section, you'll learn how to set up the OWASP Juice Shop vulnerable application on Kali Linux to safely learn how to perform active reconnaissance on web applications. Additionally, you will learn how to set up Metasploitable 3 Windows VM as a virtual machine within our lab environment.

Setting up a vulnerable web application

To get started with setting up OWASP Juice Shop on Kali Linux, please use the following steps:

1. Firstly, power on the Kali Linux virtual machine within **VirtualBox Manager**.
2. Open the Terminal and use the following command to update the package repository list:

```
kali@kali:~$ sudo apt update
```

The following snippet shows the successful execution of the command:

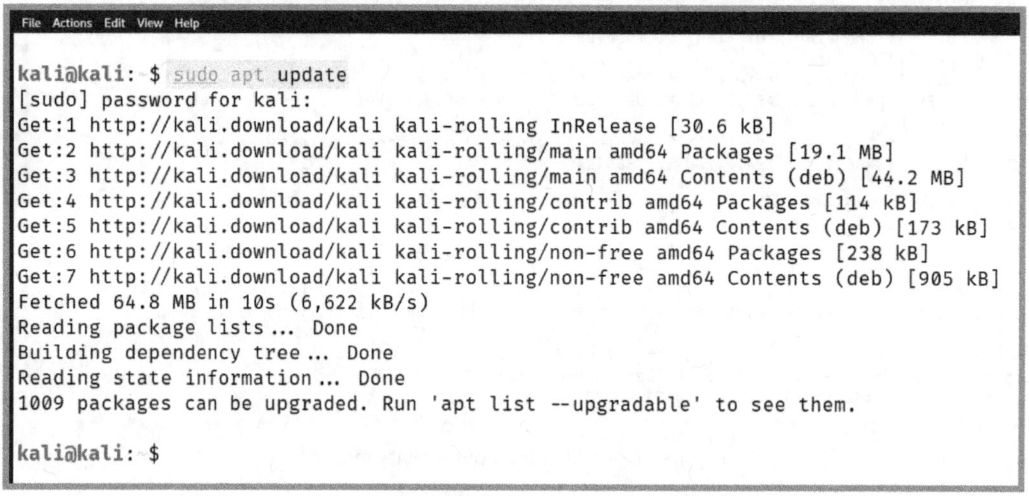

```
File Actions Edit View Help

kali@kali: $ sudo apt update
[sudo] password for kali:
Get:1 http://kali.download/kali kali-rolling InRelease [30.6 kB]
Get:2 http://kali.download/kali kali-rolling/main amd64 Packages [19.1 MB]
Get:3 http://kali.download/kali kali-rolling/main amd64 Contents (deb) [44.2 MB]
Get:4 http://kali.download/kali kali-rolling/contrib amd64 Packages [114 kB]
Get:5 http://kali.download/kali kali-rolling/contrib amd64 Contents (deb) [173 kB]
Get:6 http://kali.download/kali kali-rolling/non-free amd64 Packages [238 kB]
Get:7 http://kali.download/kali kali-rolling/non-free amd64 Contents (deb) [905 kB]
Fetched 64.8 MB in 10s (6,622 kB/s)
Reading package lists... Done
Building dependency tree... Done
Reading state information... Done
1009 packages can be upgraded. Run 'apt list --upgradable' to see them.

kali@kali: $
```

Figure 2.39 – Updating package list

3. Next, use the following command to install the Docker repository on Kali Linux:

```
kali@kali:~$ printf '%s\n' "deb https://download.docker.com/
linux/debian bullseye stable" |
  sudo tee /etc/apt/sources.list.d/docker-ce.list
```

The following snippet shows the preceding command when executed on the Terminal:

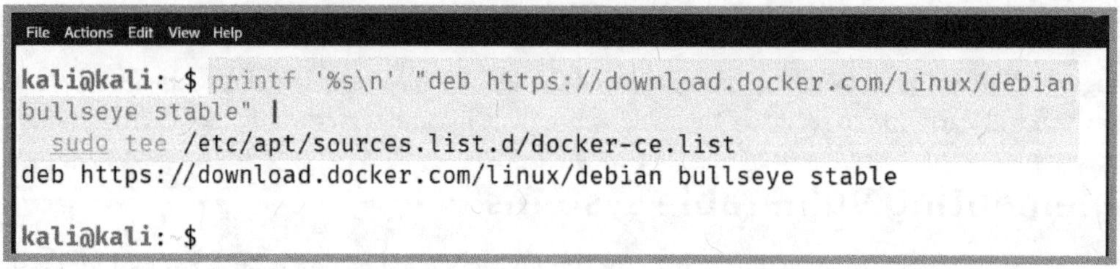

```
File Actions Edit View Help

kali@kali: $ printf '%s\n' "deb https://download.docker.com/linux/debian
bullseye stable" |
  sudo tee /etc/apt/sources.list.d/docker-ce.list
deb https://download.docker.com/linux/debian bullseye stable

kali@kali: $
```

Figure 2.40 – Adding the Docker repository

4. Next, use the following commands to import the **GNU Privacy Guard** (**GPG**) key on Kali Linux:

```
kali@kali:~$ curl -fsSL https://download.docker.com/linux/
debian/gpg |
  sudo gpg --dearmor -o /etc/apt/trusted.gpg.d/docker-ce-
archive-keyring.gpg
```

The following snippet shows the execution of the preceding commands:

```
File Actions Edit View Help

kali@kali:~$ curl -fsSL https://download.docker.com/linux/debian/gpg |
  sudo gpg --dearmor -o /etc/apt/trusted.gpg.d/docker-ce-archive-keyring.
gpg

kali@kali:~$
```

Figure 2.41 – Importing the GPG key

5. Next, re-update the package repository list on Kali Linux:

```
kali@kali:~$ sudo apt update
```

The following snippet shows the package list was updated again:

```
File Actions Edit View Help

kali@kali:~$ sudo apt update
Get:2 https://download.docker.com/linux/debian bullseye InRelease [43.3 k
B]
Get:3 https://download.docker.com/linux/debian bullseye/stable amd64 Pack
ages [16.6 kB]
Get:4 https://download.docker.com/linux/debian bullseye/stable amd64 Cont
ents (deb) [1,322 B]
Hit:1 http://kali.download/kali kali-rolling InRelease
Fetched 61.3 kB in 1s (66.4 kB/s)
Reading package lists... Done
Building dependency tree... Done
Reading state information... Done
1009 packages can be upgraded. Run 'apt list --upgradable' to see them.
```

Figure 2.42 – Package list updated

6. Next, use the following commands to install `docker-ce`:

```
kali@kali:~$ sudo apt install -y docker-ce docker-ce-cli
containerd.io
```

The following snippet shows the installation process:

```
File Actions Edit View Help

kali@kali:~$ sudo apt install -y docker-ce docker-ce-cli containerd.io
Reading package lists... Done
Building dependency tree... Done
Reading state information... Done
The following additional packages will be installed:
  docker-ce-rootless-extras docker-scan-plugin libslirp0 pigz
  slirp4netns
Suggested packages:
  aufs-tools cgroupfs-mount | cgroup-lite
The following NEW packages will be installed:
  containerd.io docker-ce docker-ce-cli docker-ce-rootless-extras
  docker-scan-plugin libslirp0 pigz slirp4netns
```

Figure 2.43 – docker-ce installation

7. Next, use the following commands to pull the OWASP Juice Shop Docker container and install it on Kali Linux:

```
kali@kali:~$ sudo docker pull bkimminich/juice-shop
```

The following snippet shows the download and setup process of OWASP Juice Shop:

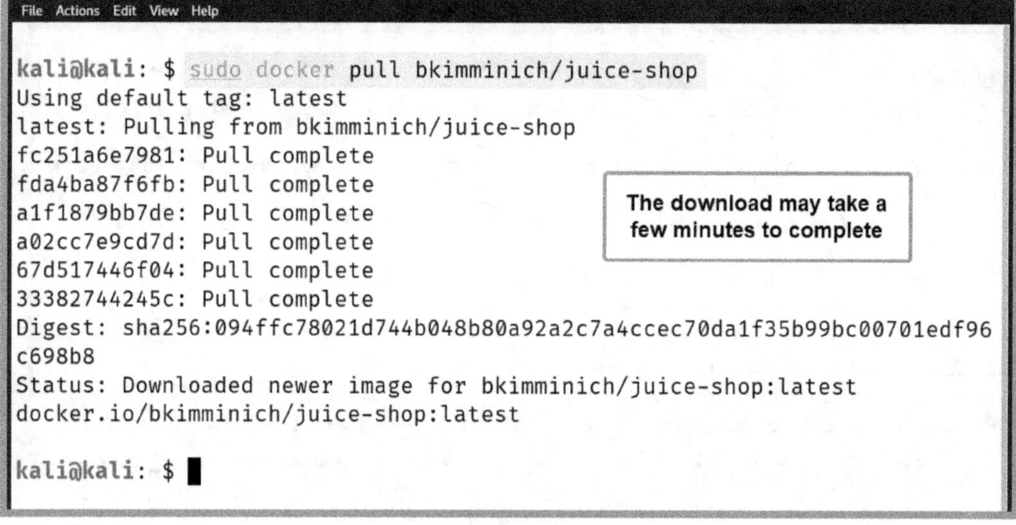

```
File Actions Edit View Help

kali@kali:~$ sudo docker pull bkimminich/juice-shop
Using default tag: latest
latest: Pulling from bkimminich/juice-shop
fc251a6e7981: Pull complete
fda4ba87f6fb: Pull complete
a1f1879bb7de: Pull complete
a02cc7e9cd7d: Pull complete
67d517446f04: Pull complete
33382744245c: Pull complete
Digest: sha256:094ffc78021d744b048b80a92a2c7a4ccec70da1f35b99bc00701edf96
c698b8
Status: Downloaded newer image for bkimminich/juice-shop:latest
docker.io/bkimminich/juice-shop:latest

kali@kali:~$
```

The download may take a few minutes to complete

Figure 2.44 – Setting up OWASP Juice Shop

8. Next, to run the OWASP Juice Shop Docker container, use the following commands:

```
kali@kali:~$ sudo docker run --rm -p 3000:3000 bkimminich/juice-
shop
```

The following snippet shows the execution of the preceding commands:

```
File Actions Edit View Help

kali@kali: $ sudo docker run --rm -p 3000:3000 bkimminich/juice-shop
info: All dependencies in ./package.json are satisfied (OK)
info: Chatbot training data botDefaultTrainingData.json validated (OK)
info: Detected Node.js version v18.12.1 (OK)
info: Detected OS linux (OK)
info: Detected CPU x64 (OK)
info: Configuration default validated (OK)
info: Entity models 19 of 19 are initialized (OK)
info: Required file server.js is present (OK)
info: Required file index.html is present (OK)
info: Required file styles.css is present (OK)
info: Required file main.js is present (OK)
info: Required file tutorial.js is present (OK)
info: Required file polyfills.js is present (OK)
info: Required file runtime.js is present (OK)
info: Required file vendor.js is present (OK)
info: Port 3000 is available (OK)
info: Server listening on port 3000
```

Figure 2.45 – Running OWASP Juice Shop

9. Lastly, open the web browser on Kali Linux and go to `http://localhost:3000/` to view the OWASP Juice Shop web application, as shown in the following screenshot:

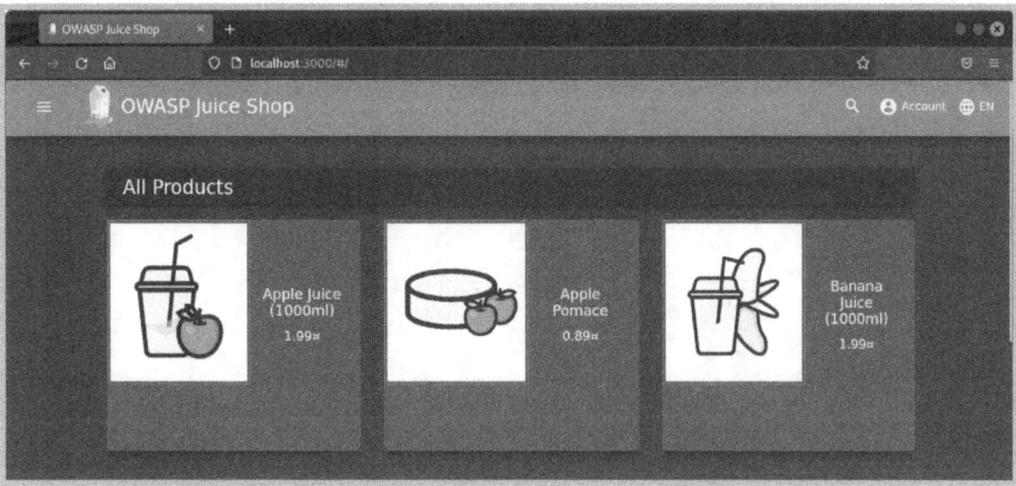

Figure 2.46 – OWASP Juice Shop web interface

> **Important note**
>
> The official instructions for setting up Docker on Kali Linux can be found at: `https://www.kali.org/docs/containers/installing-docker-on-kali/`.

Having completed this exercise, you have learned how to set up Docker and OWASP Juice Shop on Kali Linux. Next, you will learn how to set up Metasploitable 3 Windows VM in our lab environment.

Setting up a vulnerable machine

To get started on setting up Metasploitable 3 Windows VM, please use the following steps:

1. Firstly, you will need to download and install Vagrant, an open source application that enables a user to build and maintain virtual machines and applications. On your host computer, open the web browser and go to `https://developer.hashicorp.com/vagrant/downloads`, then download the **Vagrant 2.3.3 AMD64** version on your computer:

Figure 2.47 – Vagrant download page

2. After downloading Vagrant, double-click on the installer package to install the application on your host computer. After the installation is complete, you'll be prompted to restart your computer for the changes to take effect.

3. After restarting your computer, open the Windows Command Prompt and use the following commands to reload and install additional Vagrant plugins:

```
C:\Users\Glen> vagrant plugin install vagrant-reload
C:\Users\Glen> vagrant plugin install vagrant-vbguest
```

The following snippet shows the results of running the preceding commands:

```
C:\Users\Glen> vagrant plugin install vagrant-reload
==> vagrant: A new version of Vagrant is available: 2.3.4 (installed version: 2.3.3)!
==> vagrant: To upgrade visit: https://www.vagrantup.com/downloads.html

Installing the 'vagrant-reload' plugin. This can take a few minutes...
Installed the plugin 'vagrant-reload (0.0.1)'!

C:\Users\Glen> vagrant plugin install vagrant-vbguest
Installing the 'vagrant-vbguest' plugin. This can take a few minutes...
Installed the plugin 'vagrant-vbguest (0.31.0)'!

C:\Users\Glen>
```

Figure 2.48 – Installing Vagrant plugins

4. Next, use the following commands to download the Metasploitable 3 Windows VM files from the Vagrant repository:

```
C:\Users\Glen> vagrant box add rapid7/metasploitable3-win2k8
```

Ensure you enter `1` to specify VirtualBox as the provider, as shown in the following screenshot:

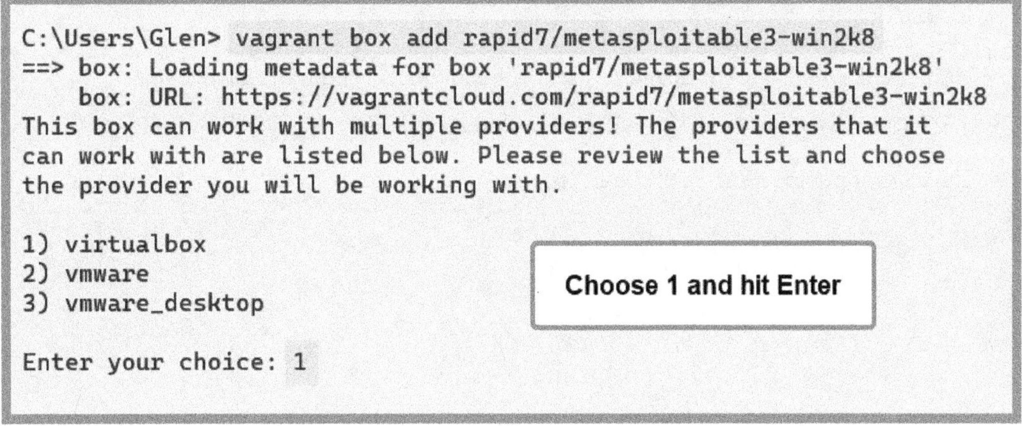
C:\Users\Glen> vagrant box add rapid7/metasploitable3-win2k8
==> box: Loading metadata for box 'rapid7/metasploitable3-win2k8'
box: URL: https://vagrantcloud.com/rapid7/metasploitable3-win2k8
This box can work with multiple providers! The providers that it
can work with are listed below. Please review the list and choose
the provider you will be working with.
1) virtualbox
2) vmware
3) vmware_desktop
Enter your choice: 1
Choose 1 and hit Enter

Figure 2.49 – Selecting the provider type

Vagrant will begin to download the Metasploitable 3 Windows VM files, as shown in the following screenshot:

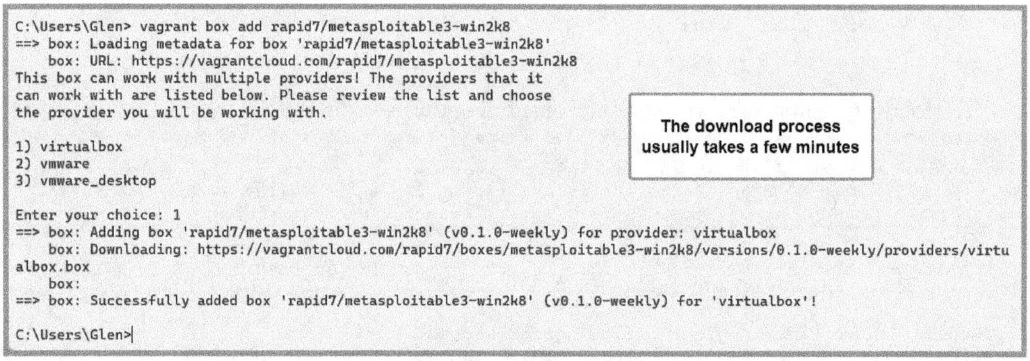

```
C:\Users\Glen> vagrant box add rapid7/metasploitable3-win2k8
==> box: Loading metadata for box 'rapid7/metasploitable3-win2k8'
    box: URL: https://vagrantcloud.com/rapid7/metasploitable3-win2k8
This box can work with multiple providers! The providers that it
can work with are listed below. Please review the list and choose
the provider you will be working with.

1) virtualbox
2) vmware
3) vmware_desktop

Enter your choice: 1
==> box: Adding box 'rapid7/metasploitable3-win2k8' (v0.1.0-weekly) for provider: virtualbox
    box: Downloading: https://vagrantcloud.com/rapid7/boxes/metasploitable3-win2k8/versions/0.1.0-weekly/providers/virtu
albox.box
    box:
==> box: Successfully added box 'rapid7/metasploitable3-win2k8' (v0.1.0-weekly) for 'virtualbox'!

C:\Users\Glen>
```

The download process usually takes a few minutes

Figure 2.50 – Download status

5. Next, open **Windows Explorer** and navigate to `C:\Users\yourusername\.vagrant.d\boxes`, where you will find the `rapid7-VAGRANTSLASH-metasploitable3-win2k8` folder. Rename the `rapid7-VAGRANTSLASH-metasploitable3-win2k8` folder to `metasploitable3-win2k8`.
6. Next, on the Windows Command Prompt, use the following commands to change the working directory and initialize the build configurations for the Metasploitable 3 Windows VM:

```
C:\Users\Glen> cd .vagrant.d\boxes
C:\Users\Glen\.vagrant.d\boxes> vagrant init metasploitable3-
win2k8
```

7. Next, use the following command to start the build process of the virtual machine:

```
C:\Users\Glen\.vagrant.d\boxes> vagrant up
```

The following snippet shows the build process:

```
C:\Users\Glen\.vagrant.d\boxes> vagrant up
Bringing machine 'default' up with 'virtualbox' provider...
==> default: Checking if box 'metasploitable3-win2k8' version '0.1.0-weekly' is up to date...
==> default: Clearing any previously set forwarded ports...
==> default: Clearing any previously set network interfaces...
==> default: Preparing network interfaces based on configuration...
    default: Adapter 1: nat
==> default: Forwarding ports...
    default: 3389 (guest) => 3389 (host) (adapter 1)
    default: 22 (guest) => 2222 (host) (adapter 1)
    default: 5985 (guest) => 55985 (host) (adapter 1)
    default: 5986 (guest) => 55986 (host) (adapter 1)
==> default: Running 'pre-boot' VM customizations...
==> default: Booting VM...
==> default: Waiting for machine to boot. This may take a few minutes...
    default: WinRM address: 127.0.0.1:55985
    default: WinRM username: vagrant
    default: WinRM execution_time_limit: PT2H
    default: WinRM transport: negotiate
==> default: Machine booted and ready!
[default] GuestAdditions versions on your host (6.1.40) and guest (6.0.8) do not match.
Copy iso file C:\Program Files\Oracle\VirtualBox\VBoxGuestAdditions.iso into the box $env:TEMP/VBoxGuestAdditions.iso
The term 'Mount-DiskImage' is not recognized as the name of a cmdlet, function, script file, or operable program. Check
the spelling of the name, or if a path was included, verify that the path is correct and try again.
At line:1 char:1
+ Mount-DiskImage -ImagePath $env:TEMP/VBoxGuestAdditions.iso
+ ~~~~~~~~~~~~~~~
    + CategoryInfo          : ObjectNotFound: (Mount-DiskImage:String) [], CommandNotFoundException
    + FullyQualifiedErrorId : CommandNotFoundException
```

Figure 2.51 – Vagrant build process

> **Important note**
>
> If the `vagrant up` command gives an error, execute it again.

This process usually takes a few minutes to complete.

8. Once the process is completed, open **VirtualBox Manager**, where you will find the Metasploitable 3 Win VM running in the background. Select the virtual machine and click on **Show**:

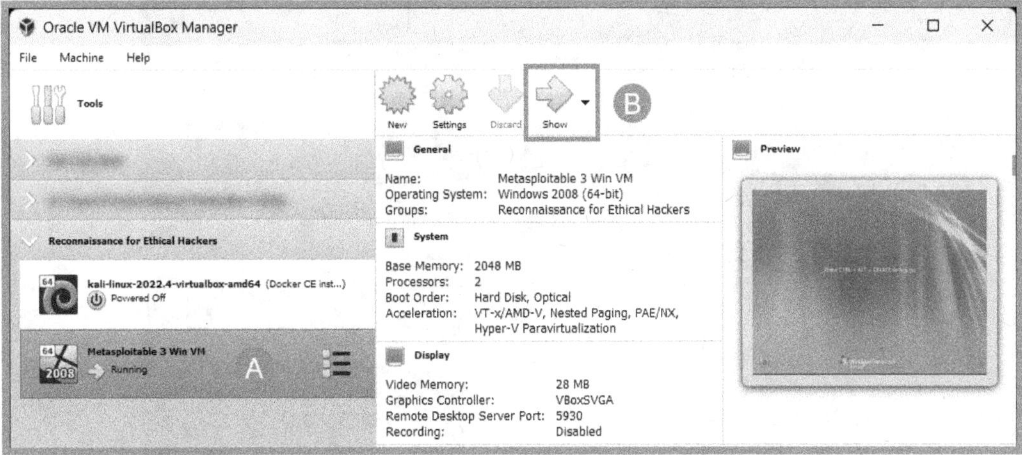

Figure 2.52 – Metasploitable 3 Windows virtual machine

9. Once the virtual machine window appears, click on **Input** > **Keyboard** > **Insert Ctrl-Alt-Del**, as shown in the following screenshot:

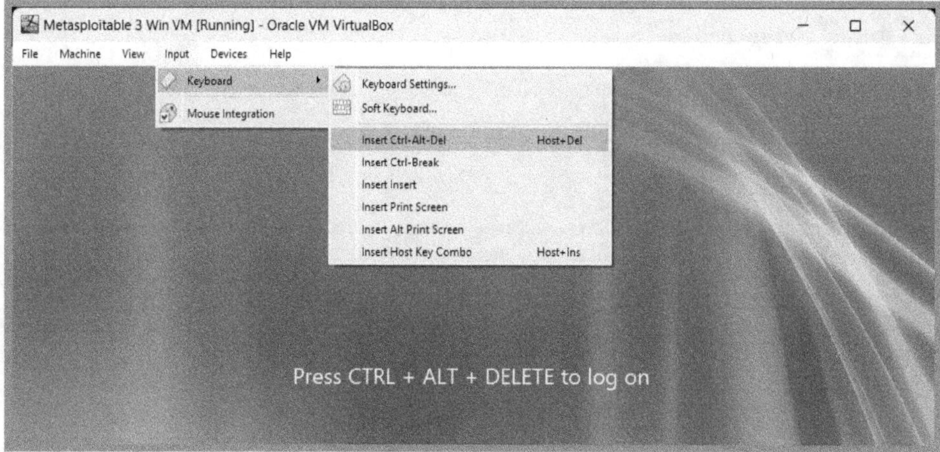

Figure 2.53 – Boot screen

10. Next, select the **Administrator** account and use the default password (`vagrant`) to log in, as shown in the following screenshot:

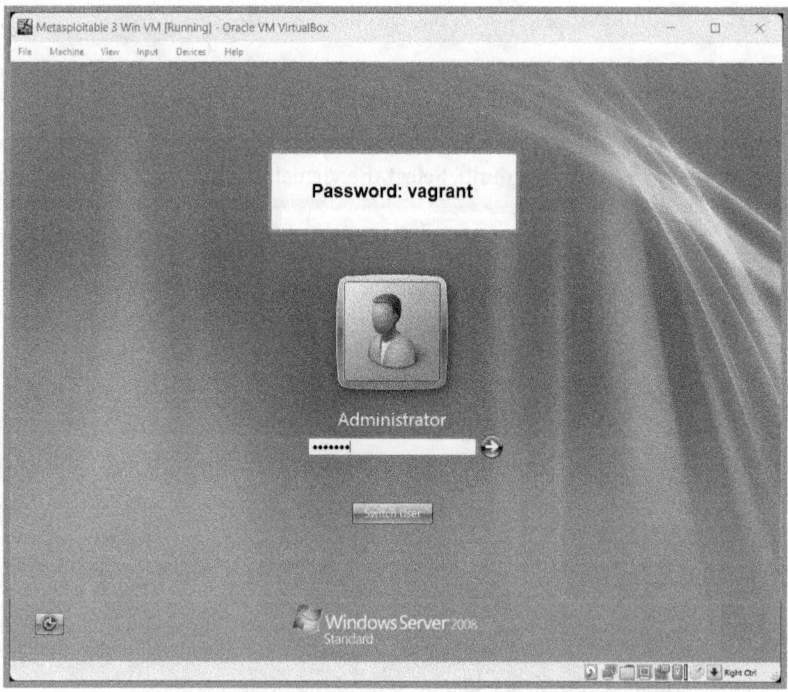

Figure 2.54 – Login window

Once you're logged in, simply close all the windows that appear and do not activate the operating system.

11. Next, click on the Windows icon in the bottom-left corner and shut down the virtual machine, as shown in the following screenshot:

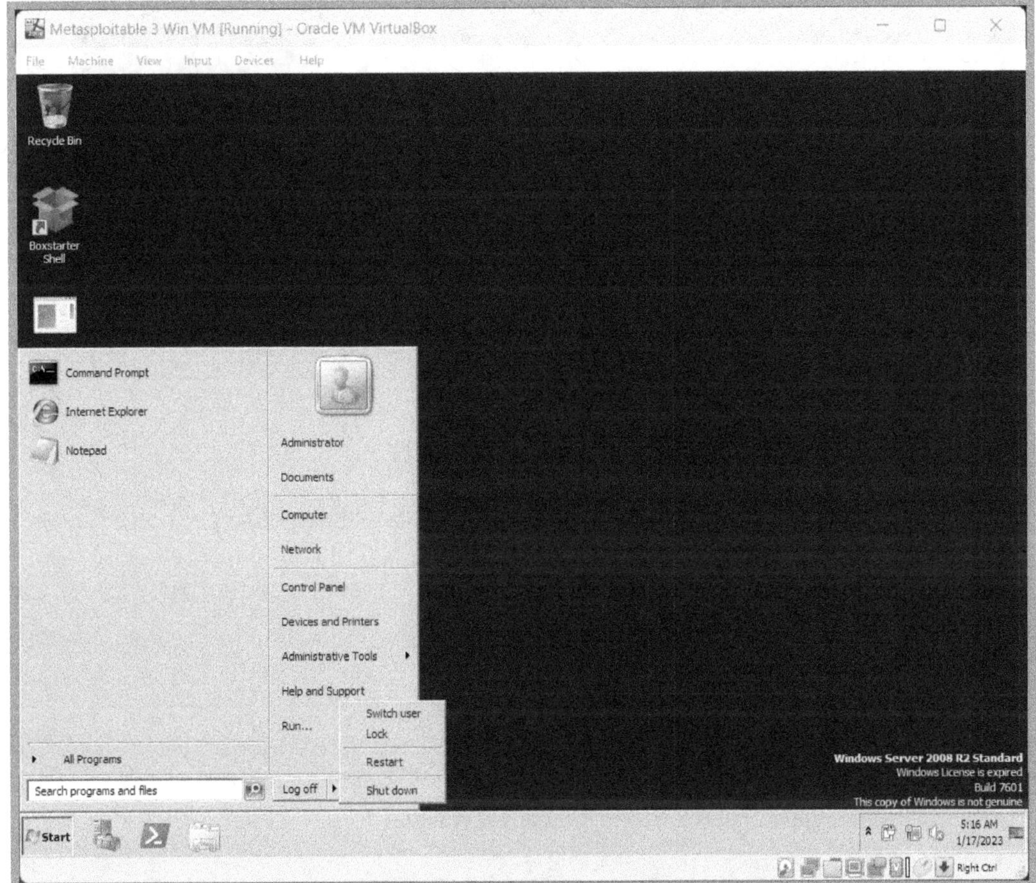

Figure 2.55 – Powering off the virtual machine

12. Once the virtual machine is powered off, select it and click on **Settings**:

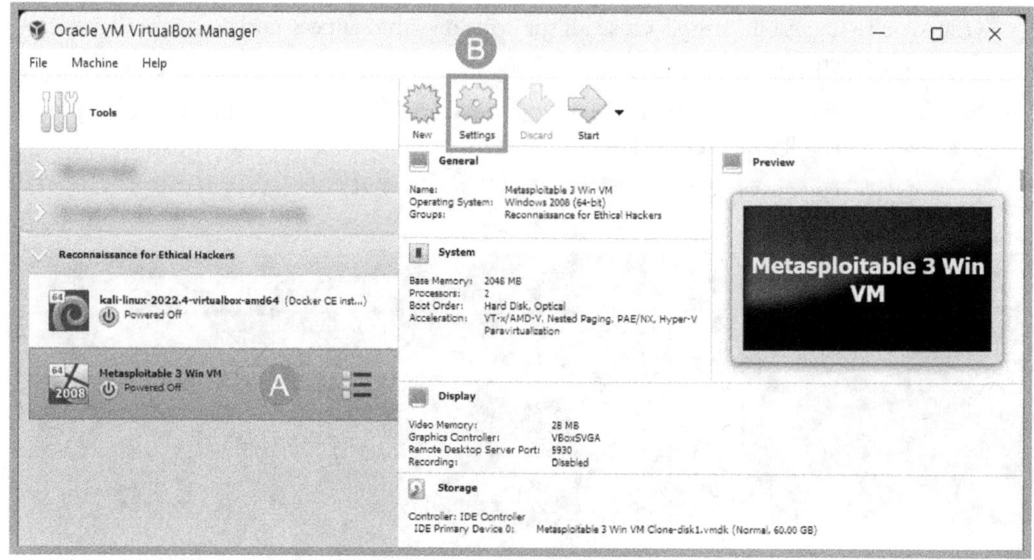

Figure 2.56 – VirtualBox Manager

13. Next, select **Network** and use the following configurations on **Adapter 1**:
    - **Enable Network Adapter**: Check the box to enable
    - **Attached to**: **Internal Network**
    - **Name**: Manually type `PentestNet` into the field
    - **Cable Connected**: Yes – check the box
    - **Promiscuous Mode**: **Allow All**

    The following snippet shows the configurations of **Adapter 1**:

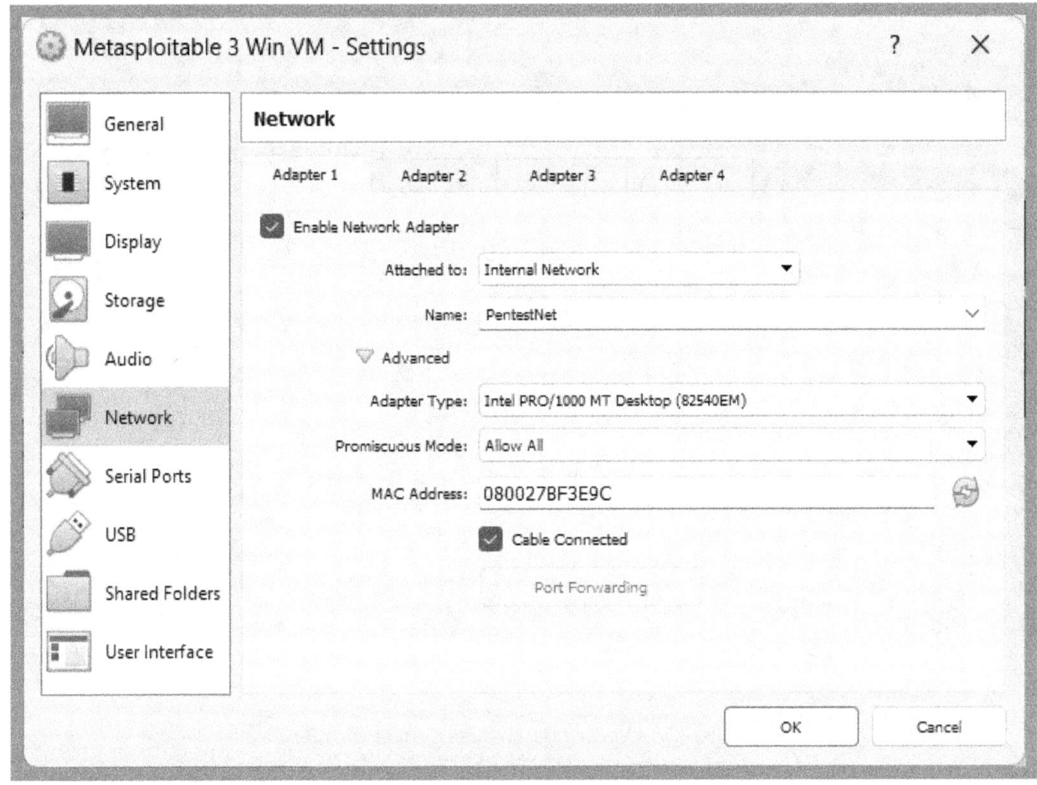

Figure 2.57 – Network adapter settings

14. Next, click on **OK** to save the configurations.
15. Lastly, power on and log in to the Metasploitable 3 Win VM with the **Administrator** user account, open the Command Prompt, and use the `ipconfig` command to verify the virtual machine is assigned an IP address from the `172.30.1.0/24` network, as shown:

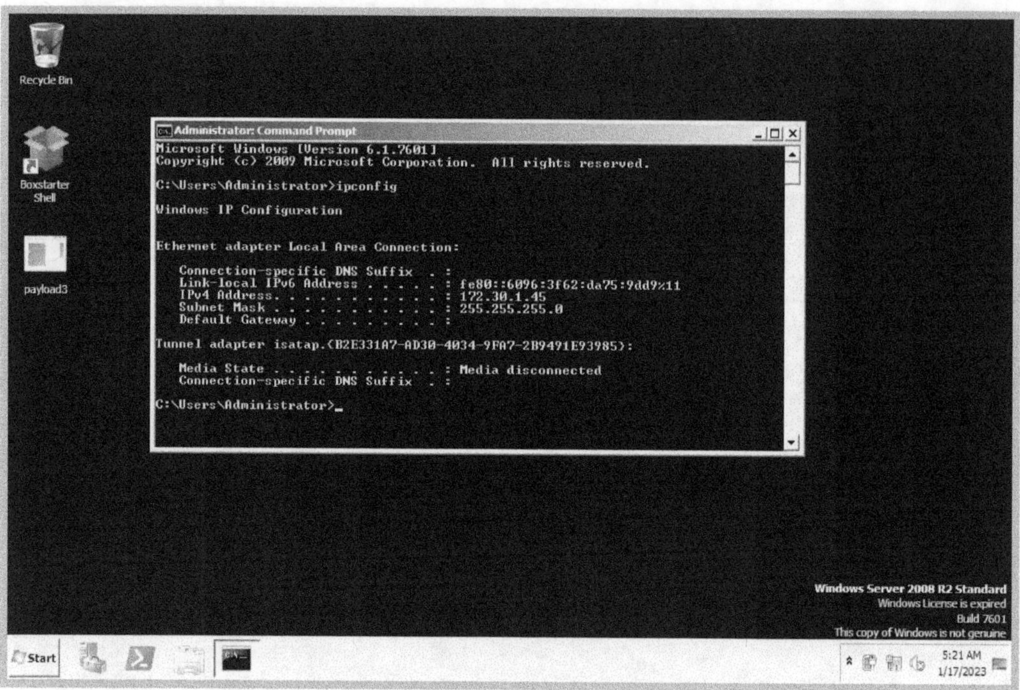

Figure 2.58 – Verifying network connectivity

16. Lastly, power off the virtual machine.

Having completed this section, you have learned how to set up the Metasploitable 3 Windows VM as a vulnerable server within the lab environment.

Summary

During the course of this chapter, you have discovered the importance and benefits of building your personal lab environment for learning and gaining hands-on experience when learning about various cybersecurity topics. Additionally, you have learned how to set up a virtualized environment and network, how to deploy Kali Linux and OSINT VM, and how to set up vulnerable systems for security testing.

I hope this chapter has been informative for you and helpful in your journey in the cybersecurity industry. In the next chapter, *Understanding Passive Reconnaissance*, you will learn about the fundamentals and importance of using passive reconnaissance techniques to discover organizations' assets that are exposed on the internet.

Further reading

- Kali Linux official documentation: `https://www.kali.org/docs/`
- VirtualBox documentation: `https://www.virtualbox.org/wiki/Documentation`
- Trace Labs OSINT VM documentation: `https://github.com/tracelabs/tlosint-live/wiki`
- OWASP Juice Shop documentation: `https://owasp.org/www-project-juice-shop/`
- Metasploitable 3 documentation: `https://github.com/rapid7/metasploitable3/wiki`

3

Understanding Passive Reconnaissance

As more people and organizations are connecting to the largest network in the world, the internet is becoming a massive storage medium. Lots of people are uploading various types of data on many websites and online platforms. While the internet is known as the *information super-highway*, enabling people around the world to share information and collaborate, it has become a medium for storing data about people, organizations, systems, and networks. Such information can be leveraged by adversaries with malicious intent to plan future cyber-attacks on a target. As an aspiring ethical hacker, it's important to understand how adversaries are able to discover and leverage publicly available information to create a profile of their target and improve their strategies and techniques in their cyber operations.

During the course of this chapter, you will learn about the fundamentals and importance of passive reconnaissance and the best practices that are used by industry experts such as ethical hackers and penetration testers, as well as how these techniques are aligned with adversaries. Furthermore, you will explore the fundamentals of **Open Source Intelligence** (**OSINT**) and how it helps organizations to reduce their attack surface. Additionally, you will also learn how to efficiently conceal your online identity as an ethical hacker when performing passive reconnaissance and how to anonymize your internet traffic.

In this chapter, we will cover the following topics:

- Exploring passive reconnaissance
- Fundamentals of OSINT
- Concealing your online identity
- Anonymizing your network traffic

Let's dive in!

Technical requirements

To follow along with the exercises in this chapter, please ensure that you have met the following hardware and software requirements:

- Kali Linux: `https://www.kali.org/get-kali/`
- Instructions on how to set up the **The Onion Router** (**TOR**) service and browser on Kali Linux: `https://www.kali.org/docs/tools/tor/`

Exploring passive reconnaissance

As time goes on, the internet is becoming a more valuable resource every day to many people and organizations, helping everyone to collaborate and share ideas with others beyond traditional borders. The internet was originally referred to as an information super-highway, enabling a user with an internet connection and a computer to easily visit websites, participate in online forums and communities, and perform e-commerce transactions without visiting a traditional *brick-and-mortar* store.

In today's world, smartphones with a data plan enable a user to easily access the internet and its resources within a few seconds. For instance, imagine you're looking to advance your career in the cybersecurity industry and want to identify any organization that's hiring professionals within your region. Using the web browser on your smartphone and Google Search, you can quickly research the top cybersecurity organizations within your region, check their career portal, and explore their hiring process. Additionally, you can further determine the mission, vision, core values, interviewing process and panel, and job responsibilities and requirements of a suitable candidate. However, some organizations leak too much data in their job posts, which is leveraged by threat actors in planning future cyber-attacks.

Every day, many people around the world use the internet and upload documents, audio, video, and other file types. This data is not automatically erased from the internet when a user decides to delete it. For instance, the **Internet Archive** is a public digital library that indexes and keeps a record of everything it knows on the internet for the past 20 years. To put it simply, if you upload or create anything on the internet and delete it after, it's not erased from the internet because it can be retrieved using the **Wayback Machine** on the Internet Archive to view historical data on the internet. Hence, threat actors leverage data collected from the Internet Archive about their targets to improve their attacks.

The following figure shows the Wayback Machine, which allows anyone to insert a **Uniform Resource Locator** (**URL**) within the search field:

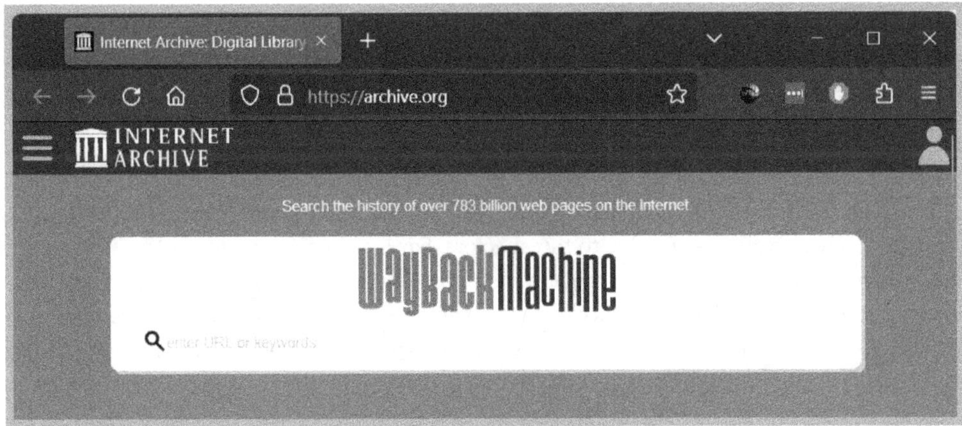

Figure 3.1 – The Wayback Machine

Once a valid URL is entered, the Wayback Machine will check its archival records for any snapshots of the website and provide a timeline, allowing a user to select an available date to view the web page at that point in time.

> **Important note**
>
> To learn more about the Internet Archive, the Wayback Machine, and its resources, please visit `https://archive.org/`.

Data privacy is a major concern around the world as more people, organizations, and governments are becoming more aware of the impact of confidential data being available to anyone with malicious intentions, such as adversaries and other threat actors. There are many data privacy concerns when visiting websites, such as whether users' data is being collected and how it's going to be used by the website owner. For instance, some websites do not have any security protection for their users and will transmit data between the web server and the user's web browser in plaintext, enabling an adversary to intercept and capture any sensitive data. For instance, websites insert cookies into the visitor's web browser, which enables the website to track the visitor's activities and information about the user while improving the user experience. If a threat actor were to capture and analyze the data found within cookies, the adversary will be able to retrieve **session IDs** and determine the user's web activities. If the user is authenticated on the website, the session ID enables the threat actor to gain unauthorized access to the victim's account on the same website without knowing the victim's user credentials.

With the growth of the internet came many social media websites, which enable people to digitally connect and share updates with their family, friends, and co-workers using an online digital platform. Social media platforms have grown a lot over the past decade as many people and organizations use these platforms to start online communities, advertise their products and services, and share news and updates. While social media websites encourage everyone to connect with each other, it's important to understand these websites create a **web of people** who are linked to each other with a mutual connection. For instance, when you visit someone's profile on a social media platform as a logged-in user, the platform shows your mutual connections or friends. If you click on a mutual contact, you're redirected to the person's profile where you may see additional mutual connections or friends, which creates a digital web of people who are interconnected in some way.

While social media platforms help people stay connected digitally, some people's online profiles may be less secure than others, leaking your data to threat actors. For instance, if you do not use social media and others do, you may appear in pictures that are posted by your family, friends, or co-workers with less secure profile. This provides the opportunity for a threat actor to use facial recognition to determine the identity of a person from a picture and use Google Images to perform **reverse image searches** to identify any websites that contain similar pictures to determine the identity or location of someone or something.

> **Tip**
>
> To learn more about advanced Google searching, please refer to *Chapter 4, Domain and DNS Intelligence*. To learn more about reverse image searching, please refer to *Chapter 6, Imagery, People, and Signals Intelligence.*

The following figure shows the **Search by Image** feature on Google Images:

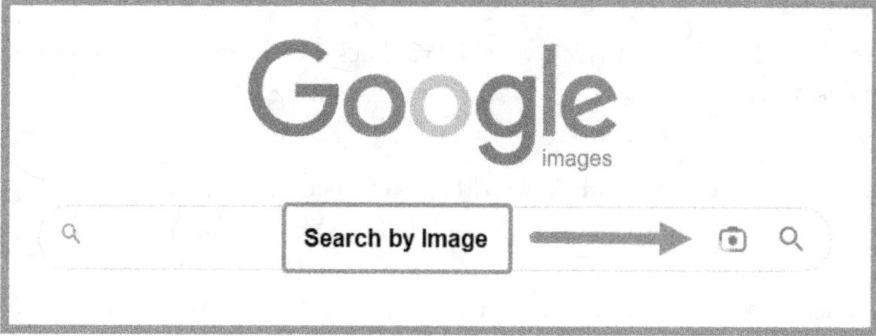

Figure 3.2 – Google Images

> **Important note**
>
> Google Images leverages **Google Lens**, an advanced image recognition technology, to help identify location, text, and objects.

There are many data sources that threat actors and ethical hackers use to collect information about organizations and people. **Data brokers** are organizations that buy, collect, analyze, and sell data about people and organizations without first gaining permission. Data brokers often sell data to anyone such as advertising agencies, people search engines, financial institutions, and marketing companies around the world. Data brokers collect data from many sources, including tracking information found within website cookies that are placed on users' web browsers, usage information collected from mobile applications that are installed on smartphones and tablets, and public information that is found on the internet.

Data brokers analyze and process the collected data to create profiles and dossiers on millions of people around the world, including their activities, preferences and dislikes, travel and health status, and residential location. The availability of such public information can be leveraged by a threat actor who is targeting a user, group of people, or organization.

Thus, so far, you have gained an understanding of how the internet is currently being used as a massive storage medium for various types of data and anyone with internet connectivity will be able to access public data. This means threat actors can spend a lot of time fine-tuning their internet searches to collect specific data from various online sources, with the intent of creating a profile and understanding the attack surface of their target.

Reconnaissance is the tactics and techniques used by threat actors and ethical hackers to collect information that can be used to discover and exploit security vulnerabilities on a target. The following are the two categories of reconnaissance:

- **Passive reconnaissance**
- **Active reconnaissance**

Passive reconnaissance is when a threat actor or ethical hacker does not directly interact with or connect to the target, and instead relies heavily upon publicly available information to collect intelligence that can be used to compromise the target.

The following are the key advantages of performing passive reconnaissance:

- **Increasing stealth**: The techniques used during passive reconnaissance ensure a threat actor or ethical hacker does not directly interact with the target. It reduces the likelihood of being detected by security systems.
- **More data sources**: The internet contains a massive wealth of information that can be found on websites, social media platforms, and internet search engines and used for passive reconnaissance by ethical hackers.
- **Reducing legal concerns**: Since passive information gathering involves collecting data from public sources, this process does not have the same risks as active reconnaissance and it's generally legal.
- **Reducing cost**: Since ethical hackers collect publicly available information from the internet to create a profile of their target, passive reconnaissance is often less expensive and does not require the same resources as compared to active reconnaissance.
- **Improving safety**: Since the ethical hacker is collecting publicly available information from the internet, they are at less risk of detection. Therefore, using passive reconnaissance is a safer option as compared to active reconnaissance.

While passive information gathering is commonly used by both threat actors and ethical hackers, there are some disadvantages, such as the following:

- **Limited data**: Since passive reconnaissance leverages publicly available information, the data found may be outdated or inaccurate, and access may be limited.
- **Inaccurate data**: Using passive reconnaissance to collect information can lead to gathering inaccurate or misleading data about the target, which can further lead to discovering fake websites and making inaccurate conclusions. Hence, ethical hackers depend on the availability and accuracy of information collected from the internet.
- **Time-consuming**: The internet contains massive amounts of data. An ethical hacker may require a lot of time to locate and gather specific data about the target from various data sources.

However, active reconnaissance enables a threat actor or ethical hacker to directly interact with or connect to the targeted system to retrieve specific details that are unlikely to be collected during passive reconnaissance. It usually involves ethical hackers using specialized tools and techniques to send probes to determine the technical architecture of a system, operating system, running services and open ports, and security vulnerabilities. Unlike passive reconnaissance, active reconnaissance techniques are more intrusive and have a higher risk of being detected and triggering alerts by security systems.

> **Important note**
>
> Since active information-gathering techniques are intrusive, as they collect sensitive details about systems, it's generally illegal to perform scanning on systems without first obtaining legal permission.

Ethical hackers and penetration testers use the same techniques as real hackers to efficiently determine the attack surface and identify security vulnerabilities and their associated security risks within organizations. Isn't it illegal to use the same techniques to compromise systems and networks that are owned by organizations? The simple answer is yes as hacking is generally illegal. However, ethical hackers and penetration testers need to obtain legal permission from the authorities before starting any work on their client's systems and networks.

The following are the benefits of why ethical hackers use the same techniques as real hackers:

- **Efficiently testing the security of a system**: Using the same techniques as threat actors, ethical hackers are able to efficiently test whether the security of a system is working as expected or contains possible security flaws.
- **Efficiently discovering security vulnerabilities**: By simulating real-world cyber-attacks on an organization's systems, ethical hackers are able to efficiently discover hidden security vulnerabilities and recommend countermeasures to prevent or mitigate a real cyber-attack.

- **Improving security awareness**: Many organizations do not fully understand the impact of a cyber-attack on their systems and network. By simulating a real-world cyber-attack, the ethical hacker will be able to demonstrate the impact and severity if a real cyber-attack were to occur. Overall, it raises the importance of cybersecurity to decision-makers of an organization.
- **Providing evidence**: Since ethical hackers use the same techniques as real hackers to test the security of a system and discover hidden security vulnerabilities, the ethical hacker can collect solid evidence of security vulnerabilities and understand their severity risk levels, which can be used to help the organization's security team prioritize resources to improve their security posture.
- **Keeping up to date**: Since there are new cyber-attacks and threats every day, ethical hackers and penetration testers always need to stay up to date on the latest techniques that are used by real hackers to efficiently test for hidden security vulnerabilities on systems.
- **Discovering data leaks**: Many organizations leak data about themselves, which can be leveraged by real hackers to plan future cyber-attacks on their targets.

To better understand what type of data should be collected and how it can be processed to create meaningful information, let's take a look at the concept of footprinting.

Understanding footprinting

Using the internet to find and collect sensitive data about a person or organization is quite fascinating for many, especially aspiring ethical hackers and penetration testers. However, collecting tons of data about a target is meaningless if we don't understand what we are looking for and how to process and convert it into information that's meaningful for us to develop a profile about a target.

Footprinting is the technique used by threat actors and ethical hackers to gather specific details about a target, such as the host operating system, open ports, and running services. The information collected during footprinting can be leveraged by an ethical hacker to identify the number of security vulnerabilities that can be exploited to gain unauthorized access to a system, network, or organization. Additionally, footprinting helps ethical hackers to gain a better understanding of the security posture of an organization and their infrastructure.

The following diagram displays the link between footprinting and reconnaissance:

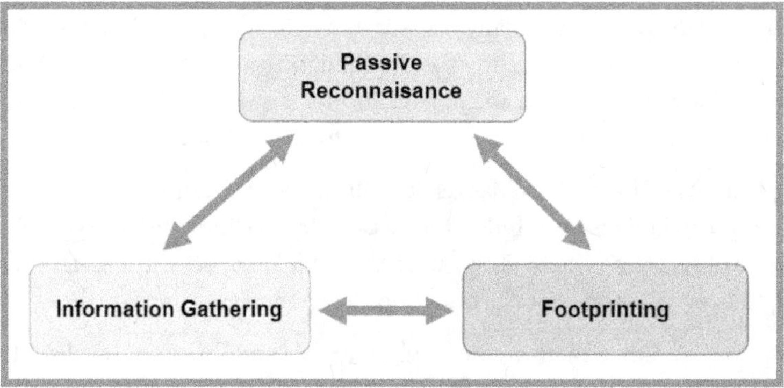

Figure 3.3 – Footprinting

Data that's collected during footprinting is generally classified into the following categories:

- System information
- Network information
- Organizational information

System information helps ethical hackers to determine what's running on the device, such as the host operating system, running services and applications, and usernames and passwords. The ethical hacker can use the operating system information to research and identify whether the operating system is vulnerable and can be exploited. Within many organizations, there are outdated and vulnerable applications and services that are running on hosts, which creates the opportunity for an ethical hacker to identify and exploit security vulnerabilities on running services to gain unauthorized access to the target. Furthermore, ethical hackers may be able to retrieve valid user credentials, which can be used to gain unauthorized remote access to a targeted system.

Network information helps ethical hackers to identify whether there are any vulnerable network protocols and services running between systems on the network. Many common network protocols were not designed with security in mind, thus enabling threat actors to exploit their security weaknesses to perform cyber-attacks. An attacker can intercept and identify sensitive network information, such as **Domain Name System** (**DNS**), **Link-Local Multicast Name Resolution** (**LLMNR**), **Address Resolution Protocol** (**ARP**), domain names, sub-domain names, and firewall rules. For instance, if a threat actor were to capture DNS queries and replies, the attacker may be able to identify the organization's DNS servers and attempt to retrieve internal hostnames and IP addresses.

Important note

DNS is an application-layer protocol that enables a system to resolve the hostname to an IP address. **LLMNR** is a layer 2 protocol that enables systems to resolve hostnames on a **Local Area Network** (**LAN**). **ARP** is a network protocol that allows systems to resolve the IP address to a **Media Access Control** (**MAC**) address.

Vulnerable network protocols, such as LLMNR and ARP, are susceptible to various layer 2-based attacks, which enables an attacker to intercept and inject unsolicited packets into hosts on a network. Furthermore, identifying the domain name of an organization provides focus to enumerate sub-domains. Some organizations will enforce additional security controls on their primary domain name but neglect some of their sub-domains. The attacker can identify whether compromising a vulnerable sub-domain will lead to gaining unauthorized access to the organization's network.

Organizational information helps ethical hackers to identify who the employees of the company are, their contact information, such as telephone numbers and email addresses, and an organizational chart that displays high-profile employees with high-privilege user accounts. The organizational information can be leveraged when planning various social engineering attacks, such as spear-phishing email campaigns that are targeted toward the employees of the organization.

While there are various methodologies used by ethical hackers and threat actors, the following is a common set of guidelines for getting started with footprinting a target:

- Use internet search engines
- Perform advanced Google searches (**Google hacking/dorking**)
- Collect information from social media websites
- Collect information from the target's website
- Perform footprinting on email messages originating from the target
- Use an online domain registry to retrieve domain information
- Perform DNS information gathering
- Perform network footprinting
- Performing social engineering

> Tip
>
> Ethical hackers can use the Google Dork `"login" site:microsoft.com` to find login pages that are associated with the `microsoft.com` domain or `site:microsoft.com AND intitle:"login"` to find pages that contain the keyword `login`.

You will explore the preceding techniques in the next chapter, which focuses on advanced intelligence gathering, to develop the industry skills that are needed by ethical hackers and penetration testers. Having completed this section, you have learned the fundamentals of passive information gathering and how the internet can be leveraged to collect data about a target. In the next section, you will learn the fundamentals of OSINT and how it's used to help ethical hackers to profile their targets.

Fundamentals of OSINT

OSINT is simply the process and techniques that are used by ethical hackers to search for, collect, and analyze data that's found on public sources such as the internet. While many people post and upload various types of data every day on the internet, ethical hackers can leverage the data on the internet to create a profile and better understand their target. While OSINT may seem to be as simple as using a search engine such as Google Search to find answers to questions, many people typically use the information and resources that are displayed on page 1 of the search results as it's generally the most related to the search criteria. However, not many people will check the remaining pages of the search results, such as pages 2–10, to determine whether the answers to the search criteria are accurate and provide comprehensive details to the user.

Ethical hackers spend a lot of time collecting data from various public sources to ensure sufficient data is gathered about their targets; however, this data does not always provide meaning and context and needs to be thoroughly analyzed to create meaning, enabling ethical hackers to better understand the system details and network and organizational information about the target. Hence, it's essential, when using internet search engines to find data, to go beyond page 1 to collect more data about the target. The idea is to collect everything, then analyze the collected data to create meaning and provide context as it relates to the target.

The following are common types of data that are found and available from public sources:

- **Media**: This includes audio, video, and pictures that are uploaded to the internet
- **Text**: These are documents, articles, and blogs that are posted online
- **Maps**: Geolocation data about a person or organization

These common data types are usually posted directly either by the target or by others about the target and they contain useful data to better understand and profile the target. For instance, if a target posts photos about their vacation on social media, an ethical hacker can observe the details of the pictures, such as the date posted and background. The date helps the ethical hacker to determine when and how often the target visited the location, specific times when the target took a vacation, and whether there are any other patterns that can be leveraged to track the target's movement. Furthermore, the background within each photo reveals identifying information about the physical location of where the photo was taken and enables the ethical hacker to use online search engines, such as Google Images, to perform a reverse image lookup to determine the geolocation of the target.

Additionally, if an ethical hacker is assigned to perform a penetration test on the target's wireless network infrastructure, identifying the geolocation helps the ethical hacker to identify the physical location of the target's premises. Furthermore, videos and vlogs that are posted online contain information about the target's environment, such as their office locations and spaces, where they've been, and how often they have visited a location.

Collecting data from documents, articles, and blogs enables an ethical hacker to identify the technologies, projects, and work that were done by the target. For instance, many people on LinkedIn post specific details about technical projects and systems they've implemented within their organization. Such information helps an attacker or ethical hacker to determine the type of infrastructure and security controls that exist within the company.

Hence, it's important for organizations to be aware of the risk of leaking data online that can lead to a possible cyber-attack and how this data is retrieved and exploited by cyber-criminals. As an ethical hacker, the most valuable data about your target is simply data that's directly posted by the target and others about them.

The following diagram shows the various sources of data when performing OSINT operations:

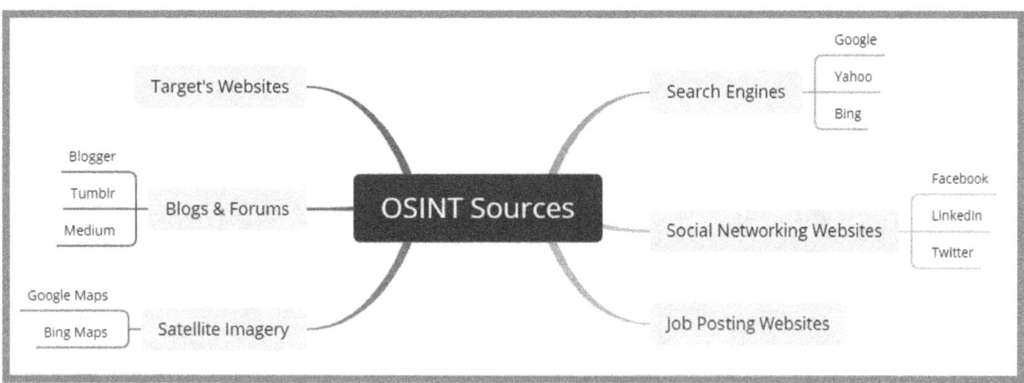

Figure 3.4 – OSINT data sources

As shown in the preceding diagram, there are various online sources that anyone can use to collect data about a target, then analyze the collected data to provide meaning and context. Using multiple internet search engines to research the target is essential as one search engine may provide additional data as compared to another. Many organizations and people use social media to connect and share information about themselves and their company with others. Sometimes, people leak sensitive data without realizing the impact if the data is exploited. Many organizations will post technical details and technologies within the description of a job post for hiring a new employee. Some of this information contains specific details, such as the operating systems of their servers, server types and architectures, cloud computing services, and even networking devices.

The OSINT life cycle

The OSINT life cycle helps ethical hackers to better understand the various phases that are involved in collecting, analyzing, and reporting data during the reconnaissance phase of ethical hacking and penetration testing.

The OSINT life cycle generally consists of the following phases:

1. Understanding the requirements for information gathering
2. Performing data collection and retrieval
3. Conducting data analysis of collected data
4. Pivoting and reporting the analysis

The following diagram shows the continuous processes when performing OSINT:

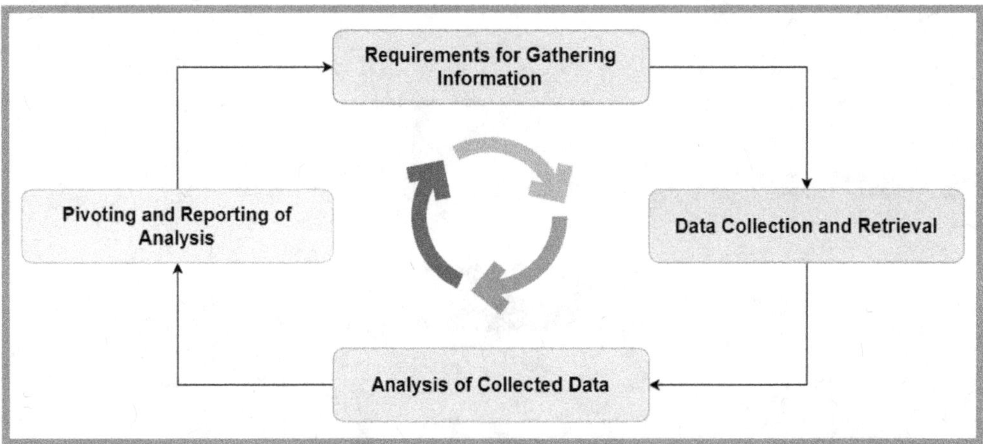

Figure 3.5 – OSINT life cycle

During phase 1, *Requirements for Gathering Information*, it's important for the ethical hacker to have a clear understanding of what the organization wants. Therefore, the following are some questions that should be asked to the client to gain clarity:

- What are the deliverables at the end of the OSINT penetration test?
- Is the organization interested in identifying whether their data is being leaked online?
- How will an attacker identify the points of entry (attack surface) and security vulnerabilities within their organization?
- What would the risk level and impact be if an attacker were to leverage OSINT to compromise the organization's systems and network?

In phase 2, *Data Collection and Retrieval*, the ethical hacker uses **Tactics, Techniques, and Procedures** (**TTPs**) that are similar to that of real hackers to collect data about the target from publicly available sources. During this phase, it's important to collect all sufficient data, such as photos, videos, documents, and other media types, from various data sources on the internet that can provide context about the targeted organization. However, it's essential the ethical hacker identifies what data is considered to be relevant information and at what point sufficient data has been collected.

During phase 3, *Analysis of Collected Data*, the ethical hacker carefully analyzes the collected data to better understand how it will be applicable to meet the customer's requirements while sorting it into both meaningful and useful categories. During this phase, the ethical hacker sometimes discovers something interesting and decides to go deeper into collecting more data. For instance, if an ethical hacker discovers the targeted organization is hiring an IT professional to manage their self-hosted servers, the ethical hacker may attempt to find additional information from current and past job posts for any technical details that may reveal the operating systems of the target's internal servers. Diving deeper, the ethical hacker may use social media, such as LinkedIn, to collect intelligence on any IT systems that were implemented by past and current employees of the target.

The collected data is processed and converted into meaningful information by determining the following:

- Is the collected data accurate?
- Was the data collected from credible sources?
- Is the collected data factual or subjective?
- Is more data needed to better understand a specific area of the target?

In phase 4, *Pivoting and Reporting of Analysis*, the ethical hacker may pivot into collecting additional data in a different area about the target. For instance, if sufficient data is collected and analyzed about the organizational structure of the targeted organization, pivoting into collecting data about the social media presence of each employee may be beneficial in finding additional data leaks about the target. For instance, pivoting can lead to discovering the social media accounts of IT professionals of the targeted organization, and enable you to identify the technologies that are used within the company if they are posted online.

Once enough/sufficient data is analyzed about the target, the ethical hacker then moves into the reporting phase, which contains the relevant data, information, and the analysis that was performed. It's important to notify the client/customer about the evidence collected during OSINT and why it should be important to them. Hence, provide awareness on how the data was leaked by employees and how it can be leveraged by real hackers to plan a possible future cyber-attack on the company.

When collecting data about a target, sometimes the process becomes a bit overwhelming and determining whether to pivot into another direction creates uncertainty. To help provide a clear visual map and interconnection of OSINT data, **Your OSINT Graphical Analyzer (YOGA)** is a free online resource that helps cybersecurity professionals to easily visualize how they can move on to another data point and collect more data during the OSINT process.

The following figure shows the YOGA user interface and data points:

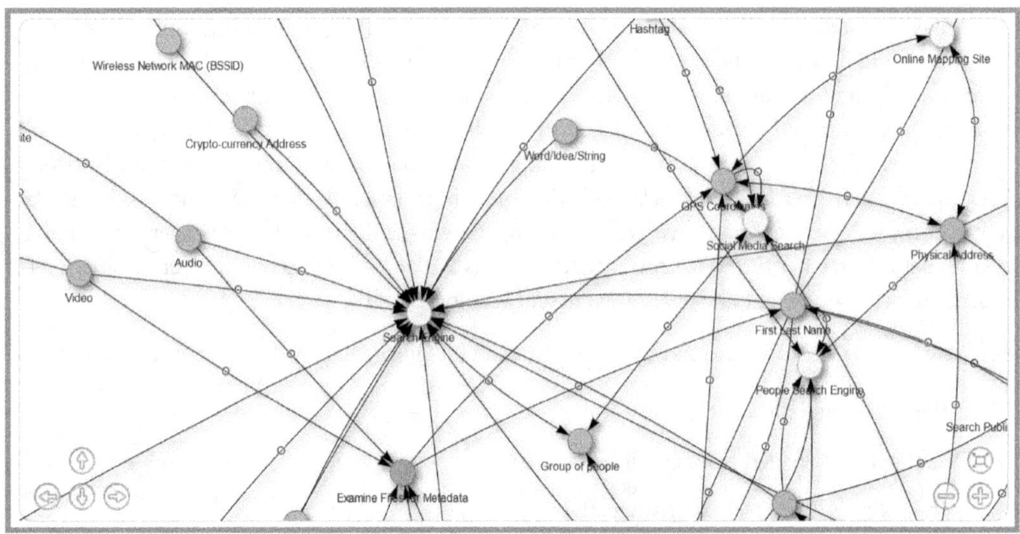

Figure 3.6 – YOGA data points and map

As shown in the preceding figure, YOGA shows how one data point is interconnected with another, which enables the ethical hacker to easily follow the map to another data point to collect additional intelligence about the target.

> **Important note**
>
> To learn more about the YOGA interactive map for OSINT data points, please see `https://yoga.myosint.training/`.

Next, you will discover how OSINT helps law enforcement discover new criminal activities and increase awareness, and how it can be leveraged by ethical hackers.

Benefits of using OSINT

Having knowledge and understanding of how to use OSINT in the right way makes a person very powerful. Similarly, there's a lot of publicly available data on the internet but understanding how to collect and analyze such data to create meaning and context on a topic can be used with both good and bad motives and intentions. As ethical hackers and penetration testers, we'll focus on using OSINT with good intentions, such as helping organizations to reduce their attack surface and improve their security posture.

OSINT is commonly used for both good and lawful purposes. For instance, law enforcement leverages the data from OSINT to determine the location and additional details about *persons of interest* and wanted persons. A wanted person may use social media to brag about their evasion from being caught by the local law enforcement agency and make an online post showing they are having fun at a party. Law enforcement professionals can monitor their target and any posts that are made about the target to determine their location and activities. Photos and video footage that are posted on social media help law enforcement to discover new crimes. Furthermore, there are people who break the law and overshare their activities on social media, which leads to self-incrimination and helps law enforcement become aware of unlawful acts.

As previously mentioned, social media platforms were created to help people stay connected and share updates with each other. However, social media platforms are commonly used by law enforcement as modern-day wanted posters to help raise awareness of suspects and criminal activities with the intention of reaching a larger audience.

There are many missing people around the world. Using OSINT can help you and law enforcement to find missing people by leveraging the information collected from the internet about the missing person. For instance, a missing person may have a social media account that contains data about their daily life and activities, friends, and groups. The daily activities can be used to identify common patterns of the missing person and whether something unusual occurred recently. The victim's friend list can help law enforcement determine who was able to view the victim's activities, posts, photos, and videos prior to the incident, and whether any of these people had a motive to do something unlawful. What if a stalker was on the victim's friend list, masking as a law-abiding citizen but secretly monitoring their daily activities and using the information to plan criminal activities? Overall, cybersecurity professionals who work with law enforcement can use OSINT to find missing people and catch unlawful persons.

As an ethical hacker, it's important to conceal your identity while performing OSINT operations on a target. In the next section, you will learn about the importance of hiding your online identity when collecting information about your target.

Concealing your online identity

To better understand the importance of concealing your identity as an ethical hacker when performing OSINT, let's imagine your organization was compromised by a threat actor who sent a phishing email message to a high-profile employee of the company who was tricked into clicking an obfuscated link that looked safe. Upon clicking the obfuscated link, malware was downloaded and executed on the victim's system, which enabled the threat actor to gain unauthorized access and exfiltrate confidential data from the organization's servers and other devices. Afterward, the threat actor sent an email message to the organization that indicated the type of data that was exfiltrated from the servers and requested the company send a payment of cryptocurrency within 24 hours, or else the threat actor would sell the data on the *dark web* to the highest bidder.

Within the organization, Bob is the security professional who manages and oversees the company's security operations and is the go-to person for security-related concerns. The security breach was brought to Bob's attention quickly as the leadership team was concerned about their intellectual property and other confidential data being sold on the dark web to other hackers. Bob decided to inspect the source code of the email message that was sent by the threat actor to identify any information that could be used to find the sender of the email and the location of the attacker. For instance, the origin IP address can be used to determine the geolocation and network provider of the sender.

> **Important note**
>
> Within the header of email messages, the origin IP address is sometimes unmasked, which enables the recipient to identify the real IP address of the sender. However, many email providers mask their sender's real IP addresses on outbound messages, hence increasing the complexity of tracing an email back to its real source. However, you can attempt to social engineer the hacker by creating and sending a **canary token** via email to the threat actor. Canary tokens (`https://canarytokens.org/generate`) are used to help cybersecurity professionals determine whether their systems or networks are breached and trick an adversary into announcing themselves and revealing their geolocation.

However, Bob was unable to find the threat actor's real IP address from the email header and decided to perform an online search using the sender's name and alias that was found within the message. Performing a Google search on the threat actor's alias may provide some results or none at all. Searching on social media websites such as LinkedIn for any profile that has the sender's name will trigger an alert to the profile owner, therefore cybersecurity professionals need to be mindful about alerting their targets. For instance, if you use your personal account to search and view someone's profile on LinkedIn, the target will be notified and given the identity of who viewed their profile. Imagine if Bob used his personal account to find the attacker, who may be using a fake profile on LinkedIn. Then, the threat actor will be notified that someone from the victim organization is performing an investigation for the perpetrator of the cyber-attack. To put it simply, Bob should use a different account that cannot be traced back to him or his company to prevent the target from being aware of the investigation.

Fundamentals of sock puppets

Whether you're performing an OSINT investigation on a cyber-criminal or as an ethical hacker for an organization, it's important to hide your real identity to avoid the target identifying who is collecting information and tracing it back to you. Hence, it's always recommended to never use personal accounts for work.

Ethical hackers use a **socket puppet** to conceal or hide their identity when performing investigations and OSINT operations on a target. A sock puppet is one or more alternate social media accounts that are set up and operated by the ethical hacker for the purpose of performing OSINT, such as collecting intelligence from various online data sources. The concept of using a sock puppet is to be pseudo-anonymous and it should never be traced back to you, the ethical hacker. It provides privacy while gathering intelligence.

Furthermore, sock puppets enable ethical hackers to create multiple custom social media accounts for the purpose of collecting intelligence about the target. However, the sock puppet accounts should have a realistic persona, making it convincing to anyone that the account is real and not fake, including the target. The sock puppet account should be an online personality that will acquire followers and encourage others to connect, enabling the ethical hacker to connect with the target and gather more information. Additionally, the sock puppet excellently conforms to **Operational Security** (**OPSEC**) practices as the account does not trace back to you. Remember, when you're performing OSINT operations or investigating a target, it's important that the target is unaware you are investigating or collecting intelligence about them.

Setting up a sock puppet

Before setting up a sock puppet, it's important the ethical hacker determines whether anonymity and/or persistence is required. If anonymity is required, this implies you do not want the target to find or trace who you are while performing intelligence gathering on them. To ensure anonymity, you will be required to carry out a lot more work when setting up a sock puppet. If persistence is required on the sock puppet, this means the ethical hacker wants to ensure the fake persona will interact with others on social media, following and connecting with others, and frequently posting new statuses and updates regularly to ensure the sock puppet has a real persona and is not identified as a fake account. Persistence also means the ethical hacker will need to infiltrate social networks, such as groups, for weeks, months, or years to build a trusted online presence. However, if you're not looking for persistence, then the sock puppet will be easier to set up and maintain compared to a persistent account.

> **Important note**
>
> Never use personal accounts for work-related activities, such as OSINT operations, investigations, ethical hacking, or penetration testing.

The following are guidelines and recommendations for creating a sock puppet:

- When creating a sock puppet, your origin details are important. Do not use your own IP address when creating and setting up a social media account. Consider visiting a local coffee shop with free wireless internet.
- Do not use **Virtual Private Network** (**VPN**) or TOR services as social media platforms can detect when origin traffic is being sent through a proxy or VPN and will require you to further validate your identity.
- Ensure your online appearance on social media looks like a normal, regular person to avoid being flagged as a fake user account.
- Use a *burner* email address when signing up for a social media account. Consider using **Proton Mail** (`https://proton.me/`) as it provides additional privacy and data encryption. However, you can also use a typical email service, such as Gmail, Yahoo Mail, or Hotmail. Ensure the email address is very vanilla (basic).

- When using Facebook, use the mobile version of the website (`https://m.facebook.com/`) as it performs fewer security checks and validations as compared to the desktop version of the website.
- Ensure you share content regularly, such as status updates, photos, and videos, and interact and connect with people on the social media platform.
- Do not use someone else's pictures. It's very easy to perform a reverse image search to determine whether a picture is fake or not.

The following are useful resources when creating a sock puppet:

- **Fake Name Generator**: `https://www.fakenamegenerator.com/`
- **This Person Does Not Exist**: `https://www.thispersondoesnotexist.com/`
- **Privacy Cards**: `https://privacy.com/`
- **Remote browser**: `https://webgap.io/`

Fake Name Generator enables an ethical hacker to quickly generate a fake identity that can be used when creating a sock puppet. The website allows you to customize the identity, such as gender, name set, and country, to create a believable persona. **This Person Does Not Exist** is a website that uses **Artificial Intelligence** (**AI**) to create fake pictures of people who do not actually exist. **Privacy Cards** is a website that enables you to create a proxy for your real credit card such that you do not need to insert your real credit card details on an e-commerce website when making purchases; a proxy credit card is created that can be used instead. Therefore, the e-commerce merchant will not know your real credit card number and details. **WEBGAP** provides remote browser isolation services that enable anyone to browse the internet using a remote browser for additional security and privacy.

The following are additional resources that may be needed based on the OSINT operations:

- A *burner* telephone number from a service provider who does not recycle the numbers each month. This *burner* telephone number can be used for account verification purposes.
- A proxy credit card can be used to purchase the burner phone and additional items if needed.
- Virtual machines for performing OSINT operations and investigations.
- A VPN service if needed.
- Always use a dedicated computer when performing OSINT operations as you do not want the target to trace anything back to you. Consider inexpensive microcomputers, such as Raspberry Pi.

Having completed this section, you have learned the importance and how to get started with using a sock puppet to hide your online identity when performing OSINT operations on a target. Next, you will learn how to anonymize your internet traffic.

Anonymizing your network traffic

As an ethical hacker who is performing OSINT operations, it's important to ensure the target is not able to trace who is collecting data about them. Skilled cybersecurity and network professionals are able to capture and analyze the data found within network packets and logs to determine the source of the traffic; hence, it's recommended to anonymize your network traffic while using the internet to mask your real geolocation and source public IP address.

The following are common techniques that are used to anonymize internet-based traffic:

- **VPNs**
- **ProxyChains**
- **TOR**

The following sub-sections will describe the advantages and how each of the preceding technologies works to anonymize your internet-based traffic.

VPNs

A VPN enables a user to establish a secure communication channel over an unsecure network, thus providing confidentiality and mitigating eavesdropping attacks over a network. VPNs are commonly used by organizations to establish secure and remote connectivity between their remote offices and the main office without requiring a managed **Wide Area Network** (**WAN**) from a telecommunication service provider. These are commonly referred to as **site-to-site VPNs**. There are also **remote access VPNs**, which are implemented by organizations to enable their remote workers to securely connect and access resources on the corporate network.

Ethical hackers can either set up their own remote access VPN server on the cloud or subscribe to a commercial VPN provider. Setting your own remote access VPN on the cloud requires some technical skills, such as knowing how to set up a virtual machine on your preferred cloud service provider environment and some networking skills. The advantage of owning the VPN server is that no one else sees your VPN traffic logs but you, hence, there's added privacy. However, it's very important that you secure the operating system of your server on the cloud to prevent threat actors from gaining unauthorized access to it. Additionally, when setting up your virtual machine on the cloud, you'll need to select a hosting location. This location will be the exit point for all traffic that uses your VPN channel. Therefore, if you're located in the **United States** (**US**) and your VPN server is hosted in the **United Kingdom** (**UK**), then all your internet-based traffic that's routed through your VPN will exit in the UK and the source public IP address will be shown as a UK-based address.

The following diagram shows the concept of using a VPN to anonymize your traffic:

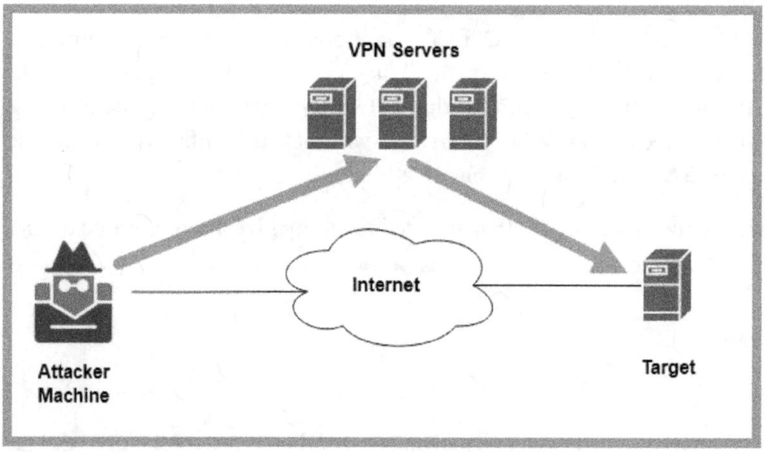

Figure 3.7 – VPN concept

As shown in the preceding diagram, the attacker machine is being used by the ethical hacker while it connects to the VPN server on the cloud. All internet-based traffic is sent through the VPN tunnel and exits at the hosting location (country) before it's sent to the target. Therefore, if the target captures and analyzes the network packets, it'll be traced back to the exit node (the VPN server) and not to the ethical hacker's machine.

> **Tip**
>
> **OpenVPN** provides a self-hosted Access Server option that enables you to connect up to two devices for free. This enables you to set up your own OpenVPN Access Server on your own device or on the cloud. Please refer to the following link to learn more: `https://openvpn.net/access-server/`.

Some ethical hackers will prefer to use a commercial VPN service provider, which reduces the need for configuring and setting up a virtual server on the cloud. One of the major benefits of using a subscription-based VPN service is that it enables the subscriber to connect to multiple VPN servers around the world in various countries that are all managed by the same commercial VPN provider. Therefore, you can easily switch between different countries and choose the location of the exit-node that will be routing your traffic to the Internet.

The following are general considerations for using VPNs:

- Whether you're hosting a VPN server on the cloud or using a commercial VPN service, there's a cost that's attached to it.
- Ensure your preferred VPN service provider does not keep logs about your traffic.

- Ensure your preferred VPN service provider does not sell your data.
- Ensure your preferred VPN service provider has unmetered bandwidth usage.
- Ensure the VPN client is supported by your computer's host operating system.
- Many cloud providers, such as Microsoft Azure and Amazon AWS, have a preconfigured OpenVPN Access Server available for quick deployment.
- When using a VPN, ensure your DNS traffic is not leaked outside the VPN, or else your real geolocation will be revealed. Consider testing your DNS leakage by using **DNS leak test** at `https://www.dnsleaktest.com/`.
- If your VPN service does not support IPv6, ensure IPv6 is disabled on your machine to prevent leakage of your geolocation.

While there are many commercial VPN service providers, the list of the top VPN service providers changes from time to time based on their services, features, performance, and reviews. Take some time to research which commercial VPN provider is most suitable for your needs or whether a self-hosted solution is better. Next, you will learn about Proxychains and how to set it up.

Proxychains

Proxychains is a common technique and tool used by ethical hackers to mask the origin of their network traffic while increasing privacy and security. Proxychains enables network traffic to be routed through multiple proxy servers on the internet, creating a chaining effect, and is useful when more than one proxy server is used to improve the desired level of anonymity and security.

The following are the advantages of using proxy servers and Proxychains for OSINT operations:

- **Providing anonymity**: Proxychains simply allows your internet-based traffic to be routed through more than one proxy server. This enables an ethical hacker to hide their real source IP address and identity during OSINT operations.
- **Bypassing geo-restrictions**: There are many networks and organizations that restrict access to specific websites and services. Using proxy chaining enables ethical hackers to bypass these geolocations and gain access to the target.
- **Circumventing security measures**: There are security systems that are implemented by organizations that are designed to detect and prevent malicious traffic based on their source address and geolocation. Using proxy chaining enables an ethical hacker to make it more difficult for these security systems to detect and restrict their traffic.

The Proxychains tool enables an ethical hacker to configure a list of proxy servers that will be used to route network traffic through them in a chain-like format. This is commonly referred to as **proxy chaining**. When the ethical hacker sends traffic to a target, the packets pass through the first proxy server in the chain, then onto another until the packet reaches the target. When the target responds, the returning packets travel in reverse order through the chain of proxy servers back to the ethical hacker's machine.

The following diagram shows the chaining of proxy servers:

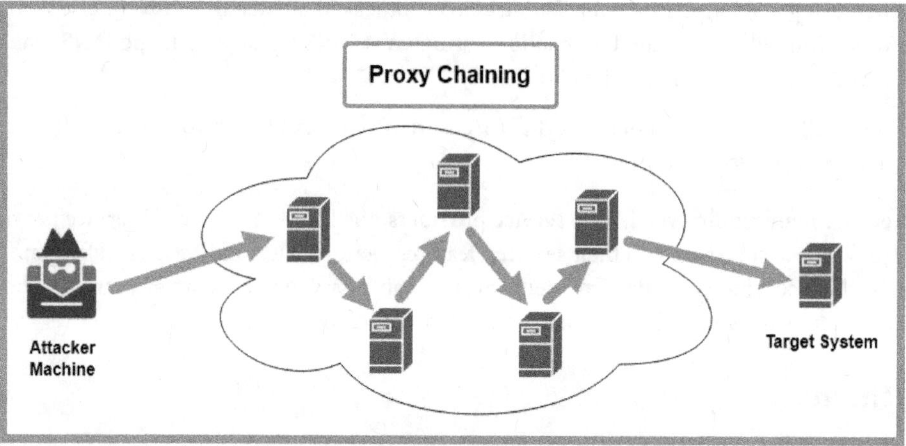

Figure 3.8 – Proxy chaining

Each proxy server within the chain of proxies operates as a relay, which then forwards the packet onto the next proxy server within the list until it arrives at the target. This technique enables an ethical hacker to hide their identity, their real public IP address, and their location. The target will see the exit node or the last proxy server from the chain of proxies as the origin of the traffic.

> **Tip**
>
> There are many websites on the internet that provide a list of free proxy servers. Consider checking out `https://spys.one/en/`.

The `proxychains4.conf` file on Kali Linux enables you to configure the Proxychains tool and add specific servers to the list of preferred proxy servers to create the proxy chaining effect. Additionally, Proxychains supports the following types of proxy servers:

- HTTP
- HTTPS
- SOCKS4
- SOCKS5

Each of these specific types of proxy servers has its own set of security features. Proxy servers that use HTTP to relay and forward traffic do not provide data encryption, while proxy servers that use HTTPS will encrypt the traffic. It is important to conduct additional research on the type of proxy servers that are the best fit for your needs prior to your OSINT operation on a target.

To get started setting up Proxychains, please use the following instructions:

1. Open VirtualBox Manager and power on the Kali Linux virtual machine.
2. Once the Kali Linux virtual machine is powered on, log in and open the Terminal and use the following commands to update the filename database and search for the Proxychains configuration file:

   ```
   kali@kali:~$ sudo updatedb
   kali@kali:~$ locate proxychain
   ```

   As shown in the following figure, the `proxychains4.conf` file is found:

   ```
   kali@kali: $ locate proxychain
   /etc/proxychains4.conf
   /etc/alternatives/proxychains
   /etc/alternatives/proxychains.1.gz
   /usr/bin/proxychains
   /usr/bin/proxychains4
   /usr/bin/proxychains4-daemon
   /usr/lib/x86_64-linux-gnu/libproxychains.so.4
   /usr/share/applications/kali-proxychains.desktop
   ```

   Figure 3.9 – Locating the Proxychains configuration file

3. Next, open your web browser and go to `https://spys.one/en/` for a list of various proxy servers.
4. After choosing a few proxy servers from the previous step, you will need to configure the `proxychains4.conf` file to use the chosen proxy servers. Use the following command to open the `proxychains4.conf` file with a command-line text editor:

   ```
   kali@kali:~$ sudo nano /etc/proxychains4.conf
   ```

5. Next, when `proxychains4.conf` is opened in Terminal, scroll down to the line that contains `dynamic_chain` and uncomment the line by removing the # character at the start of the line. Also, insert a # character at the start of the line that says `strict_chain` to make it a comment as shown:

```
# proxychains.conf  VER 4.x
#
#        HTTP, SOCKS4a, SOCKS5 tunneling proxifier with DNS.


# The option below identifies how the ProxyList is treated.
# only one option should be uncommented at time,
# otherwise the last appearing option will be accepted
#
dynamic_chain          Uncomment
#
# Dynamic - Each connection will be done via chained proxies
# all proxies chained in the order as they appear in the list
# at least one proxy must be online to play in chain
# (dead proxies are skipped)
# otherwise EINTR is returned to the app
#
#strict_chain          Comment
#
# Strict - Each connection will be done via chained proxies
# all proxies chained in the order as they appear in the list
# all proxies must be online to play in chain
# otherwise EINTR is returned to the app
#
```

Figure 3.10 – Enabling dynamic chaining

As shown in the preceding figure, uncommenting a line of code simply enables the feature, such as dynamically chaining multiple proxy servers, and commenting a line disables the strict chaining feature.

6. Next, scroll down to the end of the `proxychains4.conf` file and insert a # character at the beginning of `socks4 127.0.0.1 9050` to comment it out from being read, and insert your additional proxies at the bottom of the list, as shown:

```
[ProxyList]
# add proxy here ...
# meanwile
# defaults set to "tor"
#socks4         127.0.0.1 9050
socks5 104.236.45.251 31226
socks5 174.138.33.62 59166
```

Figure 3.11 – Adding proxies

7. Next, to save the configuration file, press *Ctrl* + *X* on your keyboard, then hit *Y* to confirm the filename, and hit *Enter* to exit and return to Terminal.

8. To use Proxychains, use the following command to launch a Firefox browsing instance that will route internet-based traffic via the chain of proxy servers:

```
kali@kali:~$ proxychains4 firefox
```

As shown in the following screenshot, Firefox is routing outbound traffic via the list of proxies from the configuration file:

```
kali@kali: $ proxychains4 firefox
[proxychains] config file found: /etc/proxychains4.conf
[proxychains] preloading /usr/lib/x86_64-linux-gnu/libproxychains.so.4
[proxychains] DLL init: proxychains-ng 4.16
[proxychains] DLL init: proxychains-ng 4.16
[proxychains] DLL init: proxychains-ng 4.16
[proxychains] DLL init: proxychains-ng 4.16
[proxychains] Dynamic chain  ...  104.236.45.251:31226 [proxychains] DLL init: proxychains-ng 4.16
Missing chrome or resource URL: resource://gre/modules/UpdateListener.jsm
Missing chrome or resource URL: resource://gre/modules/UpdateListener.sys.mjs
[proxychains] DLL init: proxychains-ng 4.16
[proxychains] DLL init: proxychains-ng 4.16
[proxychains] DLL init: proxychains-ng 4.16
 ...  timeout
[proxychains] Dynamic chain  ...  174.138.33.62:59166  ...  contile.services.mozilla.com:443  ...  OK
[proxychains] Dynamic chain  ...  174.138.33.62:59166  ...  www.google.com:443  ...  OK
[proxychains] Dynamic chain  ...  174.138.33.62:59166  ...  push.services.mozilla.com:443  ...  OK
[proxychains] Dynamic chain  ...  174.138.33.62:59166  ...  ocsp.pki.goog:80  ...  OK
[proxychains] Dynamic chain  ...  174.138.33.62:59166  ...  ocsp.digicert.com:80  ...  OK
[proxychains] Dynamic chain  ...  174.138.33.62:59166  ...  www.gstatic.com:443  ...  OK
[proxychains] DLL init: proxychains-ng 4.16
[proxychains] Dynamic chain  ...  174.138.33.62:59166  ...  adservice.google.com:443  ...  OK
[proxychains] Dynamic chain  ...  174.138.33.62:59166  ...  googleads.g.doubleclick.net:443  ...  OK
```

Figure 3.12 – Using Proxychains

9. Next, to determine whether you have anonymized your network traffic from Firefox, using the same web browser, go to `https://whatismyipaddress.com/` to verify your public IP address, as shown in the following screenshot:

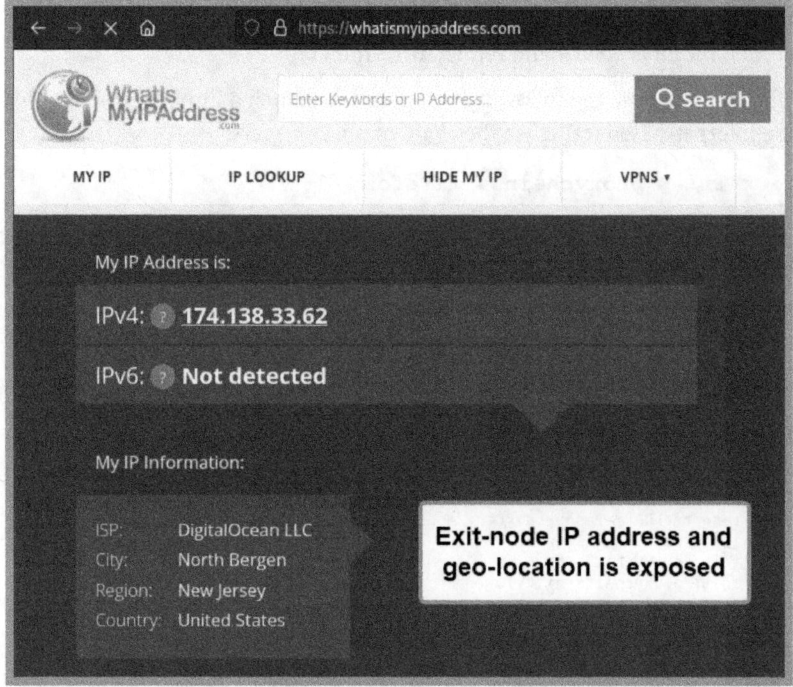

Figure 3.13 – Verifying public IP address for anonymity

As shown in the preceding figure, the public IP address is detected as the exit node, which is the last proxy server on the list from the `proxychains4.conf` file and not the real source IP address.

10. Lastly, always remember that whenever you want to use Proxychains, simply ensure the proxy servers are online and insert the `proxychains4` command at the start of Terminal, as shown in *step 8*.

Next, you will learn how to anonymize your internet-based traffic through the dark web using TOR.

TOR

TOR is an open source project that enables anyone to anonymously browse the internet. The TOR network routes a user's internet-based traffic through multiple nodes, which enables the user to hide their identity and geolocation while using the internet. This is very useful for ethical hackers. When a user sends traffic through the TOR network, it will encrypt the network traffic and wrap it in multiple layers of encryption. As the encrypted traffic is passed onto another node, the receiving node will remove a layer of encryption to reveal the next node to forward the traffic toward, and this process is repeated until the traffic arrives at the target. At this point, the target receives the encrypted data and will not be able to trace the real origin of the packets back to the sender.

The following diagram shows the general concept of using TOR:

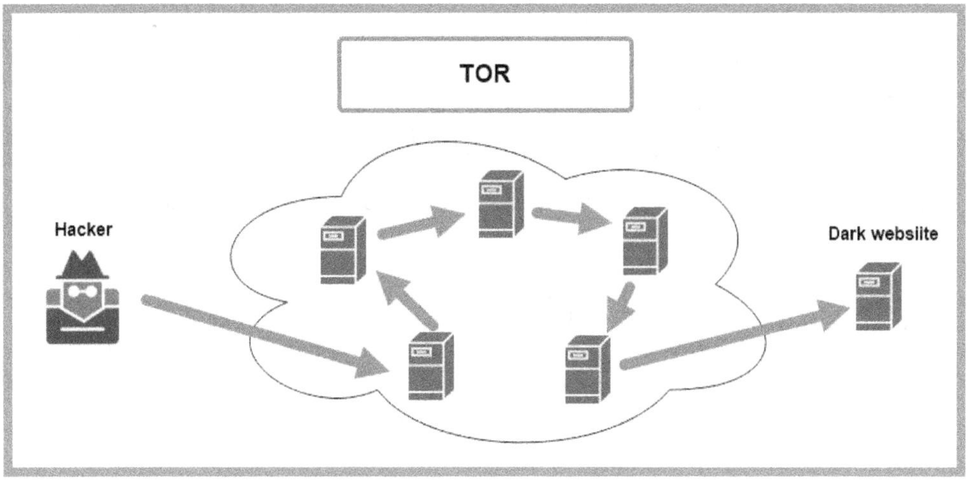

Figure 3.14 – Routing using TOR

Within the TOR network, each node only knows the previous and the next node to create a chaining effect similar to proxy chaining. However, the nodes do not know the sender (source) or the final destination (target) of the traffic. This creates a lot of complexity for any node within the TOR network or anyone to determine the real origin and destination of the traffic that's routed over the TOR network. As an additional layer of security, the traffic is encrypted, decrypted, and re-encrypted multiple times to prevent anyone who is intercepting the traffic to identify the original data or the source of the traffic. Using TOR enables users to browse the internet and access dark web websites that use the `.onion` domain.

The following is the process of what happens when you connect to the TOR network:

1. The ethical hacker will connect to TOR and open the TOR Browser.
2. The TOR browser will encrypt the outbound traffic to the first relay node within the TOR network.
3. The first node will receive the traffic, decrypt the first layer of encryption to determine the next node, and forward the traffic to the next node.
4. This process is repeated along the way as each node will decrypt one layer of encryption on the traffic, then forward the traffic to the next until it reaches the final (exit) node on the TOR network.
5. The final or exit node in TOR will forward the request to the target.
6. When the target responds, the returning traffic is sent back through the TOR network using the same path in reverse order. Each node will encrypt the traffic before forwarding it to the next. This process is repeated until it arrives at the final (exit) node in reverse order.
7. The final (exit) node will forward the encrypted traffic back to your TOR browser, which will decrypt and display the returning data from the target.

To get started setting up TOR and the TOR browser, please use the following instructions:

1. Open VirtualBox Manager and power on the Kali Linux virtual machine.
2. Once the Kali Linux virtual machine is powered on, open Terminal and use the following command to update the package repository list:

```
kali@kali:~$ sudo apt update
```

The following figure shows the complete execution of the preceding command:

```
kali@kali:~$ sudo apt update
[sudo] password for kali:
Get:2 https://download.docker.com/linux/debian bullseye InRelease [43.3 kB]
Get:3 https://download.docker.com/linux/debian bullseye/stable amd64 Packages [17.2 kB]
Get:1 http://kali.download/kali kali-rolling InRelease [30.6 kB]
Get:4 http://kali.download/kali kali-rolling/main amd64 Packages [19.2 MB]
Get:5 http://kali.download/kali kali-rolling/main amd64 Contents (deb) [44.0 MB]
Get:6 http://kali.download/kali kali-rolling/contrib amd64 Packages [111 kB]
Get:7 http://kali.download/kali kali-rolling/contrib amd64 Contents (deb) [164 kB]
Get:8 http://kali.download/kali kali-rolling/non-free amd64 Packages [237 kB]
Get:9 http://kali.download/kali kali-rolling/non-free amd64 Contents (deb) [922 kB]
Fetched 64.7 MB in 10s (6,419 kB/s)
Reading package lists... Done
Building dependency tree... Done
Reading state information... Done
1179 packages can be upgraded. Run 'apt list --upgradable' to see them.
```

Figure 3.15 – Updating the package repository

3. Next, use the following command to download and install both TOR and the TOR browser on Kali Linux:

```
kali@kali:~$ sudo apt install -y tor torbrowser-launcher
```

The following figure shows the expected output when executing the preceding commands:

```
kali@kali:~$ sudo apt install -y tor torbrowser-launcher
Reading package lists... Done
Building dependency tree... Done
Reading state information... Done
The following additional packages will be installed:
  tor-geoipdb torsocks
Suggested packages:
  mixmaster apparmor-utils nyx obfs4proxy
The following NEW packages will be installed:
  tor tor-geoipdb torbrowser-launcher torsocks
0 upgraded, 4 newly installed, 0 to remove and 1179 not upgraded.
Need to get 3,626 kB of archives.
After this operation, 17.4 MB of additional disk space will be used.
```

Figure 3.16 – Setting up TOR

4. Next, use the following command to launch the TOR browser:

```
kali@kali:~$ torbrowser-launcher
```

The following figure shows the TOR browser is set up before launch:

```
kali@kali: $ torbrowser-launcher
Tor Browser Launcher
By Micah Lee, licensed under MIT
version 0.3.6
https://github.com/micahflee/torbrowser-launcher
Creating GnuPG homedir /home/kali/.local/share/torbrowser/gnupg_homedir
Downloading Tor Browser for the first time.
Downloading https://aus1.torproject.org/torbrowser/update_3/release/Linux_x86_64-gcc3/x/ALL
Latest version: 12.0.2
Downloading https://dist.torproject.org/torbrowser/12.0.2/tor-browser-linux64-12.0.2_ALL.tar.xz.asc
Downloading https://dist.torproject.org/torbrowser/12.0.2/tor-browser-linux64-12.0.2_ALL.tar.xz
Verifying Signature
Downloading latest Tor Browser signing key ...
Key imported successfully
```

Figure 3.17 – Tor Browser services

5. Next, the TOR browser will automatically appear. Click on **Connect** to establish a connection from the TOR browser to the TOR network, as shown in the following screenshot:

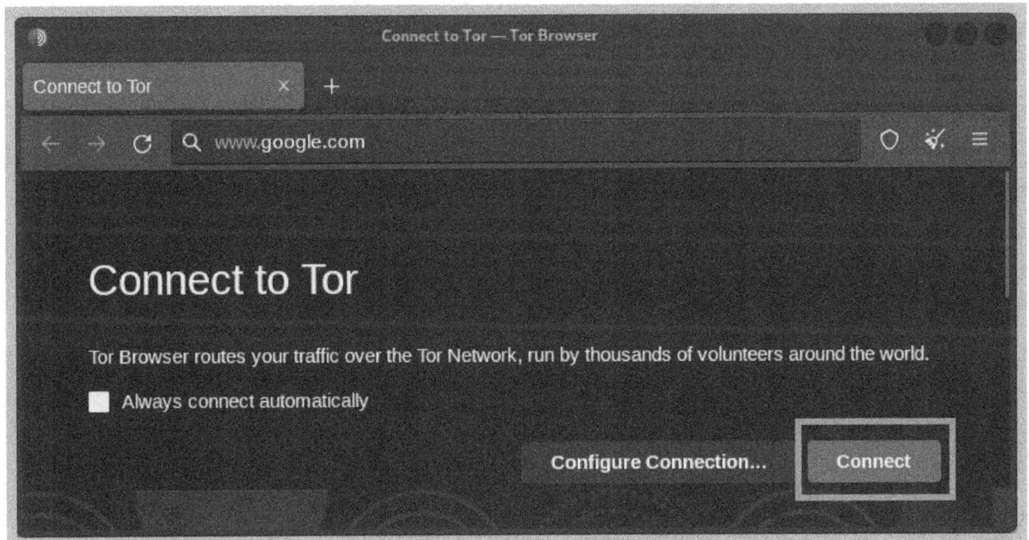

Figure 3.18 – TOR browser

Then, you will see the connection is being established to the TOR network, as shown in the following screenshot:

Figure 3.19 – Connecting to the dark web

6. Once the connection is established, go to `https://whatismyipaddress.com/` to verify whether your real address and geolocation are anonymized, as shown in the following screenshot:

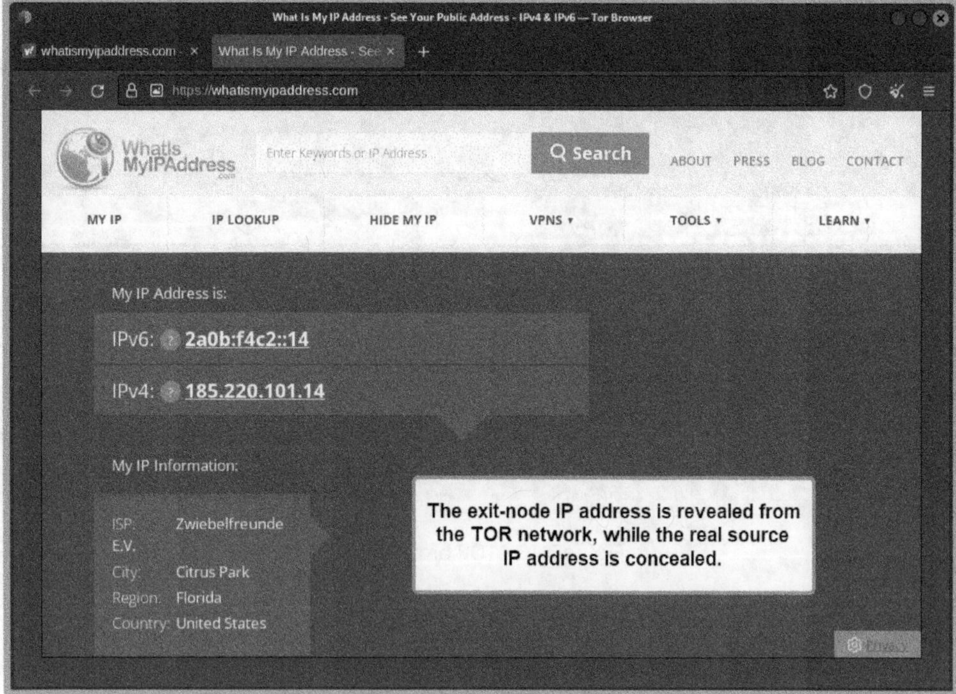

Figure 3.20 – Verify anonymity

As shown in the preceding figure, the real source address and geolocation were not revealed. However, only web traffic that is originating from the TOR browser is anonymized via the TOR network.

> **Important note**
>
> If you want to visit any website on the dark web that has a `.onion` address, ensure you use only TOR and the TOR browser. Do not download anything, do not trust anything, and do not trust anyone. Only visit at your own risk. Be warned.

7. Next, close the TOR browser to terminate the connection to the TOR network.

To route traffic from other applications and tools from Kali Linux through the TOR network, you'll need to modify the `proxychains4.conf` file by using the following instructions:

1. On Kali Linux, open Terminal and use the following command to open the `proxychains4.conf` file using `nano` as the command-line text editor:

```
kali@kali:~$ sudo nano /etc/proxychains4.conf
```

2. Next, when the `proxychains4.conf` file appears, scroll down to the end, uncomment `socks4 127.0.0.1 9050`, and insert comments (#) at the beginning of each line that contains a proxy server, as shown in the following screenshot:

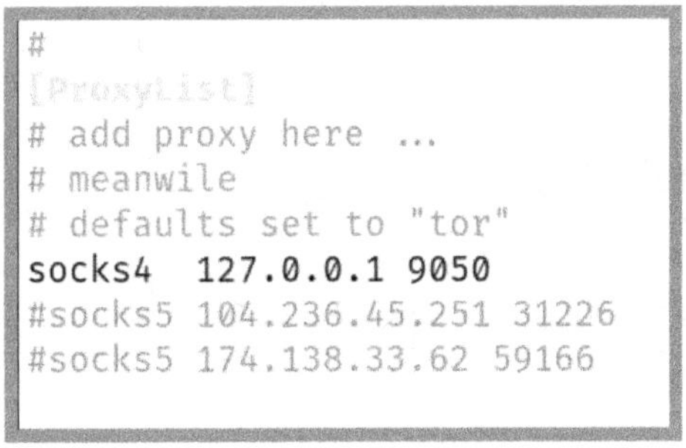

```
#
[ProxyList]
# add proxy here ...
# meanwile
# defaults set to "tor"
socks4  127.0.0.1 9050
#socks5 104.236.45.251 31226
#socks5 174.138.33.62 59166
```

Figure 3.21 – Modifying the Proxychains file

3. Next, to save the configuration file, press *Ctrl* + *X* on your keyboard, then *Y* to use the current filename, and *Enter* to confirm and return to Terminal.

4. Next, use the following commands on Terminal to start the TOR service and verify its status:

```
kali@kali:~$ sudo systemctl start tor
kali@kali:~$ sudo systemctl status tor
```

5. Next, use the following commands to connect to `https://ifconfig.co/` and verify whether your internet-based traffic is anonymized:

```
kali@kali:~$ proxychains4 curl ifconfig.co
```

The following figure shows the public IP address that was detected by the website. Since the internet-based traffic was routed through the TOR network, it was anonymized as the address shown is the IP address of the exit node on the TOR network and not the real address:

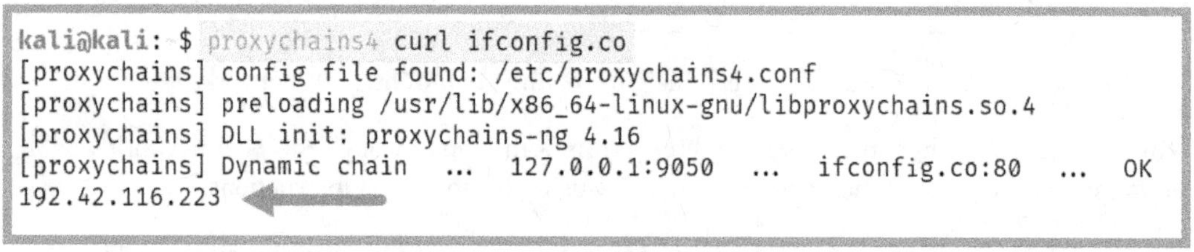

```
kali@kali: $ proxychains4 curl ifconfig.co
[proxychains] config file found: /etc/proxychains4.conf
[proxychains] preloading /usr/lib/x86_64-linux-gnu/libproxychains.so.4
[proxychains] DLL init: proxychains-ng 4.16
[proxychains] Dynamic chain  ...  127.0.0.1:9050  ...  ifconfig.co:80  ...  OK
192.42.116.223
```

Figure 3.22 – Verifying anonymity through TOR

6. Next, since the TOR service is running on Kali Linux, we can use the `proxychains4` command to launch any application while routing its traffic through the TOR network:

```
kali@kali:~$ proxychains4 firefox
```

The following figure shows Firefox is establishing a TOR circuit:

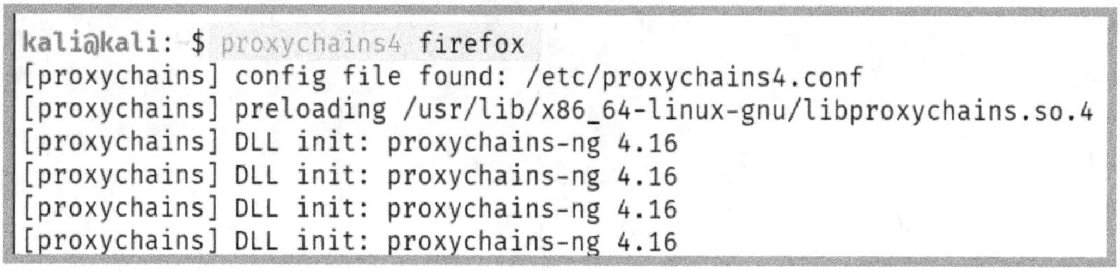

```
kali@kali: $ proxychains4 firefox
[proxychains] config file found: /etc/proxychains4.conf
[proxychains] preloading /usr/lib/x86_64-linux-gnu/libproxychains.so.4
[proxychains] DLL init: proxychains-ng 4.16
[proxychains] DLL init: proxychains-ng 4.16
[proxychains] DLL init: proxychains-ng 4.16
[proxychains] DLL init: proxychains-ng 4.16
```

Figure 3.23 – Routing application traffic through TOR

7. Next, when Firefox appears on the desktop, go to `https://whatismyipaddress.com/` to verify whether the internet-based traffic and geolocation are anonymized, as shown in the following screenshot:

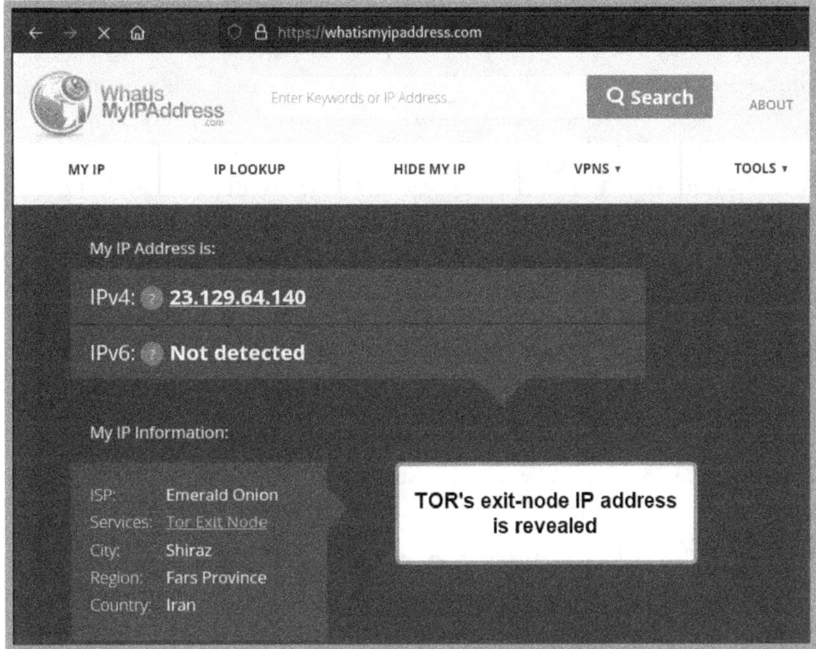

Figure 3.24 – Using TOR to anonymize traffic

As shown in the preceding figure, the IP detector website was able to identify the exit node from the TOR circuit, and not the real IP address or geolocation.

8. Next, to stop the TOR service and terminate the TOR circuit, use the following commands:

```
kali@kali:~$ sudo systemctl stop tor
kali@kali:~$ sudo systemctl status tor
```

The following figure shows the expected results of executing the preceding commands:

```
kali@kali: $ sudo systemctl stop tor

kali@kali: $ sudo systemctl status tor
o tor.service - Anonymizing overlay network for TCP (multi-instance-master)
     Loaded: loaded (/lib/systemd/system/tor.service; disabled; preset: disabled)
     Active: inactive (dead)

Jan 26 12:36:16 kali systemd[1]: Starting Anonymizing overlay network for TCP (multi-instance-master)...
Jan 26 12:36:16 kali systemd[1]: Finished Anonymizing overlay network for TCP (multi-instance-master).
Jan 26 12:43:15 kali systemd[1]: tor.service: Deactivated successfully.
Jan 26 12:43:15 kali systemd[1]: Stopped Anonymizing overlay network for TCP (multi-instance-master).
```

Figure 3.25 – Disabling the TOR service

As shown in the preceding figure, the TOR service is now inactive.

Having completed this section, you have learned how to anonymize your internet traffic to prevent your target from tracing your OSINT operations back to you.

Summary

During the course of this chapter, you learned how ethical hackers can use passive information-gathering techniques and OSINT to collect, gather, and analyze publicly available data from the internet to better understand their targets and security vulnerabilities. Additionally, you discovered how to hide your identity and anonymize your internet-based traffic as an ethical hacker using a VPN, Proxychains, and TOR.

I hope this chapter has been informative for you and is helpful in your journey in the cybersecurity industry. In the next chapter, *Chapter 4, Domain and DNS Intelligence*, you will gain the practical skills that are needed in the industry to efficiently perform OSINT operations on a target as an ethical hacker.

Further reading

- Steps of footprinting: `https://www.eccouncil.org/cybersecurity-exchange/penetration-testing/footprinting-steps-penetration-testing/`
- Understanding OSINT: `https://www.crowdstrike.com/cybersecurity-101/osint-open-source-intelligence/`
- Creating a sock puppet: `https://www.maltego.com/blog/creating-sock-puppets-for-your-investigations/`
- Installing TOR on Kali Linux: `https://www.kali.org/docs/tools/tor/`

4

Domain and DNS Intelligence

As the internet continues to grow, there's more access to information. **Open Source Intelligence** (**OSINT**) can be collected from any publicly available sources, such as the internet. Hackers collect OSINT about their target to help improve the **Tactics, Techniques, and Procedures** (**TTPs**) for their cyber-attack and operations. Hence, it's important for cybersecurity professionals such as ethical hackers and penetration testers to have a solid understanding of how an adversary can collect and leverage OSINT, with the intention of the ethical hacker to use the same TTPs to help reduce the risk of a cyber-attack or threat and improve the cyber defenses of their organization.

Throughout the course of this chapter, you will learn about and gain practical skills commonly used by ethical hackers and penetration testers to gather publicly available information from various online sources and use popular techniques and tools to efficiently profile an organization. You will learn how to leverage data collected from internet search engines, acquire domain intelligence, identify sub-domains from organizations, and perform **Domain Name System** (**DNS**) reconnaissance.

In this chapter, we will cover the following topics:

- Leveraging search engines for OSINT
- Domain intelligence
- Discovering sub-domains
- DNS reconnaissance

Let's dive in!

Technical requirements

To follow along with the exercises in this chapter, please ensure that you have met the following hardware and software requirements:

- Kali Linux: `https://www.kali.org/get-kali/`

Leveraging search engines for OSINT

The internet is the largest network in the world, and navigating and finding resources can be challenging at times, especially if we don't know the domain name or **Uniform Resource Locator** (**URL**) of a specific website. To help us quickly find the right domain name for a website, we can use an internet search engine. Internet search engines crawl and index everything on the internet, which enables a person to perform an internet search using keywords to find all relevant resources while providing safe results.

Without internet search engines, a person needs to know the right domain name, hostname, or URL for the website they want to visit. For instance, imagine a world without internet search engines; each person would need to keep an up-to-date record of each new website and its IP address. If an organization changed its domain name, perhaps due to rebranding, the information would need to be shared with all internet subscribers. However, since the world is a very large place, the update may not reach everyone immediately, and some people may attempt to connect to the old domain name, which may not resolve to the new domain or website.

Hence, internet search engines play an important role on the internet by crawling billions of domain names, websites, and their pages using web crawlers. Web crawlers (bots or spiders) enable the internet search engine provider to navigate the internet, download each web page, and follow web links to efficiently discover new websites and pages that are available on the internet. Hence, internet search engines are tools to help us find and collect intelligence about an organization. As ethical hackers, we can use various internet search engines to find and collect information about our organization and determine what the company has exposed through data leaks.

The following are common internet search engines:

- Google: `https://www.google.com/`
- Yahoo!: `https://www.yahoo.com/`
- Bing: `https://www.bing.com/`
- DuckDuckGo: `https://duckduckgo.com/`
- Yandex: `https://yandex.com/`

> **Important note**
> The DuckDuckGo internet search engine focuses on privacy and does not store the user's internet searches or track their browser. Yandex is a Russian-based internet search engine that focuses on content within the Asia and Europe regions.

A typical person may choose one of the preceding internet search engines, perform an internet search using a keyword(s), and will most likely click on a link on the first page of the search results. As an ethical hacker, it's important to understand that page 1 of the search results usually provides the most accurate data based on the search criteria; however, checking through page 2 and onward will provide additional information that may be meaningful to us about the target. Since internet search engines collect and record data, it's essential to examine all the search results for any connections to the targeted organization.

In addition, each internet search engine uses a unique algorithm for how they crawl and index websites on the internet. As an ethical hacker, using only one internet search engine may not provide the same results that another internet search engine may provide. For instance, if you're looking for specific data about a target, one internet search engine may not provide it due to its algorithm, while another may provide it to you. Hence, it's recommended to always use multiple internet search engines to collect and analyze data found on the first 10 pages of each internet search engine's result pages.

On many websites, you will find a `robots.txt` file that contains a list of directories. This file is commonly used by web administrators to allow or disallow the indexing of specific directories of a website by internet web crawlers. For instance, if a web administrator wanted to prevent internet search engines from indexing and displaying the administrator portal of their website to anyone searching the organization's domain name, they could insert a `disallow /administrator/` statement within the `robots.txt` file.

The following screenshot shows the contents of a sample `robots.txt` file:

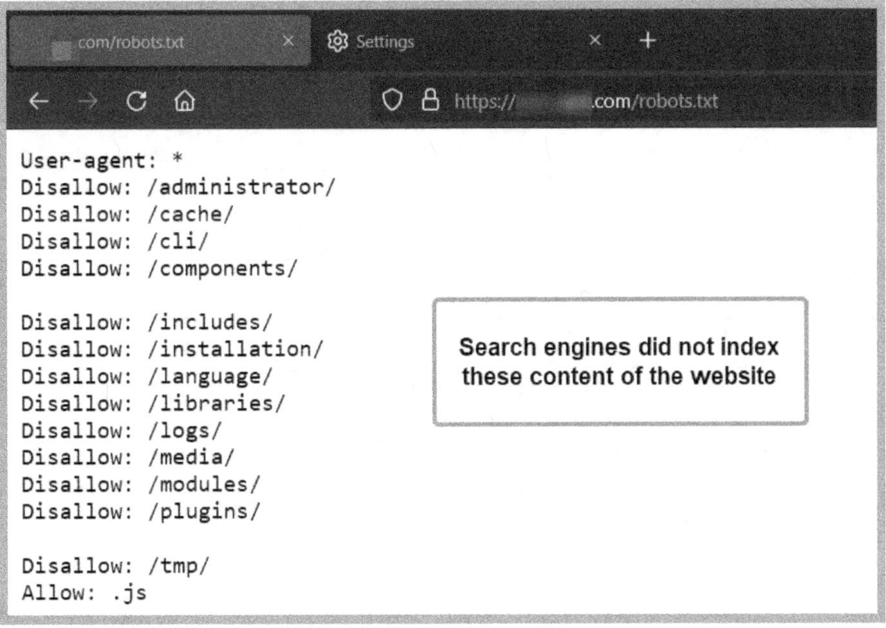

Figure 4.1 – robots.txt file

As shown in the preceding screenshot, the `robots.txt` file seems to contain a listing of sensitive directories that should not appear on internet searches. Since the `robots.txt` file is publicly available on a website, ethical hackers can collect and analyze the data found within it to discover hidden directories of a target, which can be used to plan future operations. Furthermore, visiting each hidden directory may lead to discovering additional data leaks and resources exposed by an organization.

However, since you'll be directly accessing the `robots.txt` file on the target's website, this is a form of active reconnaissance, and the target will know you've made contact. Therefore, it's always recommended to anonymize your internet traffic to help conceal your online identity and reduce your threat level during the reconnaissance phase.

Google hacking techniques

Google hacking (dorking) is a common technique used by ethical hackers to leverage Google's internet search algorithm to find sensitive information about people, places, organizations, and other things on the internet. The term *Google hacking* does not mean we're actually hacking into Google's infrastructure or systems in any way; it is simply a term that's used to describe how we are leveraging the power of Google's search engine to find specific details about our target without performing illegal activities.

For instance, searching for an organization based on a single keyword provides billions of results, as shown in the following screenshot:

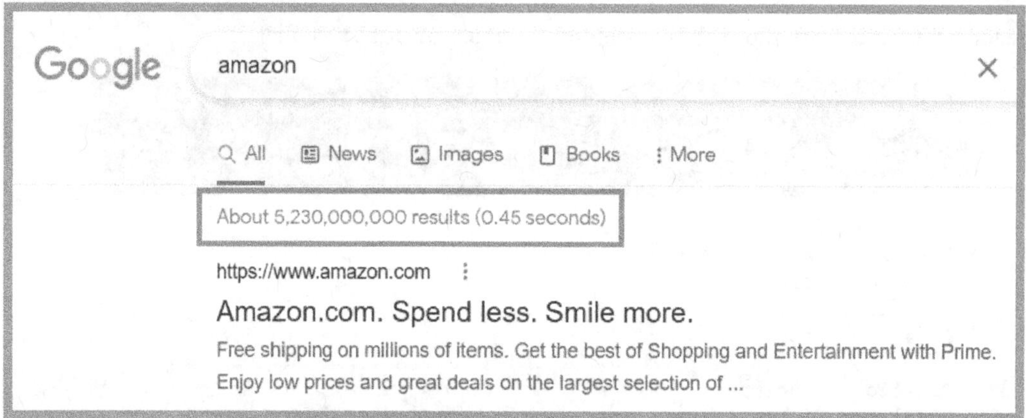

Figure 4.2 – Google search

As an ethical hacker, it's important to collect and analyze the data found not only on page 1 of the search results but all others to create a better profile of the target. However, as with any typical internet search on Google, it provides generic results that are associated with the keyword.

> **Tip**
>
> Ensure you are not signed in to your Google account or the Chrome web browser when collecting information about a target.

To perform Google hacking techniques, it's essential to understand how advanced Google search operators can help us filter specific results. The following are common Google advanced search operators:

- `-` (minus)—The dash is used to negate a term from the search results. For instance, `cybersecurity -Cisco` will provide results that contain the word *cybersecurity* but remove all entries that contain *Cisco*.
- `OR`—This is a conditional operator to provide results of either one term or the other. For example, `cybersecurity OR networking` will provide results containing either the word *cybersecurity* or *networking* in them.
- `AND`—This conditional operator is used to provide results that include two terms. For instance, `football AND england` will return results that contain both keywords.
- `intitle:`—This operator is used to search for title pages that contain a specific term(s). For instance, `intitle: administrator` will search for all websites where *administrator* is found in the title and return these results.
- `inurl:`—This operator searches for URLs that contain a specific keyword. For example, `inurl:security` provides all URLs that contain the keyword *security*.
- `intext:`—This operator enables Google to examine whether a specific keyword is found within the body of a website and provide results for those websites only. For example, `intext:cryptocurrency` provides results of web pages that contain the word *cryptocurrency* within their body.
- `ext:`/ `filetype:`—These operators are used to specify a file extension. For instance, `cybersecurity filetype:pdf` will provide results that contain the word *cybersecurity* and a PDF file.
- `site:`—This operator is used to provide results for a specific domain name. For example, `site:microsoft.com` will provide results for `microsoft.com` only.
- `before: yyyy-mm-dd`—This operator filters results before a particular time. This is useful if you're interested in finding results or data that was published before a certain time.
- `after: yyyy-mm-dd`—This operator filters results after a particular time. This is useful if you're interested in finding results or data that was published after a certain time.
- `*` (asterisk)—This operator is used as a wildcard to match any keyword. For example, `cyber *` will provide results showing anything that starts with *cyber*.

- `Keyword1 keyword2`—This search format is used to provide results with specific keywords. For instance, `cybersecurity microsoft` provides results on these related terms.
- `" "` (quotation marks)—Quotation marks are used to enclose a specific search term such as `"John Doe"`.

The following are common scenarios for using Google advanced search operators to collect intelligence on a target:

- Imagine you want to filter the Google results for a specific domain only. Using the `site:domain-name` syntax provides the following results:

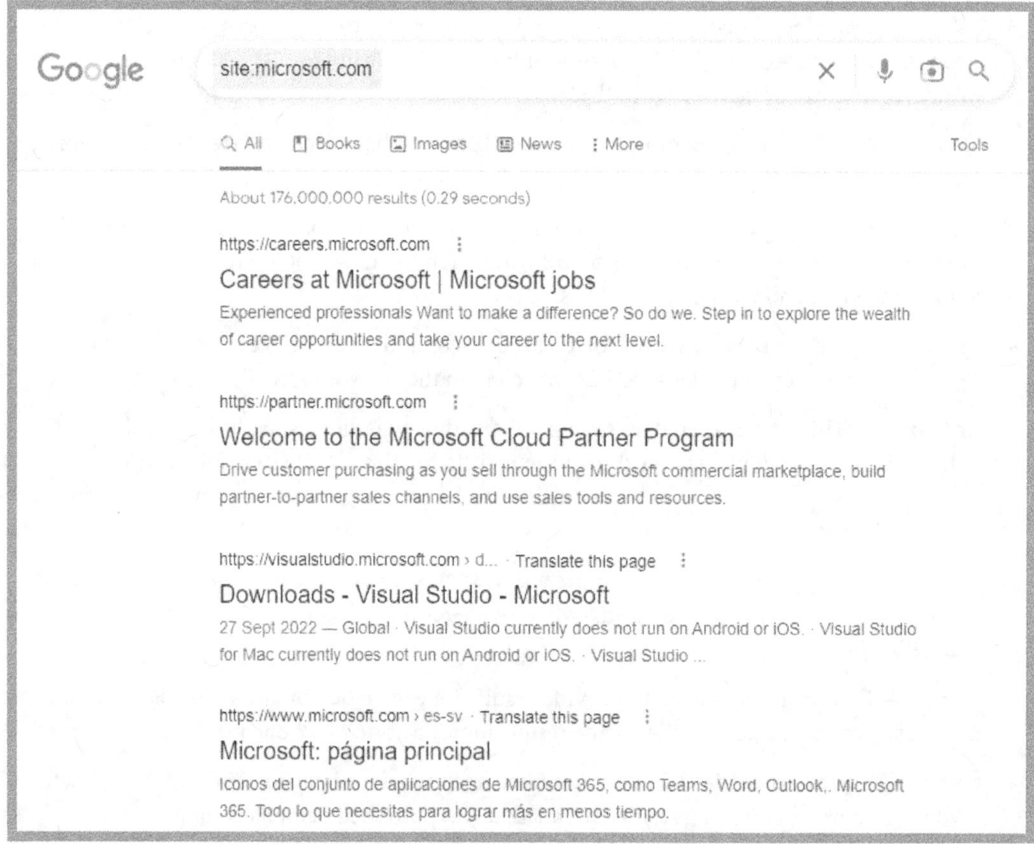

Figure 4.3 – Filtering for a specific domain

- Imagine you're interested in searching for a specific security vulnerability on a particular domain such as Microsoft's website. Using the `printnightmare site:microsoft.com` syntax will provide results for **PrintNightmare** on Microsoft's website:

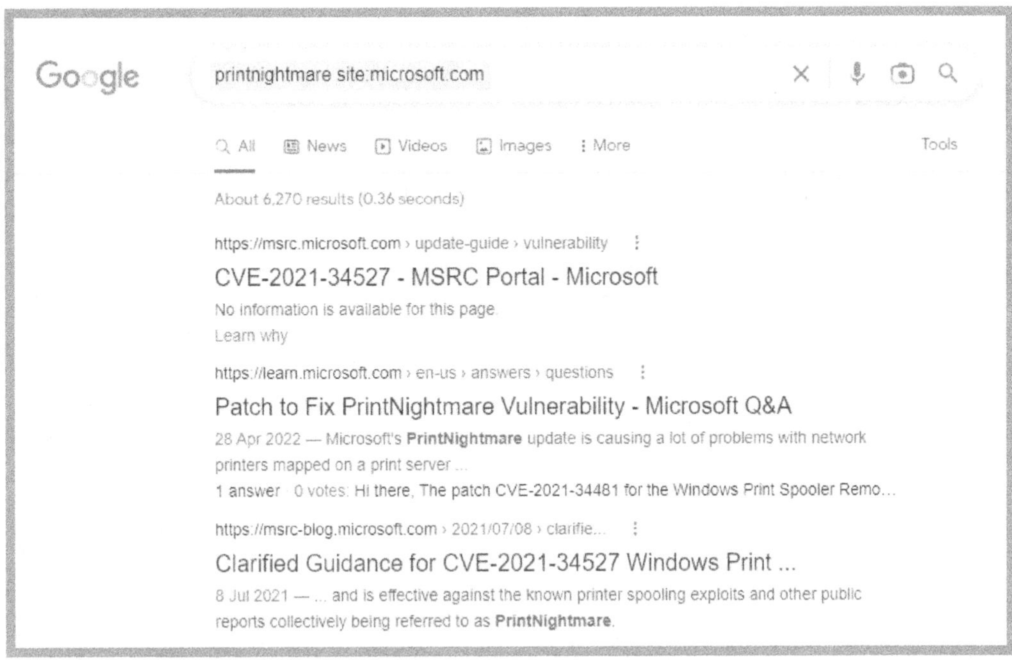

Figure 4.4 – Combining a keyword with the site operator

- If you're interested in finding the customer portals on a targeted website, using the `customer AND login site:domain-name` syntax filters the search results to look for customers and logins on a specific domain, as shown in the following screenshot:

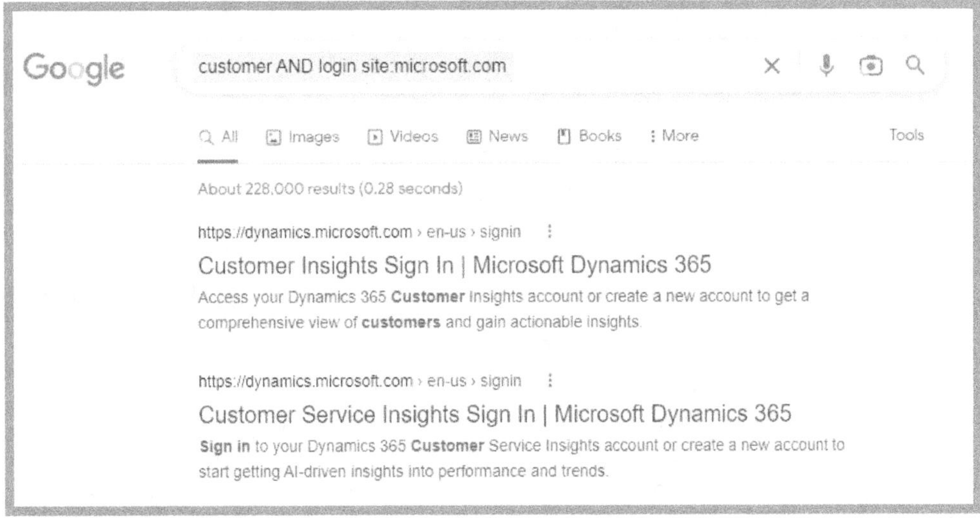

Figure 4.5 – Finding customer portals on a domain

- If you're looking for login portals on a specific domain, using the `"login" site:domain-name` syntax will filter results with the specific keyword *login* on the domain, as shown in the following screenshot:

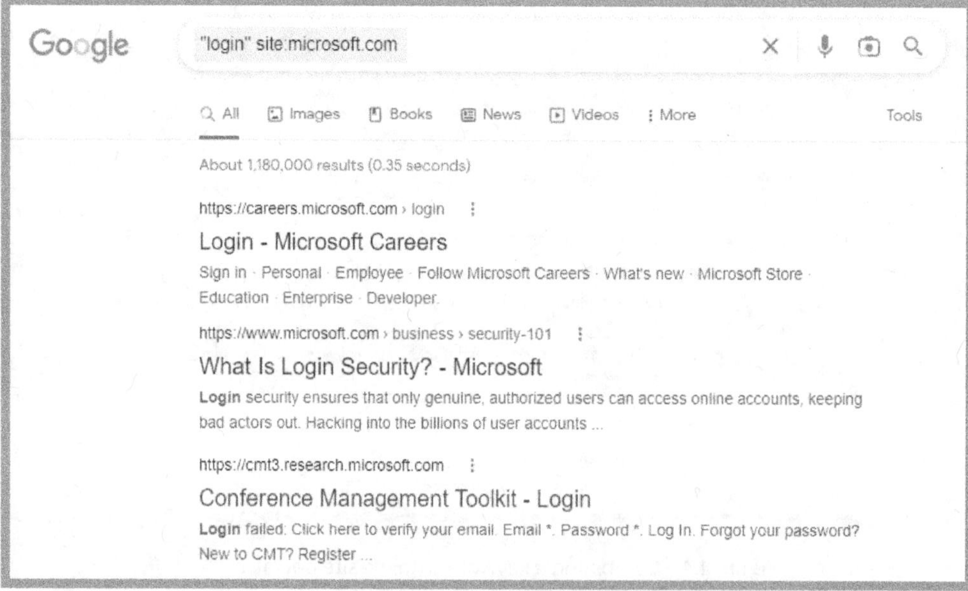

Figure 4.6 – Finding login portals

- Imagine you're interested in finding specific file types on a domain. Using `site:domain-name filetype:txt` or `site:domain-name filetype:pdf` filters the Google search results, as shown in the following screenshot:

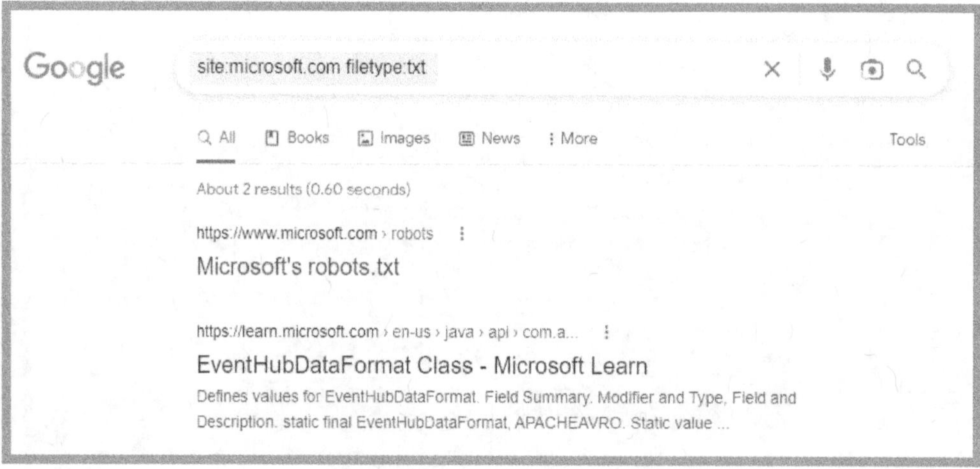

Figure 4.7 – Searching for file types on a domain

- Imagine you're interested in looking for all websites that contain the word *login* within their title for a specific domain. You can do this by using the `site:domain-name intitle:login` syntax, as shown in the following screenshot:

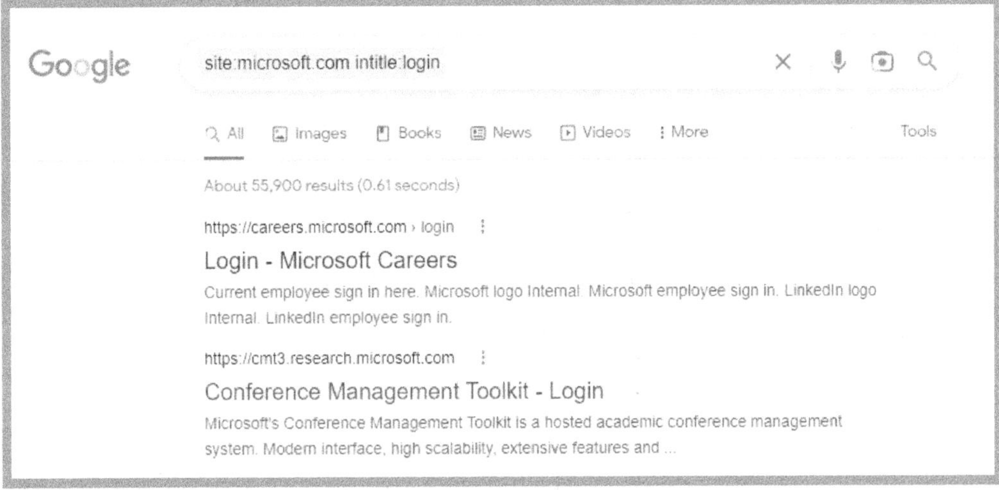

Figure 4.8 – Finding login portals

- To find sub-domains for an organization, use the `site:domain-name -www` syntax to exclude the www parameter, as shown in the following screenshot:

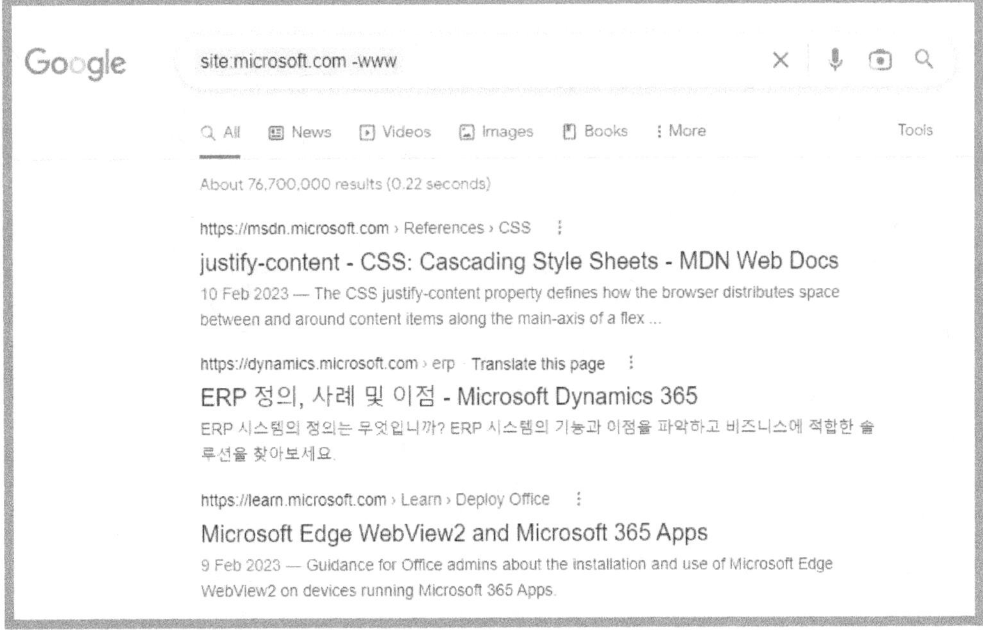

Figure 4.9 – Finding sub-domains

> **Tip**
>
> Additionally, you can use the `site:*.domain.com -site:www.domain.com` Google dork syntax to find sub-domains of a target. For instance, if you want to find sub-domains of *microsoft.com*, the `site:*.microsoft.com -site:www.microsoft.com` syntax will filter the Google search results.

Many possible combinations of Google advanced search operators will provide unique results for us. To help ease the learning curve with Google's advanced search operators, Google has included an **Advanced Search** feature on their main Google search page that reduces the need to use specific advanced search operators.

To access Google's **Advanced Search** feature, please use the following instructions:

1. Go to Google's home page at `https://www.google.com/` and click on **Settings**, as shown in the following screenshot:

Figure 4.10 – Google home page

2. Next, click on **Advanced search** as shown:

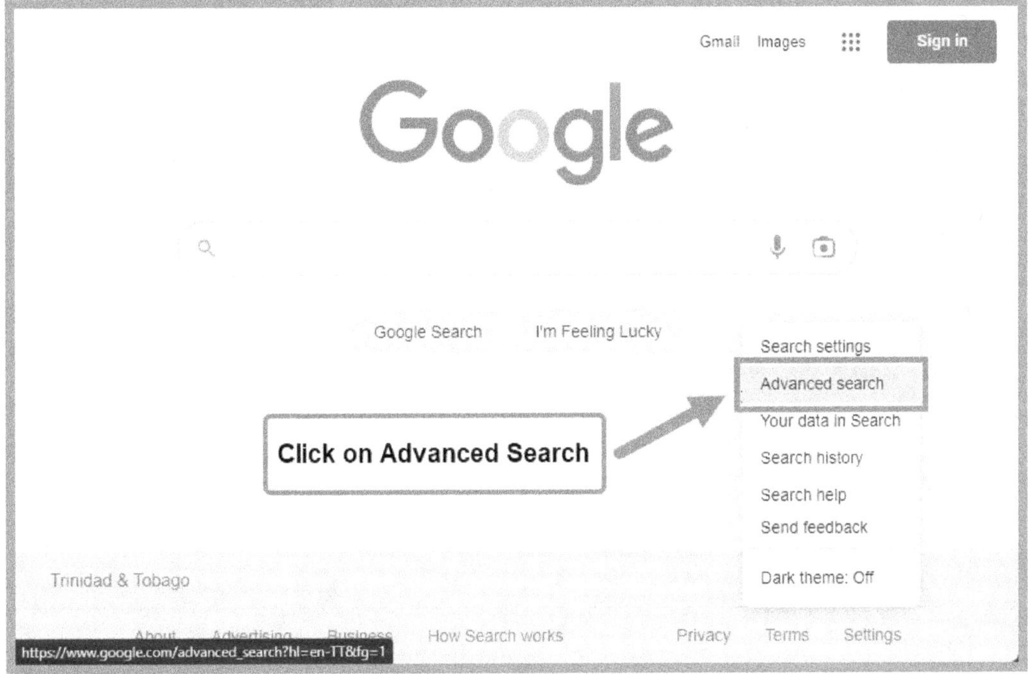

Figure 4.11 – Advanced search

3. Lastly, you'll access the **Advanced Search** menu, which provides many fields to help you perform advanced Google searches, as shown in the following screenshot:

Advanced Search

Find pages with...		To do this in the search box.
all these words:		Type the important words: tri-colour rat terrier
this exact word or phrase:		Put exact words in quotes: "rat terrier"
any of these words:		Type OR between all the words you want: miniature OR standard
none of these words:		Put a minus sign just before words that you don't want: -rodent, -"Jack Russell"
numbers ranging from:	to	Put two full stops between the numbers and add a unit of measurement: 10..35 kg, £300..£500, 2010..2011

Then narrow your results by...		
language:	any language	Find pages in the language that you select.
region:	any region	Find pages published in a particular region.
last update:	anytime	Find pages updated within the time that you specify.
site or domain:		Search one site (like wikipedia.org) or limit your results to a domain like .edu, .org or .gov
terms appearing:	anywhere in the page	Search for terms in the whole page, page title or web address, or links to the page you're looking for.
file type:	any format	Find pages in the format that you prefer.
usage rights:	not filtered by licence	Find pages that you are free to use yourself.

Advanced Search

Figure 4.12 – Google Advanced Search menu

Additionally, **Exploit Database** (`https://www.exploit-db.com/`) manages the **Google Hacking Database** (**GHDB**), a public repository of advanced Google search operators and combinations to find exposed systems and resources on the internet. The GHDB can be found at `https://www.exploit-db.com/google-hacking-database`.

The following screenshot shows some of the many Google dorks found on the GHDB:

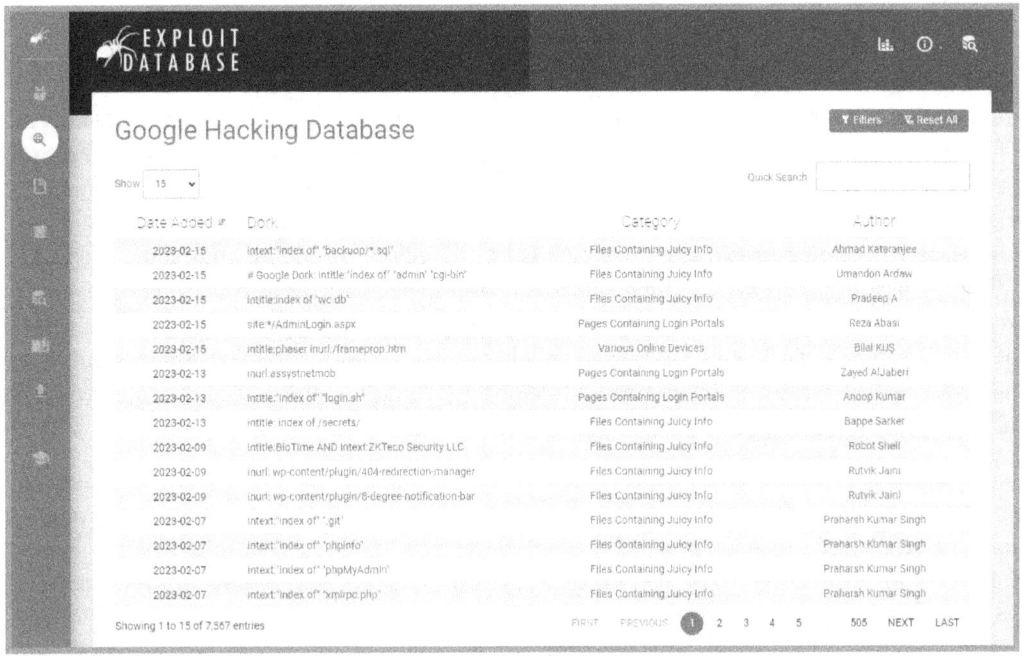

Figure 4.13 – GHDB

As shown in the preceding screenshot, each Google dork can be used to find sensitive and unintentionally exposed systems and files on the internet. Please be very mindful and careful when lurking around using Google hacking techniques. Do not use the information you find for malicious purposes or to cause harm to a system or network.

> **Important note**
>
> As an ethical hacker, Google dorks can help you find publicly exposed documents that contain metadata of username structures, which allows you to gain further insights into the way usernames are created for a targeted organization.

Having completed this section, you have learned how to use internet search engines to collect intelligence about a target. In the next section, you will learn how to collect domain intelligence.

Domain intelligence

Domain intelligence focuses on collecting and analyzing domain registry details and DNS records of a targeted organization to identify their systems, network infrastructure, and attack surface. As the internet continues to expand, there are more registered domain names on the internet as more organizations create an online presence to extend their market reach to new customers around the world. In addition, it's quite easy for anyone to register a public domain name on the internet and use the namespace for their personal usage. Domain names have been around for quite some time, and they help us to easily access websites on the internet. For instance, imagine how challenging it would be for each person to know and record the IP addresses of each website they wanted to visit. If the IP address were unknown, then the user would not know how to connect to the destination website. Similarly to a postal service, if the destination mailing address is unknown, then the courier won't be able to deliver the letter.

Domain names are easy-to-remember addresses that are unique on the internet, and they help us connect to websites such as `google.com` and `amazon.com`. However, a domain name is a string of unique characters that are usually mapped to an IP address of a device on a network. Domain names usually end with a **Top Level Domain** (**TLD**) such as `.com`, `.net`, `.org`, and so on. Different root DNS servers are managed by the **Internet Assigned Numbers Authority** (**IANA**). For instance, the `.com` root DNS server will contain the records for all domain names that have the `.com` TLD and their associated IP addresses. Therefore, if a user wants to retrieve the IP address for `www.google.com`, they will need to query the `.com` root DNS server, which searches its database for all records of `google.com`. Once the `google.com` domain is found within the DNS server, the server then searches for the www record and its associated IP address and responds to the user's computer. You can think of DNS servers as very large directories that contain many records of hostnames and their associated IP addresses, which helps many people around the world to easily connect to their destination websites and servers.

Important note

DNS is a common network protocol that allows a device to resolve a hostname to an IP address. DNS operates on **User Datagram Protocol** (**UDP**) port `53` to handle DNS queries and responses between clients and the DNS server. However, DNS servers also use **Transmission Control Protocol** (**TCP**) port `53` for the zone transfer of records between one DNS server and another.

The following diagram shows the concept of a root DNS server:

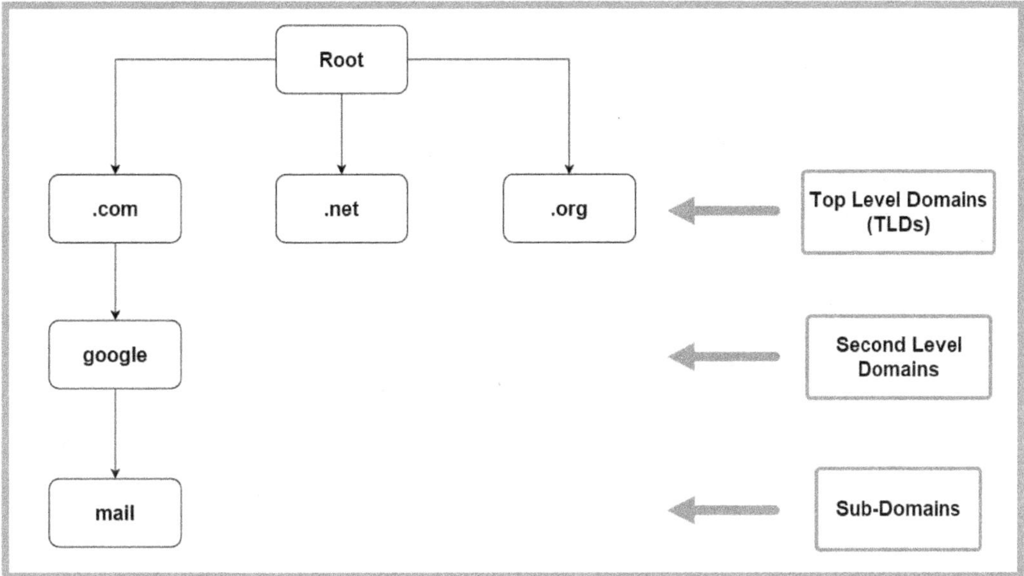

Figure 4.14 – Root DNS server structure

Prior to using DNS servers for domain name resolution, each computer used a `Hosts` file that contained a list of hostnames and their associated IP addresses. This `Hosts` file required frequent updates to ensure it had the latest records. However, managing this file on all client computers became a challenging task, and DNS servers were used to overcome this issue. However, the `Hosts` file on Windows-based operating systems can be found in the `C:\Windows\System32\drivers\etc` folder directory. However, when a computer connects to a network, the **Dynamic Host Configuration Protocol** (**DHCP**) server usually provides the network's DNS server IP addresses to the client. Therefore, the host operating system of the client device no longer queries the `Hosts` file, instead sending all DNS queries to the DNS servers.

The following screenshot shows the contents of the `Hosts` file of a Windows 11 operating system:

```
# Copyright (c) 1993-2009 Microsoft Corp.
#
# This is a sample HOSTS file used by Microsoft TCP/IP for Windows.
#
# This file contains the mappings of IP addresses to host names. Each
# entry should be kept on an individual line. The IP address should
# be placed in the first column followed by the corresponding host name.
# The IP address and the host name should be separated by at least one
# space.
#
# Additionally, comments (such as these) may be inserted on individual
# lines or following the machine name denoted by a '#' symbol.
#
# For example:
#
#      102.54.94.97     rhino.acme.com          # source server
#       38.25.63.10     x.acme.com              # x client host

# localhost name resolution is handled within DNS itself.
#     127.0.0.1       localhost
#     ::1             localhost
```

Figure 4.15 – Hosts file

As shown in the preceding screenshot, the file was originally created to map hostnames to IP addresses. However, all the lines are commented by default to prevent the operating system from reaching any of its contents, including the default entries.

The following are the general steps taken when performing a DNS query for a hostname:

1. When a user enters a domain name or hostname into the address bar of their web browser, the client device checks its local DNS cache for any previous entries that match the hostname. If no previous entries are found, the client device sends a DNS query to the preferred DNS server, as shown in the following screenshot:

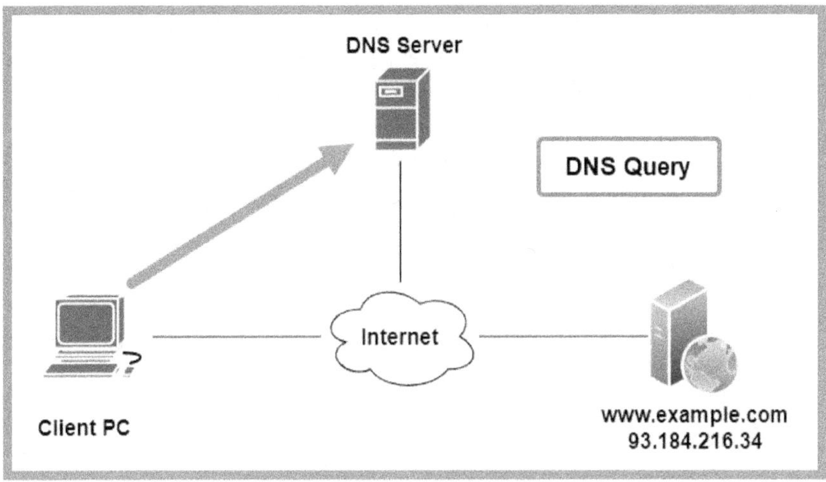

Figure 4.16 – DNS query

> **Tip**
>
> The `ipconfig /displaydns` command will display the local DNS cache on a Windows-based operating system.

2. The DNS server will check its local database for the request record and respond with a DNS reply to the client device with the IP address for the hostname, as shown in the following screenshot:

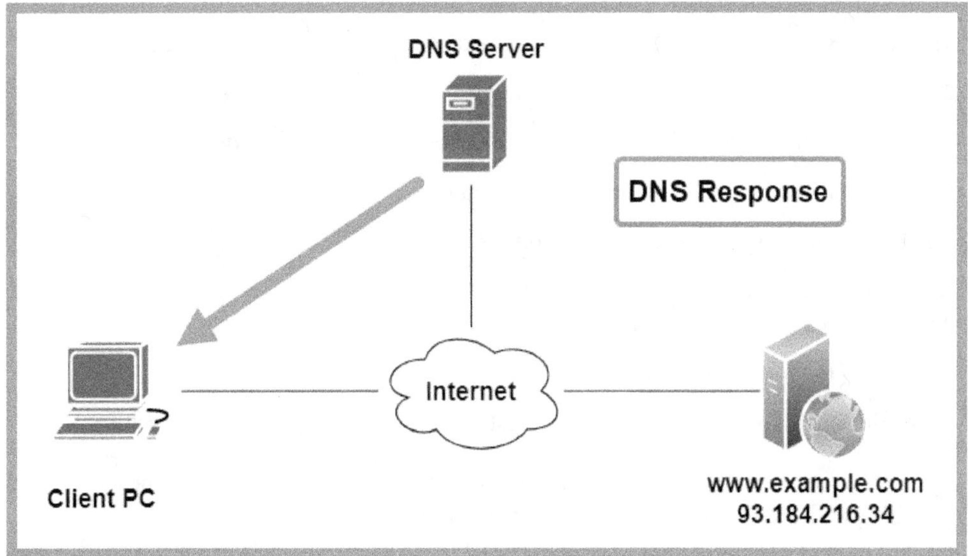

Figure 4.17 – DNS response

3. When the client device receives the IP address, it will insert it as the destination IP address within the `Layer 3` header of the IP packet and send it out on the network to establish a connection to the destination web server, as shown in the following screenshot:

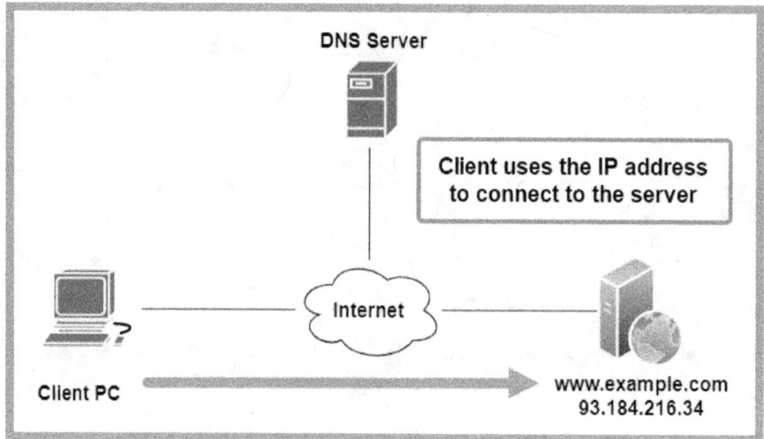

Figure 4.18 – Connection to the destination server

Domain owners and DNS server administrators can create various types of DNS records for a domain, such as the following:

- `A`: This record maps a hostname to an IPv4 address
- `AAAA`: This record maps a hostname to an IPv6 address
- `MX`: This record is used to specify the (e)mail exchange servers for the domain
- `NS`: Contains the name servers for the domain
- `CNAME`: Specifies the canonical name (alias) for the domain or a sub-domain
- `PTR`: Used to resolve an IP address to a hostname
- `RP`: Used to specify the responsible person for the domain
- `TXT`: This record allows the domain owner to specify a text record, which is commonly used to validate ownership of the domain
- `SRV`: Used to specify a service port
- `SOA`: This record is used for storing administrative information about the domain

Public DNS servers are available to anyone on the internet and usually contain a copy of the DNS records for many registered domain names. As an ethical hacker or penetration tester, you can leverage both the domain registration details and DNS records of a domain name to help you better understand the target's network infrastructure, publicly available servers, and additional assets that are owned by the target.

The following diagram shows a visual mind map of key areas for collecting domain intelligence:

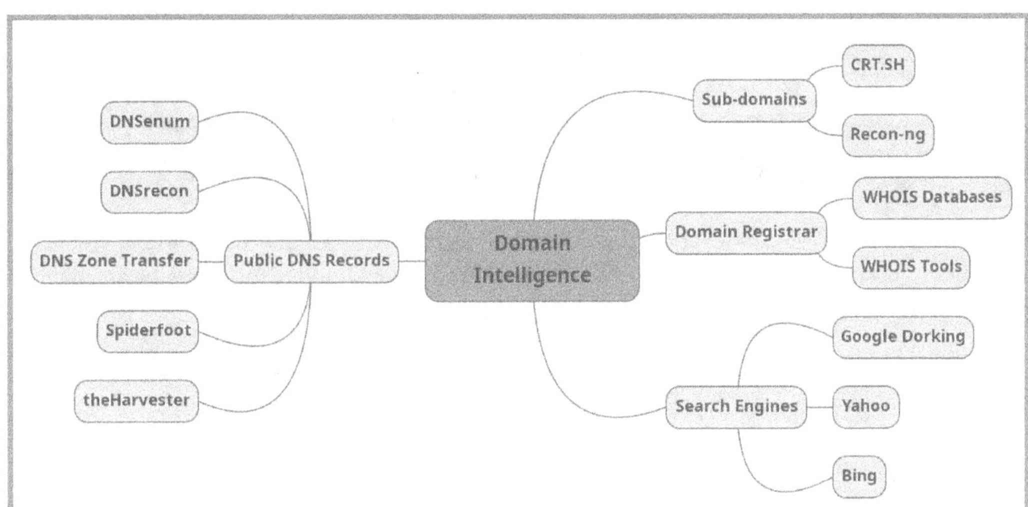

Figure 4.19 – Domain intelligence mind map

As shown in the preceding diagram, ethical hackers can collect domain registration details from public WHOIS databases, identify sub-domains using various tools and internet search engines, and retrieve DNS records from public DNS servers.

Working with WHOIS databases

Registering a domain name on the internet is quite a simple process. The user visits an online domain registrar, checks the availability of a preferred domain, and registers it for a number of years. During the registration process, the domain registrar collects a lot of **Personally Identifiable Information** (**PII**) about the new domain owner, such as the following:

- Full name
- Physical address
- Email address
- Telephone number
- Mobile number
- Organization name

By registering a domain on the internet, the registration details are publicly available by default, enabling anyone to identify the owner of the domain name and other sensitive details. However, many domain registrars provide an additional paid service to keep the registration details private for the duration of the lease. In addition, during the registration process, many people and organizations will choose to pay an additional fee to protect their privacy and keep their domain registration details private. However, not everyone chooses to pay the additional privacy charges, thus leaving their registration details exposed on the internet.

As an ethical hacker, you can leverage the information found on a WHOIS database to create a profile about a target and identify their contact details and physical location. The following are public WHOIS databases:

- `https://who.is/`
- `https://www.whois.com/whois/`
- `https://whois.domaintools.com/`

The preceding list is just a few of the many public WHOIS online databases available. When performing a WHOIS lookup, it's important to query the targeted domain on multiple WHOIS databases to ensure the information is accurate and available.

To get started with performing WHOIS lookup and retrieving domain registration information, please use the following instructions:

1. Firstly, power on your **Kali Linux** virtual machine and verify there's internet connectivity on it.
2. Next, open the Terminal on Kali Linux and use the `whois` command, like so:

   ```
   kali@kali:~$ whois apple.com
   ```

   The following screenshot shows the `whois` response for the targeted domain:

```
kali@kali:~$ whois apple.com
   Domain Name: APPLE.COM
   Registry Domain ID: 1225976_DOMAIN_COM-VRSN
   Registrar WHOIS Server: whois.corporatedomains.com
   Registrar URL: http://cscdbs.com
   Updated Date: 2023-02-16T06:14:38Z
   Creation Date: 1987-02-19T05:00:00Z
   Registry Expiry Date: 2024-02-20T05:00:00Z
   Registrar: CSC Corporate Domains, Inc.
   Registrar IANA ID: 299
   Registrar Abuse Contact Email: domainabuse@cscglobal.com
   Registrar Abuse Contact Phone: 8887802723
   Domain Status: clientTransferProhibited https://icann.org/epp#clientTransferProhibited
   Domain Status: serverDeleteProhibited https://icann.org/epp#serverDeleteProhibited
   Domain Status: serverTransferProhibited https://icann.org/epp#serverTransferProhibited
   Domain Status: serverUpdateProhibited https://icann.org/epp#serverUpdateProhibited
   Name Server: A.NS.APPLE.COM
   Name Server: B.NS.APPLE.COM
   Name Server: C.NS.APPLE.COM
   Name Server: D.NS.APPLE.COM
   DNSSEC: unsigned
   URL of the ICANN Whois Inaccuracy Complaint Form: https://www.icann.org/wicf/
>>> Last update of whois database: 2023-03-05T20:16:21Z <<<
```

Figure 4.20 – whois response

As shown in the preceding screenshot, the `whois` tool attempted to retrieve all the domain registration details, such as the domain registration period, the name servers, the owner's personal information, and their contact details. However, the domain owner has domain privacy to protect their registration details from being exposed on the internet.

3. Next, open the web browser within Kali Linux, go to `https://mxtoolbox.com/Whois.aspx`, and enter a targeted domain name, as shown in the following screenshot:

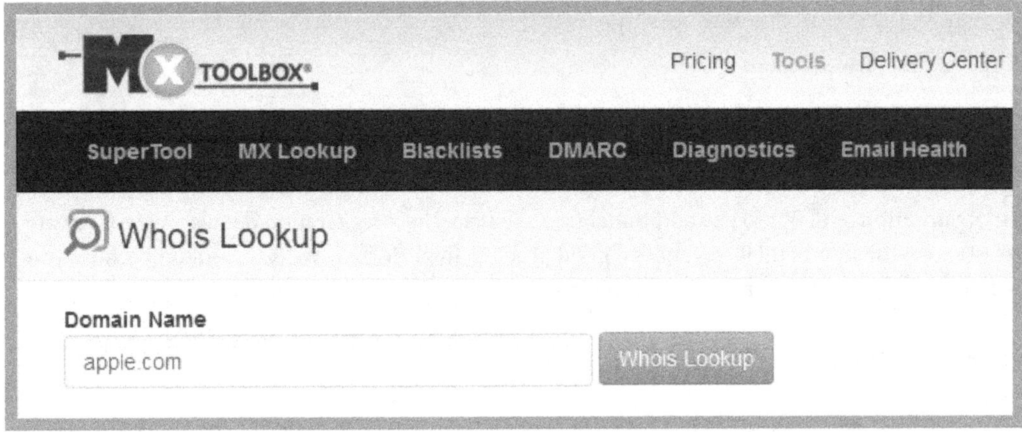

Figure 4.21 – MXToolbox Whois Lookup

As shown in the following screenshot, the MXToolbox WHOIS database provided the name servers of the domain and registration details:

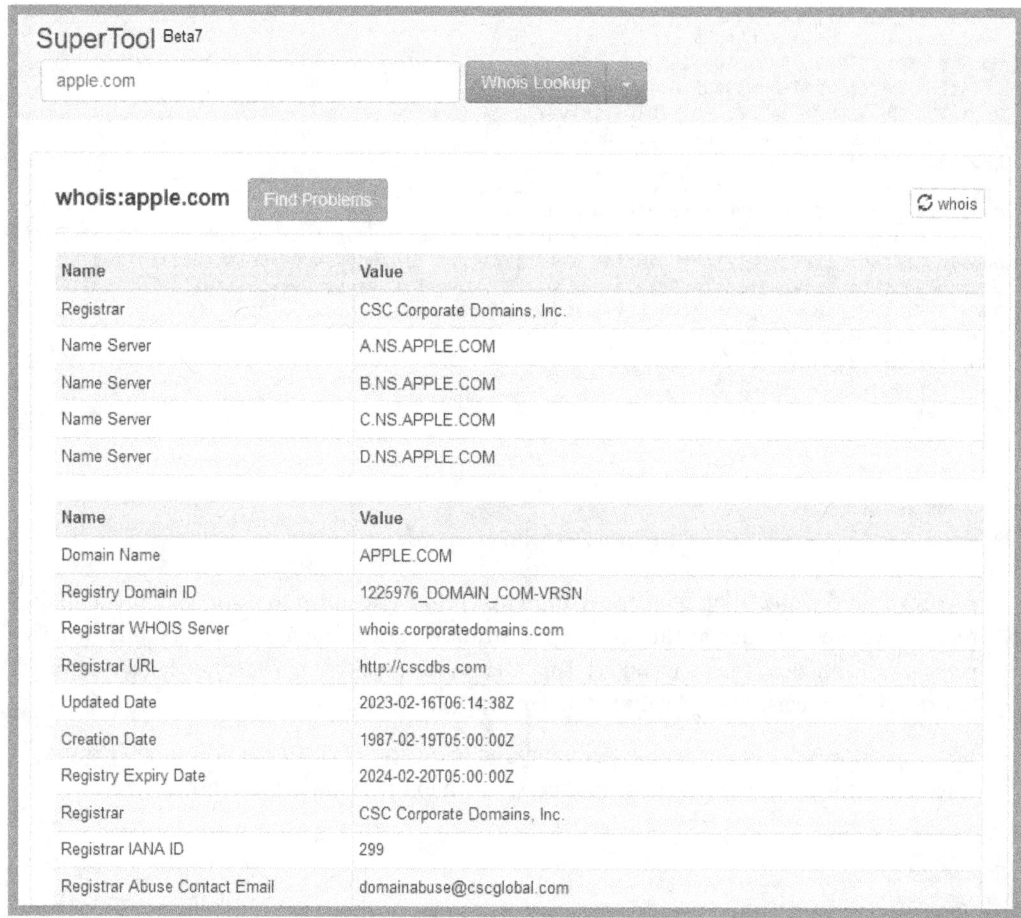

Figure 4.22 – WHOIS results

Most organizations will opt to pay additional fees to ensure their registration details are kept private from users on the internet. However, it's a good practice to perform a WHOIS lookup because you may find something interesting to help you plan future operations in ethical hacking.

Using nslookup for reconnaissance

The `nslookup` tool is a popular network administration tool that is built into many operating systems such as Windows and Linux and enables IT professionals to troubleshoot DNS issues on a network. As an ethical hacker, you can leverage `nslookup` to retrieve domain information such as DNS records from the targeted DNS server and other publicly available DNS servers on the internet.

To get started with learning how to use `nslookup`, please use the following instructions:

1. Firstly, power on the **Kali Linux** virtual machine and ensure it has internet connectivity.
2. Next, open the Terminal on Kali Linux and execute the `nslookup` command. After executing the command, the Terminal prompt will change, and all queries on the Terminal will be sent to the IP address of the DNS server that's configured on Kali Linux.
3. To change the current server to Google's DNS server, use the `server 8.8.8.8` command. This command will send all DNS queries that are performed on the Terminal to Google's DNS servers.
4. Next, enter a domain such as `amazon.com` to perform a query, as shown in the following screenshot:

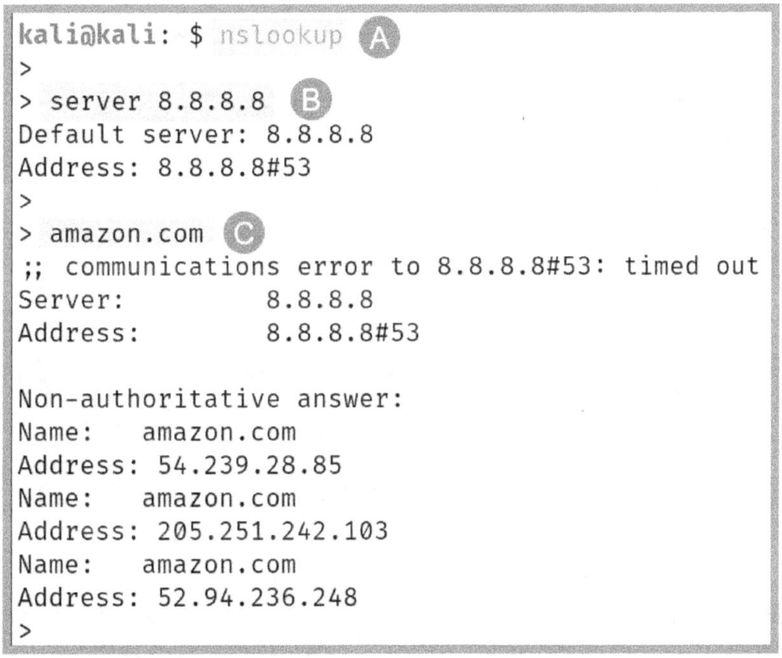

```
kali@kali: $ nslookup  A
>
> server 8.8.8.8  B
Default server: 8.8.8.8
Address: 8.8.8.8#53
>
> amazon.com  C
;; communications error to 8.8.8.8#53: timed out
Server:         8.8.8.8
Address:        8.8.8.8#53

Non-authoritative answer:
Name:    amazon.com
Address: 54.239.28.85
Name:    amazon.com
Address: 205.251.242.103
Name:    amazon.com
Address: 52.94.236.248
>
```

Figure 4.23 – nslookup tool

As shown in the preceding screenshot, `nslookup` sends the DNS query to `8.8.8.8` (Google's DNS server), retrieves the DNS `A` records for the domain name, and shows the public IPv4 addresses of their web servers.

Important note

The non-authoritative answer means the information was retrieved from a DNS server other than the server with the original DNS records. Non-authoritative DNS servers usually have a cached copy of the original DNS records.

5. Next, use the `set type=mx` command to change the query type and enter the targeted domain again to retrieve the email server information:

```
> set type=mx
>
> amazon.com
;; communications error to 8.8.8.8#53: timed out
Server:         8.8.8.8
Address:        8.8.8.8#53

Non-authoritative answer:
amazon.com      mail exchanger = 5 amazon-smtp.amazon.com.
```

Figure 4.24 – Retrieving email server information

As shown in the preceding screenshot, the hostname of the email server was found.

6. To resolve the email server hostname to an IP address, use the `set type=a` command, then enter the hostname of the email server, as shown in the following screenshot:

```
> set type=a
>
> amazon-smtp.amazon.com
;; communications error to 8.8.8.8#53: timed out
Server:         8.8.8.8
Address:        8.8.8.8#53

Non-authoritative answer:
Name:   amazon-smtp.amazon.com
Address: 35.172.144.184
>
```

Figure 4.25 – Resolving a hostname to an IP address

As shown in the preceding screenshot, the public IPv4 address of the email server was retrieved from a public DNS server. Hence, it's important to understand the type of records that are created on a DNS server and how ethical hackers can collect and analyze such information to create a profile about a target.

Discovering sub-domains

Sub-domains are used to both organize and develop a hierarchical structure for domain names, such as dividing a parent domain into smaller parts that can be easily managed by IT professionals. Furthermore, sub-domains are easily created by adding a prefix value to an existing parent domain. For instance, Google owns the `google.com` domain and created `mail.google.com` as a sub-domain that points to its online email platform. This means that sub-domains can have their own IP address, content, and hosting server, and can be managed separately from the parent domain.

As an ethical hacker, it's important to discover sub-domains of a target as they can provide valuable information during your reconnaissance phase. They can help you identify additional systems and network infrastructure and determine the security posture of the target. In addition, you can expand the attack surface, discovering security vulnerabilities that can lead to compromising the target and gaining a foothold. Furthermore, ethical hackers can spoof sub-domains that appear to be legitimate to a target's domain name during a social engineering attack, tricking employees into revealing sensitive details such as their user credentials and even downloading malware onto their company-owned systems.

Certificate searching

Certificate Search is an online database that collects information from digital certificates that are publicly available and found on the internet. Digital certificates enable a user to validate the authenticity of a system on an untrusted network such as the internet. For instance, hackers can easily set up fake websites on the internet to perform social engineering attacks on their victims, such as tricking a person into revealing their user credentials for their online account. Digital certificates are used to help a user verify the identity of a website or device while identifying the certificate issuer, the organization, and its location.

Keep in mind that digital certificates play an important role in e-commerce as they are used to help improve the security of online transactions and communication to ensure all transmitted data between the user's web browser and the server is encrypted, thus ensuring the encrypted data can only be decrypted by the intended recipient.

Digital certificates are commonly issued by a third-party, trusted organization known as a **certificate authority** (**CA**) on the internet. The CA is responsible for validating the identity of the certificate holder; therefore, the issued certificate contains the following data:

- **Subject name**—This field contains the name of the entity that will be using the certificate. The entity can be a person, website, server, or organization.
- **Public key**—This field contains the public key that will be used with the entity to perform data encryption between the client and server.
- **Issuer name**—This field contains the name of the CA that issued the certificate.
- **Validity dates**—This field indicates the validity duration of the certificate, after which the certificate will expire and become untrusted if it's not renewed by the entity/certificate holder.
- **Certificate serial number**—The serial number is a unique value assigned by the CA for identifying the certification.
- **Signature algorithm**—This field contains the cryptographic algorithms that are used by the CA to digitally sign the certificate.
- **Thumbprint**—This field contains a **unique identifier** (**UID**) value that is used for verification.

The following screenshot shows general information on a digital certificate for `www.amazon.com`:

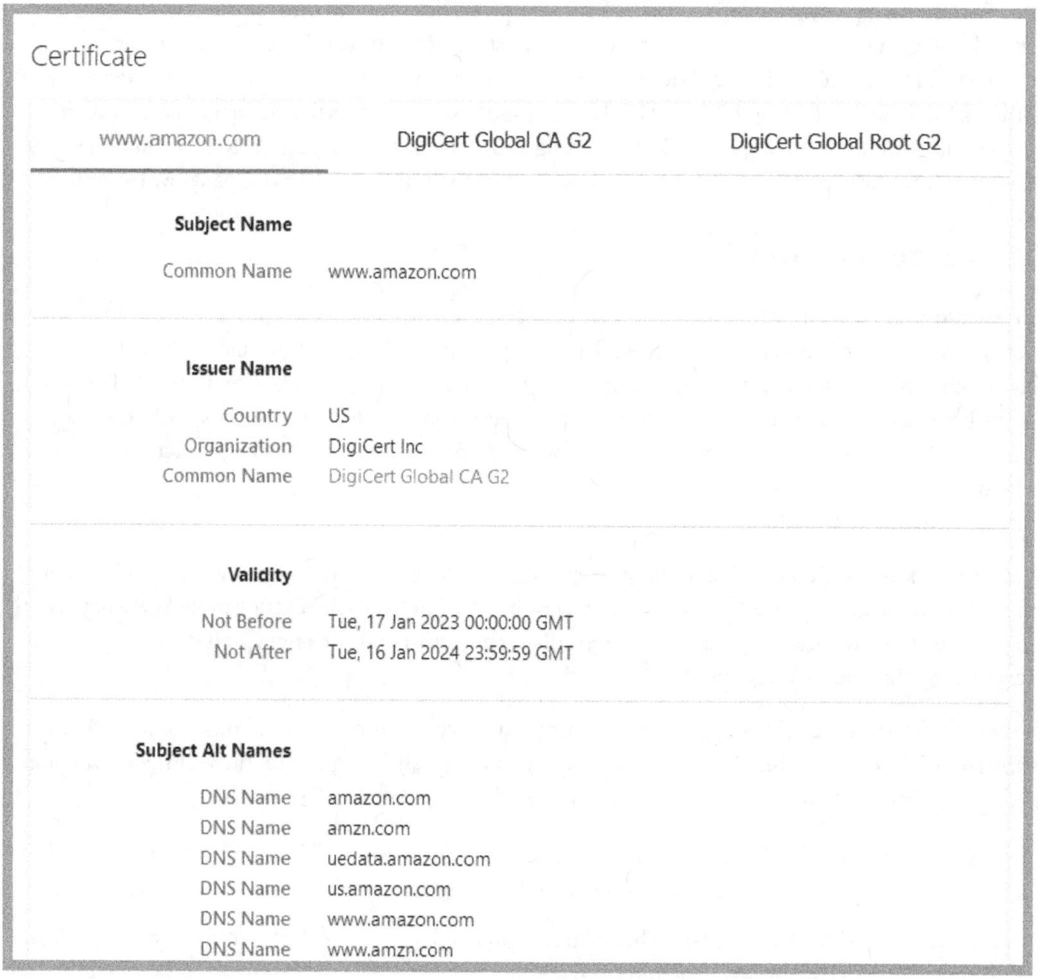

Certificate

www.amazon.com | DigiCert Global CA G2 | DigiCert Global Root G2

Subject Name

Common Name www.amazon.com

Issuer Name

Country US
Organization DigiCert Inc
Common Name DigiCert Global CA G2

Validity

Not Before Tue, 17 Jan 2023 00:00:00 GMT
Not After Tue, 16 Jan 2024 23:59:59 GMT

Subject Alt Names

DNS Name amazon.com
DNS Name amzn.com
DNS Name uedata.amazon.com
DNS Name us.amazon.com
DNS Name www.amazon.com
DNS Name www.amzn.com

Figure 4.26 – Certificate details

As shown in the preceding screenshot, anyone can identify the CA as DigiCert Global Root CA, the intermediate CA as DigiCert Global CA GA, the holder of the certificate as `www.amazon.com`, and its details. Additionally, the **Subject Alt Names** field indicates this digital certificate can be used with the additional sub-domains listed within this field. As an ethical hacker, this field helps us to easily discover additional sub-domains of a targeted organization, which can lead to additional points of entry into a target.

Certificate Search can help us to efficiently discover the digital certificates of an organization using passive reconnaissance. To get started with using Certificate Search as an online OSINT tool to discover the sub-domains of an organization, please use the following instructions:

1. Open the web browser on your computer or Kali Linux virtual machine, and go to the official website of Certificate Search at `https://crt.sh/`.
2. Once the website loads, enter a public domain and click on the **Search** button, as shown in the following screenshot:

Figure 4.27 – Certificate Search

As shown in the preceding screenshot, Certificate Search allows anyone to enter an identity such as a domain name or organizational name, or a certificate fingerprint.

3. Next, Certificate Search will return all results from its database showing all digital certificates that contain `apple.com` as the domain, and sub-domains:

crt.sh Identity Search Group by Issuer

Criteria Type: Identity Match: ILIKE Search: 'apple.com'

crt.sh ID	Logged At	Not Before	Not After	Common Name	Matching Identities	
2382386751	2020-01-27	2013-08-09	2015-08-09	Matt Martin-MPKI SSL-Premium	mattmartin@apple.com	C=US, O="VeriSign, Inc.", OU=VeriSign Trust Netwo
2382089966	2020-01-27	2013-09-23	2015-09-24	food.apple.com	food.apple.com	C=US, O="Entrust, Inc.", OU=www.entrust.net/rpa is L1C
2381996484	2020-01-27	2014-07-22	2015-07-22	ecommerce-qa.apple.com	ecommerce-qa.apple.com	C=US, O="Entrust, Inc.", OU=www.entrust.net/rpa is L1C
2381996432	2020-01-27	2014-05-05	2015-05-05	b2b-test.apple.com	b2b-test.apple.com	C=US, O="Entrust, Inc.", OU=www.entrust.net/rpa is L1C
2381422119	2020-01-27	2014-09-11	2015-09-12	afsportal2.euro.apple.com	afsportal2.euro.apple.com	C=US, O="Entrust, Inc.", OU=www.entrust.net/rpa is L1C
2381408260	2020-01-27	2011-04-11	2015-06-08	b2btest.apple.com	b2btest.apple.com	C=US, O="Entrust, Inc.", OU=www.entrust.net/rpa is L1C
2381374674	2020-01-27	2013-05-01	2015-05-02	hrweb-maint.apple.com	hrweb-maint.apple.com	C=US, O="Entrust, Inc.", OU=www.entrust.net/rpa is L1C
2381374708	2020-01-27	2014-07-25	2015-07-24	wdg02-uat.apple.com	sso-uat-nc.corp.apple.com wdg02-uat.apple.com	C=US, O="Entrust, Inc.", OU=www.entrust.net/rpa is L1C
2381374659	2020-01-27	2014-04-23	2015-04-24	hrweb-qa.apple.com	hrweb-qa.apple.com	C=US, O="Entrust, Inc.", OU=www.entrust.net/rpa is L1C
2380977932	2020-01-26	2014-03-04	2015-03-04	applechinawifi.apple.com	applechinawifi.apple.com	C=US, O="Entrust, Inc.", OU=www.entrust.net/rpa is L1C
2380927914	2020-01-26	2012-11-02	2014-11-02	ray.apple.com	ray.apple.com	C=US, O="Entrust, Inc.", OU=www.entrust.net/rpa is L1C

Figure 4.28 – Certificate results

As shown in the preceding screenshot, the results contain the `crt.sh` ID, the logged date, validity period, **Common Name** (**CN**), matching identities, and additional details. The **Matching Identities** field displays the sub-domains of the organization; these hostnames can be easily resolved to IP addresses using the DNS protocols.

4. Next, click on any `crt.sh` ID to view the details within a digital certificate:

Figure 4.29 – Certificate details

As shown in the preceding screenshot, the digital certificate contains the locale information about the organization, which can be used to find the physical location of the company. Such information is important when conducting a black-box penetration test on a target when only the organization name is given to the ethical hacker or penetration tester. It's important that the ethical hacker finds the accurate physical location of the company to identify the company's wireless networks and physical security presence.

As you have seen in the preceding steps, Certificate Search is a valuable resource to help us discover digital certificates associated with a target and displays their locale and sub-domain information, which helps ethical hackers to create a profile about their targets. Next, you will learn how to discover sub-domains using an OSINT tool on Kali Linux.

Working with Recon-ng

Recon-ng is a popular open source reconnaissance tool that helps ethical hackers and penetration testers collect and analyze OSINT from the internet. This tool can be used to identify whether an organization is leaking sensitive data that can be leveraged by potential threat actors when planning a future cyber-attack. Recon-ng provides a command-line interface and contains various modules that are designed for specific tasks and collecting OSINT from various data sources.

The following are some advantages of using Recon-ng for collecting and analyzing OSINT:

- **Automation**—Recon-ng can be automated to collect OSINT from online data sources
- **Modules**—There are a lot of built-in modules that can be customized to provide additional features and capabilities
- **Data management**—Recon-ng has tools that enable a user to both organize and manage collected data efficiently during reconnaissance
- **Reporting**—There are multiple reporting modules and features that enable the user to generate reports in formats such as **HyperText Markup Language** (HTML), **Extensible Markup Language** (XML), and **Comma-Separated Values** (CSV) files

To get started with using Recon-ng to discover and collect sub-domains of a target, please use the following instructions:

1. Firstly, power on the **Kali Linux** virtual machine and ensure it has internet connectivity.
2. Next, open the Terminal and execute the following command to start Recon-ng:

   ```
   kali@kali:~$ recon-ng
   ```

3. Next, Recon-ng has an online marketplace with additional modules that can be downloaded and installed. To install all modules from the marketplace, use the following commands:

   ```
   [recon-ng][default] > marketplace install all
   ```

   The following screenshot shows the additional modules are being downloaded and installed on Recon-ng:

```
[recon-ng][default] > marketplace install all
[*] Module installed: discovery/info_disclosure/cache_snoop
[*] Module installed: discovery/info_disclosure/interesting_files
[*] Module installed: exploitation/injection/command_injector
[*] Module installed: exploitation/injection/xpath_bruter
[*] Module installed: import/csv_file
[*] Module installed: import/list
[*] Module installed: import/masscan
[*] Module installed: import/nmap
[*] Module installed: recon/companies-contacts/bing_linkedin_cache
[*] Module installed: recon/companies-contacts/censys_email_address
[*] Module installed: recon/companies-contacts/pen
```

Figure 4.30 – Installing modules

After the modules are installed, Recon-ng will reload all modules on the application, and you will see a lot of warning messages similar to these:

```
[*] Reloading modules ...
[!] 'google_api' key not set. pushpin module will likely fail at runtime. See 'keys add'.
[!] 'hashes_api' key not set. hashes_org module will likely fail at runtime. See 'keys add'.
[!] 'bing_api' key not set. bing_linkedin_cache module will likely fail at runtime. See 'keys add'.
```

Figure 4.31 – Missing application programming interface (API) keys

As shown in the preceding screenshot, many Recon-ng modules require a valid API key to communicate with an online data source to query and retrieve data about a target. In a later step, you will learn how to add API keys for various Recon-ng modules.

4. Next, to display a list of API keys and the associated module, use the following command:

```
[recon-ng] [default] > keys list
```

As shown in the following screenshot, there are no API keys assigned to any module by default:

```
[recon-ng][default] > keys list

  +-----------------------------------+
  |        Name         | Value |
  +-----------------------------------+
  | binaryedge_api      |       |
  | bing_api            |       |
  | builtwith_api       |       |
  | censysio_id         |       |
  | censysio_secret     |       |
  | flickr_api          |       |
  | fullcontact_api     |       |
  | github_api          |       |
  | google_api          |       |
  | hashes_api          |       |
```

Figure 4.32 – Checking API keys

5. Next, to get an API key for **BuiltWith**, go to https://builtwith.com/ and create a free account. After creating an account, log in and click on **Tools** > **API Access** to get your API key.
6. Next, to get an API key for **HunterIO**, go to https://hunter.io/ and create a free account. When you're logged in, click on your profile and select **</> API** to view your API key for this platform.
7. Repeat the previous steps to get API keys from **VirusTotal** at https://www.virustotal.com/, **Censys** at https://search.censys.io/, and **Shodan** at https://www.shodan.io/. Feel free to get additional API keys for supported modules on Recon-ng.
8. Next, to add an API key to an API-supported module, use the keys add <API module name> <API key value> command on Recon-ng—for instance, keys add builtwith_api 12345.
9. Next, use the keys list command to verify the API keys were added successfully, as shown in the following screenshot:

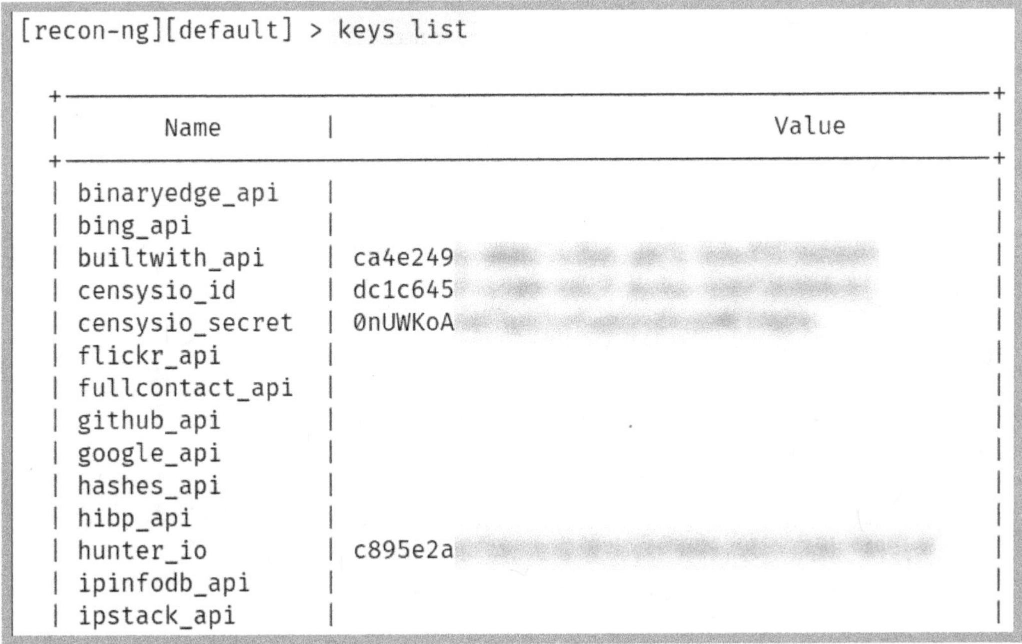

Figure 4.33 – Verifying API keys

10. Next, when collecting data about a target, it's good practice to keep your physical and digital workspace organized to easily find information when needed. Therefore, use the following commands to create a digital workspace on Recon-ng:

```
[recon-ng][default] > workspaces create target1_recon
```

11. Next, use the following command to view a list of all created workspaces within Recon-ng:

```
[recon-ng] [target1_recon] > workspaces list
```

As shown in the following screenshot, the newly created workspace is available and we're currently operating within it:

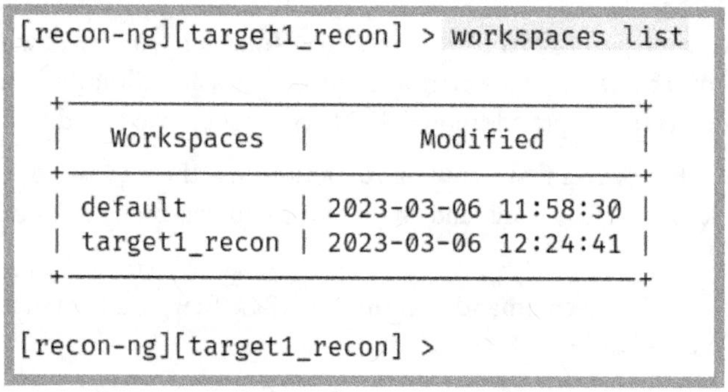

```
[recon-ng][target1_recon] > workspaces list

  +-------------------------------------------+
  |   Workspaces  |         Modified          |
  +-------------------------------------------+
  | default       | 2023-03-06 11:58:30 |
  | target1_recon | 2023-03-06 12:24:41 |
  +-------------------------------------------+

[recon-ng][target1_recon] >
```

Figure 4.34 – Viewing workspaces

12. Next, let's use the `modules search` command to find the `bing_domain` module that will help us identify the geo-location and sub-domains of a target:

```
[recon-ng] [target1_recon] > modules search bing_domain
```

The following screenshot shows two modules were identified and returned:

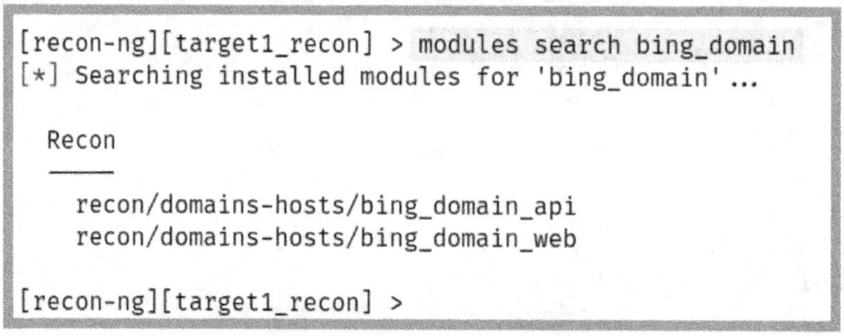

```
[recon-ng][target1_recon] > modules search bing_domain
[*] Searching installed modules for 'bing_domain'...

  Recon
  -----
    recon/domains-hosts/bing_domain_api
    recon/domains-hosts/bing_domain_web

[recon-ng][target1_recon] >
```

Figure 4.35 – Search results

13. Next, let's use the `recon/domains-hosts/bing_domain_web` module and display its description:

```
[recon-ng] [target1_recon] > modules load recon/domains-hosts/
bing_domain_web
[recon-ng] [target1_recon] [bing_domain_web] > info
```

The following screenshot shows the module's description and the required options to use the module:

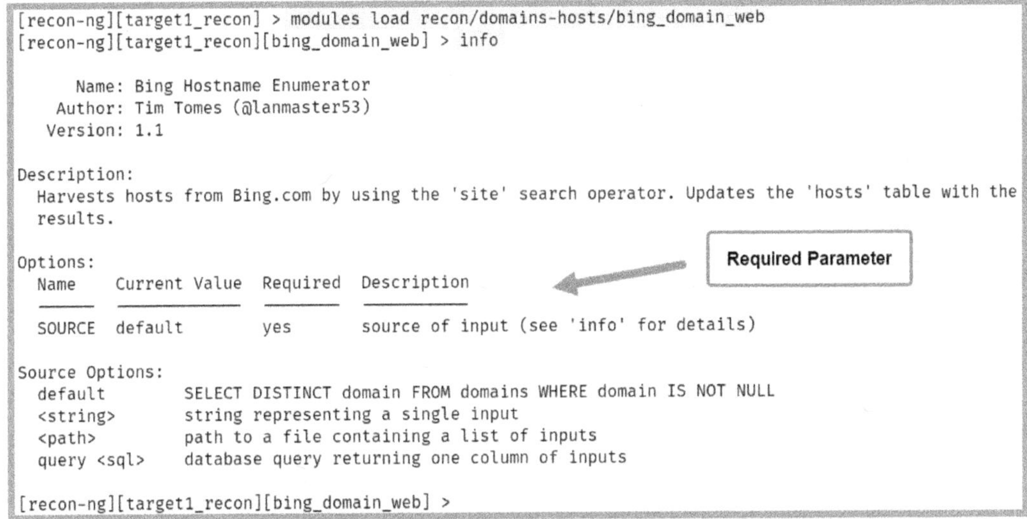

[recon-ng][target1_recon] > modules load recon/domains-hosts/bing_domain_web
[recon-ng][target1_recon][bing_domain_web] > info
Name: Bing Hostname Enumerator
Author: Tim Tomes (@lanmaster53)
Version: 1.1
Description:
Harvests hosts from Bing.com by using the 'site' search operator. Updates the 'hosts' table with the results.
Options:
Name Current Value Required Description
SOURCE default yes source of input (see 'info' for details)
Required Parameter
Source Options:
default SELECT DISTINCT domain FROM domains WHERE domain IS NOT NULL
<string> string representing a single input
<path> path to a file containing a list of inputs
query <sql> database query returning one column of inputs
[recon-ng][target1_recon][bing_domain_web] >

Figure 4.36 – Displaying module descriptions

As shown in the preceding screenshot, the SOURCE value must be set before using the module to collect OSINT from the internet.

14. Next, let's set microsoft.com as the target and execute the module using the following commands:

```
[recon-ng][target1_recon][bing_domain_web] > options set SOURCE microsoft.com
[recon-ng][target1_recon][bing_domain_web] > run
```

> **Tip**
>
> To unset a value within a module, use the options unset <parameter/value> command.

The following screenshot shows the module has started collecting OSINT about the targeted domain name:

```
[recon-ng][target1_recon][bing_domain_web] > run

---------------
MICROSOFT.COM
---------------
[*] URL: https://www.bing.com/search?first=0&q=domain%3Amicrosoft.com
[*] Country: None
[*] Host: www.microsoft.com
[*] Ip_Address: None
[*] Latitude: None
[*] Longitude: None
[*] Notes: None
[*] Region: None
[*] -------------------------------------------------
[*] Country: None
[*] Host: myaccount.microsoft.com
[*] Ip_Address: None
[*] Latitude: None
[*] Longitude: None
[*] Notes: None
[*] Region: None
[*] -------------------------------------------------
```

Figure 4.37 – Running a module

15. Once the module has finished collecting OSINT on the target, execute the `show hosts` command to view the collected data:

```
[recon-ng][target1_recon][bing_domain_web] > show hosts

  +---------------------------------------------------------------------------------------------------------+
  | rowid |                host                | ip_address | region | notes |       module      |
  +---------------------------------------------------------------------------------------------------------+
  | 1     | www.microsoft.com                  |            |        |       | bing_domain_web |
  | 2     | myaccount.microsoft.com            |            |        |       | bing_domain_web |
  | 3     | mysignins.microsoft.com            |            |        |       | bing_domain_web |
  | 4     | support.microsoft.com              |            |        |       | bing_domain_web |
  | 5     | account.microsoft.com              |            |        |       | bing_domain_web |
  | 6     | myaccess.microsoft.com             |            |        |       | bing_domain_web |
  | 7     | admin.microsoft.com                |            |        |       | bing_domain_web |
  | 8     | myapps.microsoft.com               |            |        |       | bing_domain_web |
  | 9     | appsource.microsoft.com            |            |        |       | bing_domain_web |
  | 10    | www.catalog.update.microsoft.com   |            |        |       | bing_domain_web |
  | 11    | learn.microsoft.com                |            |        |       | bing_domain_web |
  | 12    | setup.microsoft.com                |            |        |       | bing_domain_web |
  | 13    | developer.microsoft.com            |            |        |       | bing_domain_web |
  | 14    | apps.microsoft.com                 |            |        |       | bing_domain_web |
  | 15    | client.wvd.microsoft.com           |            |        |       | bing_domain_web |
```

Figure 4.38 – Viewing collected data

16. To exit the current module on Recon-ng, type `back` and hit *Enter*.

Important note

The show command can be used with show [companies] [credentials] [hosts] [locations] [ports] [pushpins] [vulnerabilities] [contacts] [domains] [leaks] [netblocks] [profiles] [repositories] to view specific information that was collected by Recon-ng.

17. Next, display a summary of all OSINT that was collected by Recon-ng:

```
[recon-ng][target1_recon] > dashboard
```

18. The following screenshot shows the summary of activities that were performed and collected by Recon-ng:

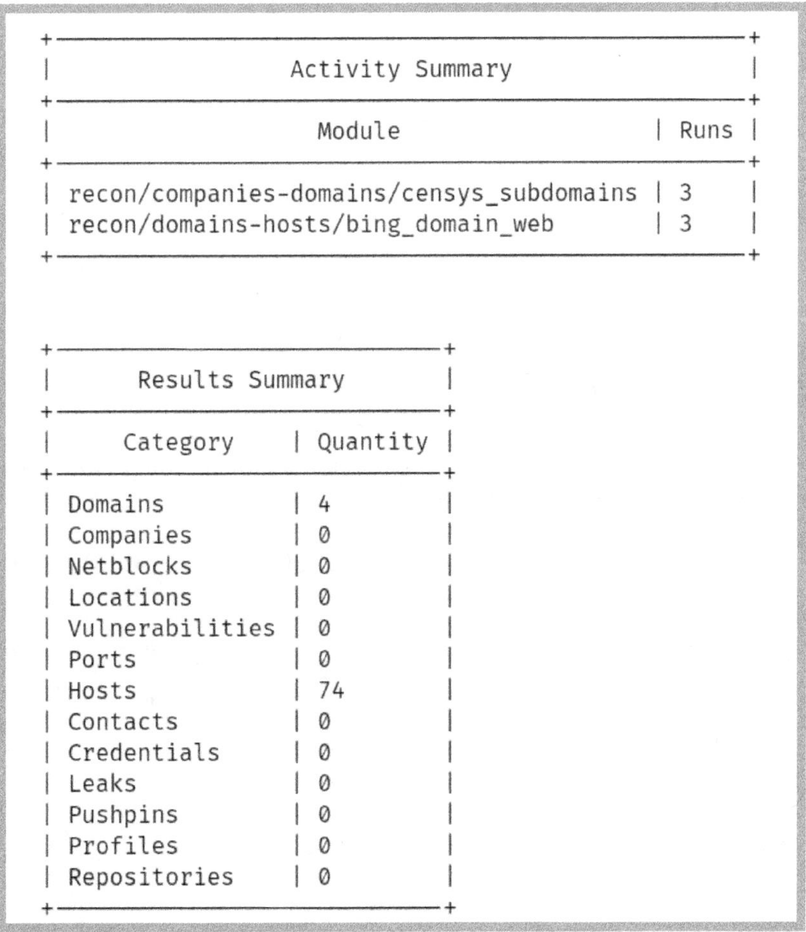

Activity Summary	
Module	Runs
recon/companies-domains/censys_subdomains	3
recon/domains-hosts/bing_domain_web	3

Results Summary	
Category	Quantity
Domains	4
Companies	0
Netblocks	0
Locations	0
Vulnerabilities	0
Ports	0
Hosts	74
Contacts	0
Credentials	0
Leaks	0
Pushpins	0
Profiles	0
Repositories	0

Figure 4.39 – Activity summary on Recon-ng

19. Next, to generate a report using an HTML reporting module, use the following commands:

```
[recon-ng][target1_recon] > modules load reporting/html
[recon-ng][target1_recon][html] > info
[recon-ng][target1_recon][html] > options set CREATOR Glen
[recon-ng][target1_recon][html] > options set CUSTOMER
NewCustomer1
[recon-ng][target1_recon][html] > options set FILENAME /home/
kali/Recon-Report1.html
[recon-ng][target1_recon][html] > run
```

The following screenshot shows the execution of the preceding commands:

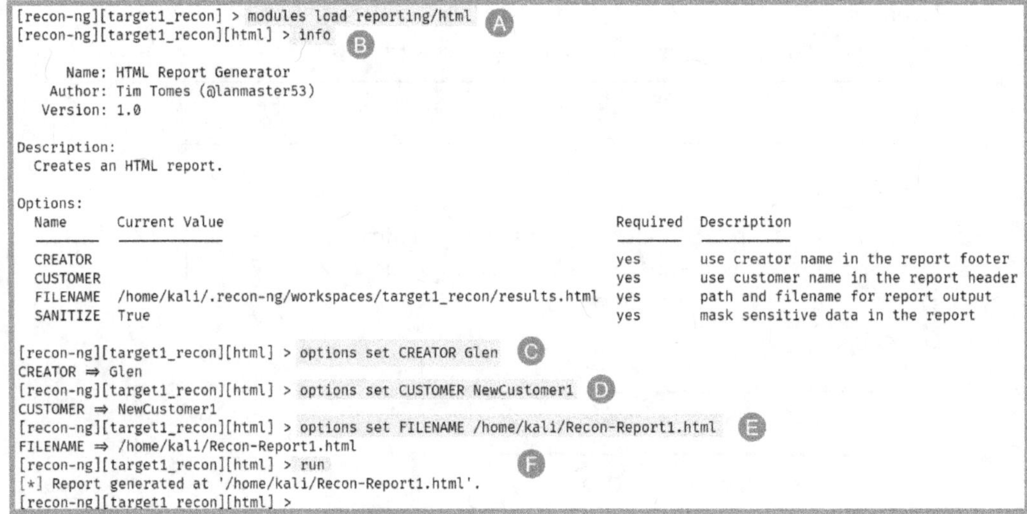

Figure 4.40 – Using the reporting module

As shown in the preceding screenshot, a report was generated and stored in the `/home/kali/` directory.

20. To view the report using a web browser, open a new Terminal and use the following commands:

```
kali@kali:~$ firefox /home/kali/Recon-Report1.html
```

The following screenshot shows the HTML version of the report with the data collected about the target domain name:

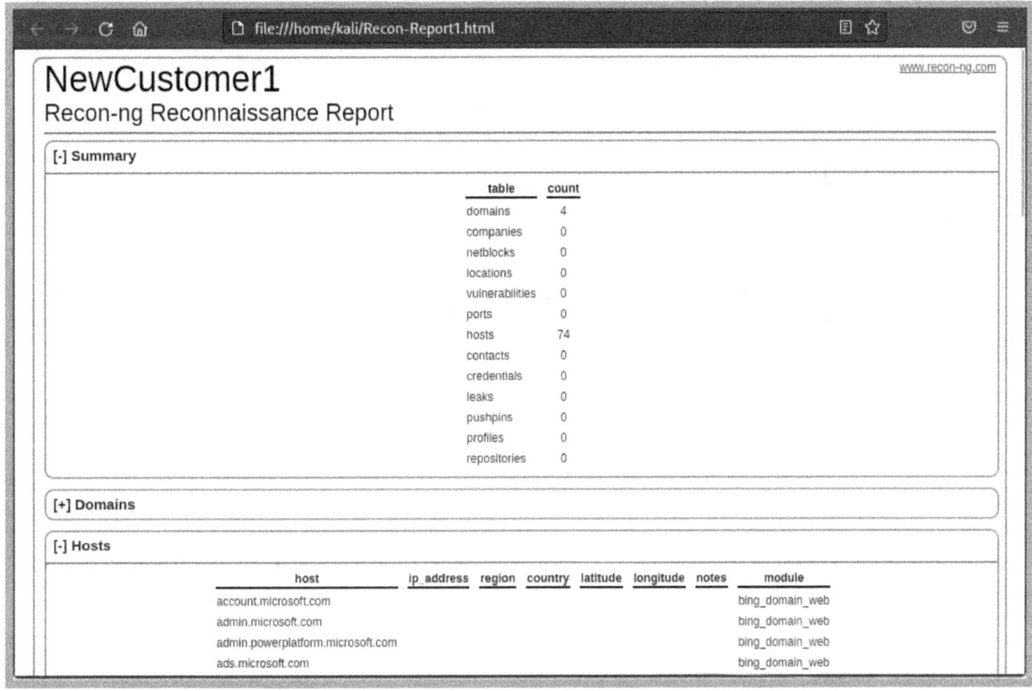

Figure 4.41 – HTML report

21. Lastly, to access the web interface of Recon-ng, open a new Terminal and execute the `recon-web` command, then open a web browser and go to `http://127.0.0.1:5000/`, as shown in the following screenshot:

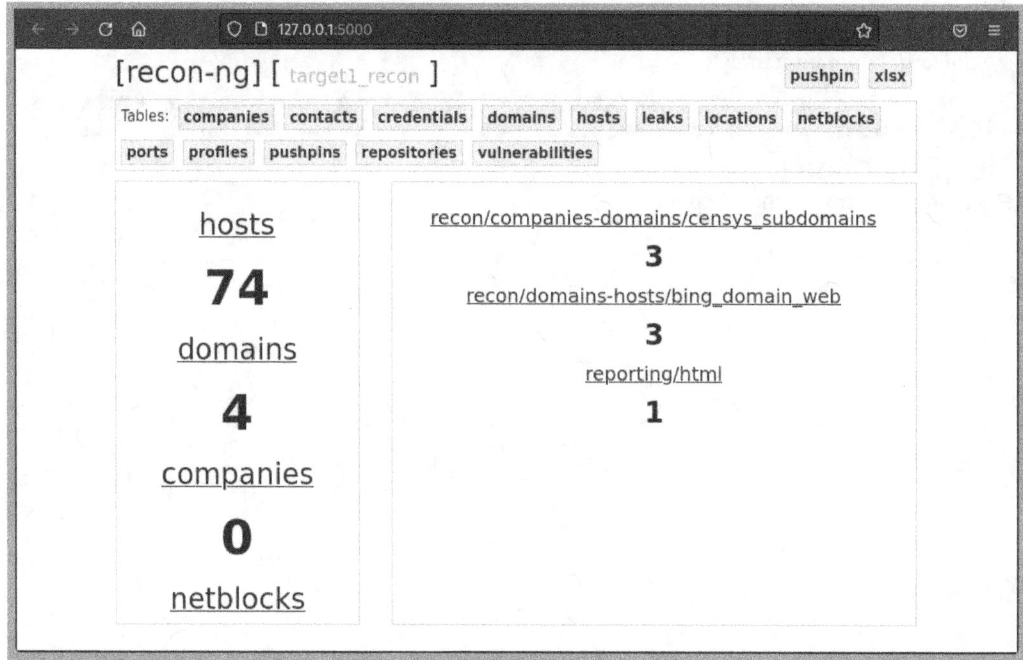

Figure 4.42 – Recon-ng web interface

Having completed this exercise, you have learned the fundamentals of using Recon-ng to collect and analyze OSINT. Next, you will learn how to perform DNS reconnaissance on a target.

DNS reconnaissance

DNS reconnaissance is a common technique used by ethical hackers and threat actors to collect information about a target by gathering publicly available information from DNS servers on the internet. If you recall, DNS is a network protocol that allows a system to resolve a hostname to an IP address, and there are many DNS servers on the internet that contain public DNS records about various registered domain names.

As an ethical hacker, DNS reconnaissance focuses on sending DNS queries to either the target-owned DNS server or a public DNS server to retrieve DNS records about the target's domain name, IP addresses, and sub-domains. The collected responses are analyzed to identify any valuable information about the target such as their systems and network architecture, potential security vulnerabilities, and application platforms.

In this section, you will learn how to use various tools and techniques to efficiently collect and analyze DNS information to create a profile about a target.

Using DNSenum

DNSenum is a common DNS enumeration tool that's used to collect OSINT on domain names. It operates by sending various types of queries to a DNS server about a targeted domain name, IP addresses, email servers, name servers, and sub-domains.

To get started with using DNSenum to collect DNS information about a targeted domain name, please use the following instructions:

1. Firstly, power on the **Kali Linux** virtual machine and ensure it has internet connectivity.
2. Next, open the Terminal and use the following command to query Google's DNS server for DNS records that belong to `apple.com`:

   ```
   kali@kali:~$ dnsenum --dnsserver 8.8.8.8 apple.com
   ```

   As shown in the following screenshot, DNSenum is attempting to retrieve all DNS record types and name servers of the target domain name:

```
kali@kali: $ dnsenum --dnsserver 8.8.8.8 apple.com
dnsenum VERSION:1.2.6

-----   apple.com   -----


Host's addresses:
__________________

apple.com.                              453      IN    A     17.253.144.10


Name Servers:
______________

b.ns.apple.com.                         21600    IN    A     17.253.207.1
a.ns.apple.com.                         21600    IN    A     17.253.200.1
c.ns.apple.com.                         21401    IN    A     204.19.119.1
d.ns.apple.com.                         20043    IN    A     204.26.57.1


Mail (MX) Servers:
___________________

mx-in.g.apple.com.                      30       IN    A     17.32.222.242
mx-in-crk.apple.com.                    841      IN    A     17.72.136.242
mx-in-mdn.apple.com.                    2226     IN    A     17.32.222.242
mx-in-rno.apple.com.                    3288     IN    A     17.179.253.242
mx-in-hfd.apple.com.                    3388     IN    A     17.57.165.2
mx-in-vib.apple.com.                    1630     IN    A     17.57.170.2
```

Figure 4.43 – DNSenum output

As shown in the preceding screenshot, DNSenum was able to retrieve the public IP addresses of the target's web server, name servers, and email servers.

3. Once DNSenum has collected general DNS record details, it will automatically attempt to perform sub-domain enumeration by using a default wordlist to identify common sub-domains of the target, as shown in the following screenshot:

```
Brute forcing with /usr/share/dnsenum/dns.txt:
_______________________________________________

access.apple.com.                        21600   IN    CNAME    www.access.apple.com.
www.access.apple.com.                    21600   IN    A        17.254.3.40
ads.apple.com.                           21600   IN    CNAME    ads.apple.com.akadns.net.
ads.apple.com.akadns.net.                600     IN    CNAME    ioshost.qtlcdn.com.
ioshost.qtlcdn.com.                      20      IN    A        61.161.1.55
ioshost.qtlcdn.com.                      20      IN    A        113.5.170.192
apps.apple.com.                          117     IN    CNAME    itunes-cdn.itunes-apple.com.akadns.net.
itunes-cdn.itunes-apple.com.akadns.net.  3577    IN    CNAME         (
itunes.apple.com.edgekey.net.            7281    IN    CNAME    e673.dsce9.akamaiedge.net.
e673.dsce9.akamaiedge.net.               20      IN    A        96.17.60.35
asia.apple.com.                          178     IN    A        17.253.144.10
autodiscover.apple.com.                  21600   IN    CNAME    mailpex.apple.com.
mailpex.apple.com.                       21600   IN    CNAME    hybridpex.v.aaplimg.com.
hybridpex.v.aaplimg.com.                 30      IN    A        17.32.214.19
av.apple.com.                            300     IN    CNAME    savant-bz.apple.com.
savant-bz.apple.com.                     300     IN    A        17.171.99.83
savant-bz.apple.com.                     300     IN    A        17.171.49.133
```

Figure 4.44 – Sub-domain enumeration

As shown in the preceding screenshot, DNSenum was able to enumerate many sub-domains of the target by performing a brute-force query to Google's DNS server.

Having completed this exercise, you have learned how to automate the collection of DNS records of a target from a public DNS server. Next, you will learn how to use DNSRecon for performing DNS reconnaissance.

Working with DNSRecon

DNSRecon is another common DNS reconnaissance tool that enables ethical hackers and penetration testers to discover DNS records about their target's domain name. This tool helps automate the processes of retrieving various DNS records found on public DNS servers about a specific domain name.

To get started with using DNSRecon, please use the following instructions:

1. Firstly, power on the **Kali Linux** virtual machine and ensure it has internet connectivity.
2. Next, open the Terminal and use the following commands to query Google's DNS server for the DNS records of `apple.com`:

```
kali@kali:~$ dnsrecon -d apple.com -n 8.8.8.8
```

As shown in the following screenshot, DNSRecon was able to retrieve the public DNS records of the target:

```
kali@kali: $ dnsrecon -d apple.com -n 8.8.8.8
[*] std: Performing General Enumeration against: apple.com...
[-] DNSSEC is not configured for apple.com
[*]      SOA usmsc2-extxfr-001.dns.apple.com 17.47.176.10
[*]      NS b.ns.apple.com 17.253.207.1
[*]      NS b.ns.apple.com 2620:149:ae7::53
[*]      NS a.ns.apple.com 17.253.200.1
[*]      NS a.ns.apple.com 2620:149:ae0::53
[*]      NS c.ns.apple.com 204.19.119.1
[*]      NS c.ns.apple.com 2620:171:800:714::1
[*]      NS d.ns.apple.com 204.26.57.1
[*]      NS d.ns.apple.com 2620:171:801:714::1
[*]      MX mx-in.g.apple.com 17.32.222.242
[*]      MX mx-in-crk.apple.com 17.72.136.242
[*]      MX mx-in-mdn.apple.com 17.32.222.242
[*]      MX mx-in-rno.apple.com 17.179.253.242
[*]      MX mx-in-hfd.apple.com 17.57.165.2
[*]      MX mx-in-vib.apple.com 17.57.170.2
[*]      A apple.com 17.253.144.10
[*]      AAAA apple.com 2620:149:af0::10
[*]      TXT apple.com 77a4a6de-da14-449c-83c4-85366e0f55f9
[*]      TXT apple.com apple-domain-verification=X5Jt76bn3Dnmgzjj
```

Figure 4.45 – DNSRecon output

As shown in the preceding screenshot, DNSRecon was able to retrieve various types of DNS records such as A, AAAA, MX, TXT, NS, and more.

3. Lastly, DNSRecon attempted to automatically enumerate the SRV record for the targeted domain, as shown in the following screenshot:

```
[*] Enumerating SRV Records
[+]      SRV _sips._tcp.apple.com gslb-b2b-ext.v.aaplimg.com 17.47.48.136 5061
[+]      SRV _sips._tcp.apple.com gslb-b2b-ext.v.aaplimg.com 17.47.49.71 5061
[+]      SRV _sip._udp.apple.com gslb-b2b-ext.v.aaplimg.com 17.47.49.71 5060
[+]      SRV _sip._udp.apple.com gslb-b2b-ext.v.aaplimg.com 17.47.49.79 5060
[+]      SRV _sip._tcp.apple.com gslb-b2b-ext.v.aaplimg.com 17.47.49.79 5060
[+]      SRV _sip._tcp.apple.com gslb-b2b-ext.v.aaplimg.com 17.47.48.136 5060
[+]      SRV _sip._tls.apple.com gslb-b2b-ext.v.aaplimg.com 17.47.49.72 5060
[+]      SRV _sip._tls.apple.com gslb-b2b-ext.v.aaplimg.com 17.47.49.70 5060
[+] 8 Records Found
```

Figure 4.46 – Enumerating SRV records

Having completed this exercise, you have learned how to enumerate DNS records using the DNSRecon tool to find sub-domains and IP addresses of a target. Next, you will learn how to perform DNS zone transfers on a misconfigured DNS server.

Performing DNS zone transfers

As you have learned so far, DNS servers have an important role in internal networks and on the internet. A DNS zone transfer is the process of copying the zone records or entire contents of a DNS zone from a primary DNS server onto another DNS server. DNS zone transfers are used to ensure DNS data is synchronized between primary and secondary DNS servers. However, the primary (master) DNS server holds the master zone record, and the secondary DNS server holds a copy of the same zone record file. Whenever there's a change to the master zone record on the primary DNS server, the secondary DNS server needs to update its copy of the zone record too, hence the need for DNS zone transfers between DNS servers.

However, if a DNS server is misconfigured on a network or on the internet, an ethical hacker can attempt to copy the master zone records of an organization to identify security vulnerabilities on the assets owned by the target. Imagine a targeted organization did not separate its internal and external namespaces on its DNS server. If a DNS zone transfer is possible, the ethical hacker will be able to retrieve information about the organization's internal systems and external systems. Keep in mind that it's important to ensure ethical hackers obtain explicit permission before attempting to perform DNS zone transfers on customers' DNS servers.

> **Important note**
> The folks at *DigiNinja* (`https://digi.ninja`) have created a vulnerable DNS server environment for learning how DNS zone transfers can impact organizations and have created the `zonetransfer.me` domain, which contains a DNS zone transfer security vulnerability.

To get started with learning how to perform a DNS zone transfer on a vulnerable DNS server, please use the following instructions:

1. Firstly, power on the **Kali Linux** virtual machine and ensure it has internet connectivity.
2. Next, open the Terminal on Kali Linux and execute the following commands to discover the DNS records of the `zonetransfer.me` domain:

   ```
   kali@kali:~$ host zonetransfer.me
   ```

   The following screenshot shows the DNS `A` and `MX` records for the targeted domain name:

```
kali@kali:~$ host zonetransfer.me
zonetransfer.me has address 5.196.105.14
zonetransfer.me mail is handled by 20 ASPMX5.GOOGLEMAIL.COM.
zonetransfer.me mail is handled by 20 ASPMX3.GOOGLEMAIL.COM.
zonetransfer.me mail is handled by 0 ASPMX.L.GOOGLE.COM.
zonetransfer.me mail is handled by 10 ALT2.ASPMX.L.GOOGLE.COM.
zonetransfer.me mail is handled by 10 ALT1.ASPMX.L.GOOGLE.COM.
zonetransfer.me mail is handled by 20 ASPMX2.GOOGLEMAIL.COM.
zonetransfer.me mail is handled by 20 ASPMX4.GOOGLEMAIL.COM.
```

Figure 4.47 – Finding DNS records

3. Next, use the following command to retrieve the name servers of the targeted domain:

   ```
   kali@kali:~$ host -t ns zonetransfer.me
   ```

   The following screenshot shows the name servers of the targeted domain were found:

   ```
   kali@kali:~$ host -t ns zonetransfer.me
   zonetransfer.me name server nsztm2.digi.ninja.
   zonetransfer.me name server nsztm1.digi.ninja.
   ```

   Figure 4.48 – Retrieving name servers

   As shown in the preceding screenshot, the following are the two name servers for the domain:

   - `nsztm2.digi.ninja`
   - `nsztm1.digi.ninja`

   As ethical hackers, we can use these two name servers to determine whether the targeted domain is leaking zone records.

4. Next, let's query the targeted domain name with one of the newly discovered name servers:

   ```
   kali@kali:~$ host -l zonetransfer.me nsztm1.digi.ninja
   ```

   The following screenshot shows the DNS records that were retrieved from the `nsztm1.digi.ninja` name server for the `zonetransfer.me` domain name:

```
kali@kali: $ host -l zonetransfer.me nsztm1.digi.ninja
Using domain server:
Name: nsztm1.digi.ninja
Address: 81.4.108.41#53
Aliases:

zonetransfer.me has address 5.196.105.14
zonetransfer.me name server nsztm1.digi.ninja.
zonetransfer.me name server nsztm2.digi.ninja.
14.105.196.5.IN-ADDR.ARPA.zonetransfer.me domain name pointer www.zonetransfer.me.
asfdbbox.zonetransfer.me has address 127.0.0.1
canberra-office.zonetransfer.me has address 202.14.81.230
dc-office.zonetransfer.me has address 143.228.181.132
deadbeef.zonetransfer.me has IPv6 address dead:beaf::
email.zonetransfer.me has address 74.125.206.26
home.zonetransfer.me has address 127.0.0.1
internal.zonetransfer.me name server intns1.zonetransfer.me.
internal.zonetransfer.me name server intns2.zonetransfer.me.
intns1.zonetransfer.me has address 81.4.108.41
intns2.zonetransfer.me has address 167.88.42.94
office.zonetransfer.me has address 4.23.39.254
ipv6actnow.org.zonetransfer.me has IPv6 address 2001:67c:2e8:11::c100:1332
owa.zonetransfer.me has address 207.46.197.32
alltcpportsopen.firewall.test.zonetransfer.me has address 127.0.0.1
vpn.zonetransfer.me has address 174.36.59.154
```

Figure 4.49 – DNS records

As shown in the preceding screenshot, there are many sensitive and interesting hostnames found within the results. Some of these hostnames and IP addresses may not be intentionally exposed by the organization; however, such information can be leveraged by an ethical hacker to expand the attack surface.

> **Important note**
>
> When performing DNS reconnaissance and zone transfers, ensure you query all name servers for a targeted domain as one server may lack proper security configuration compared with the others.

5. Next, let's use **DNSenum** to automate the DNS zone transfer process on the target domain name:

```
kali@kali:~$ dnsenum zonetransfer.me
```

The following screenshot shows that DNSenum was able to successfully perform a DNS zone transfer from the vulnerable `nsztm2.digi.ninja` name server of the target:

```
Trying Zone Transfer for zonetransfer.me on nsztm2.digi.ninja ...
zonetransfer.me.                        7200     IN    SOA          (
zonetransfer.me.                        300      IN    HINFO        "Casio
zonetransfer.me.                        301      IN    TXT          (
zonetransfer.me.                        7200     IN    MX           0
zonetransfer.me.                        7200     IN    MX           10
zonetransfer.me.                        7200     IN    MX           10
zonetransfer.me.                        7200     IN    MX           20
zonetransfer.me.                        7200     IN    MX           20
zonetransfer.me.                        7200     IN    MX           20
zonetransfer.me.                        7200     IN    MX           20
zonetransfer.me.                        7200     IN    A       5.196.105.14
zonetransfer.me.                        7200     IN    NS      nsztm1.digi.ninja.
zonetransfer.me.                        7200     IN    NS      nsztm2.digi.ninja.
_acme-challenge.zonetransfer.me.        301      IN    TXT          (
_acme-challenge.zonetransfer.me.        301      IN    TXT          (
_sip._tcp.zonetransfer.me.              14000    IN    SRV          0
14.105.196.5.IN-ADDR.ARPA.zonetransfer.me. 7200     IN    PTR     www.zonetransfer.me.
asfdbauthdns.zonetransfer.me.           7900     IN    AFSDB        1
asfdbbox.zonetransfer.me.               7200     IN    A        127.0.0.1
asfdbvolume.zonetransfer.me.            7800     IN    AFSDB        1
canberra-office.zonetransfer.me.        7200     IN    A       202.14.81.230
cmdexec.zonetransfer.me.                300      IN    TXT          ";
contact.zonetransfer.me.                2592000  IN    TXT          (
dc-office.zonetransfer.me.              7200     IN    A       143.228.181.132
deadbeef.zonetransfer.me.               7201     IN    AAAA    dead:beaf::
```

Figure 4.50 – DNS zone transfer

Furthermore, DNSenum was able to retrieve additional interesting records from the vulnerable DNS server:

```
intns1.zonetransfer.me.                 300      IN    A        81.4.108.41
intns2.zonetransfer.me.                 300      IN    A        52.91.28.78
office.zonetransfer.me.                 7200     IN    A        4.23.39.254
ipv6actnow.org.zonetransfer.me.         7200     IN    AAAA     2001:67c:2e8:11::c100:1332
owa.zonetransfer.me.                    7200     IN    A        207.46.197.32
robinwood.zonetransfer.me.              302      IN    TXT         "Robin
rp.zonetransfer.me.                     321      IN    RP            (
sip.zonetransfer.me.                    3333     IN    NAPTR         (
sqli.zonetransfer.me.                   300      IN    TXT           "'
sshock.zonetransfer.me.                 7200     IN    TXT          "()
staging.zonetransfer.me.                7200     IN    CNAME    www.sydneyoperahouse.com.
alltcpportsopen.firewall.test.zonetransfer.me. 301      IN    A        127.0.0.1
testing.zonetransfer.me.                301      IN    CNAME    www.zonetransfer.me.
vpn.zonetransfer.me.                    4000     IN    A        174.36.59.154
www.zonetransfer.me.                    7200     IN    A        5.196.105.14
xss.zonetransfer.me.                    300      IN    TXT      "'><script>alert('Boo')</script>"
```

Figure 4.51 – Zone records

The information collected from zone transfers can help ethical hackers to identify the network infrastructure of the targeted organization such as the hostname of systems and their IP addresses. Having completed this exercise, you have learned how to use perform DNS zone transfers on a vulnerable DNS server. Next, you will how SpiderFoot helps ethical hackers during their reconnaissance phase in penetration testing.

Exploring SpiderFoot

SpiderFoot is a popular OSINT tool used by cybersecurity professionals for reconnaissance, **threat intelligence** (**TI**), and investigation purposes. SpiderFoot can help ethical hackers to automate the collection of OSINT from various data sources on the internet. For instance, SpiderFoot operates by collecting data from various internet search engines, social media platforms, DNS records, WHOIS databases, and other publicly available sources. Then, SpiderFoot will both organize the collected data and present information in a way that helps ethical hackers easily analyze it.

> **Important note**
>
> To learn more about the SpiderFoot project, please visit `https://github.com/smicallef/spiderfoot`.

To get started with using SpiderFoot for reconnaissance, please use the following instructions:

1. Firstly, power on the **Kali Linux** virtual machine and ensure it has internet connectivity.
2. Next, open the Terminal on Kali Linux and execute the following command to start the SpiderFoot web interface:

   ```
   kali@kali:~$ spiderfoot -l 0.0.0.0:1234
   ```

   The following screenshot shows the web interface for SpiderFoot has started:

   ```
   kali@kali: $ spiderfoot -l 0.0.0.0:1234
   2023-03-12 20:35:09,884 [INFO] sf : Starting web server at 0.0.0.0:1234 ...

   *************************************************************
    Use SpiderFoot by starting your web browser of choice and
    browse to http://127.0.0.1:1234
   *************************************************************

   2023-03-12 20:35:09,906 [WARNING] sf :
   *********************************************************************
   Warning: passwd file contains no passwords. Authentication disabled.
   Please consider adding authentication to protect this instance!
   Refer to https://www.spiderfoot.net/documentation/#security.
   *********************************************************************
   ```

   Figure 4.52 – Starting the web interface for SpiderFoot

3. Next, open the web browser within Kali Linux and go to `http://127.0.0.1:1234` to access the web interface.
4. Next, click **New Scan**, set a **Scan Name** value, set a **Scan Target** value, select **Passive**, and click on **Run Scan Now**, as shown in the following screenshot:

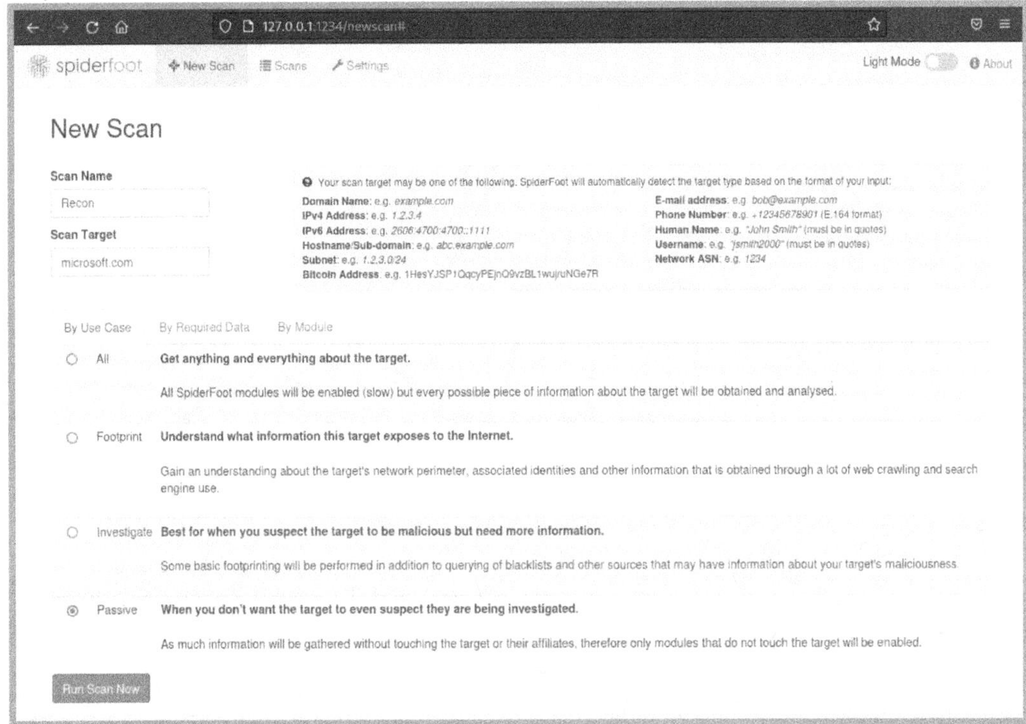

Figure 4.53 – Setting up a new scan

5. Next, the scan will start, and SpiderFoot will automatically populate the scan progress with the collected OSINT, as shown in the following screenshot:

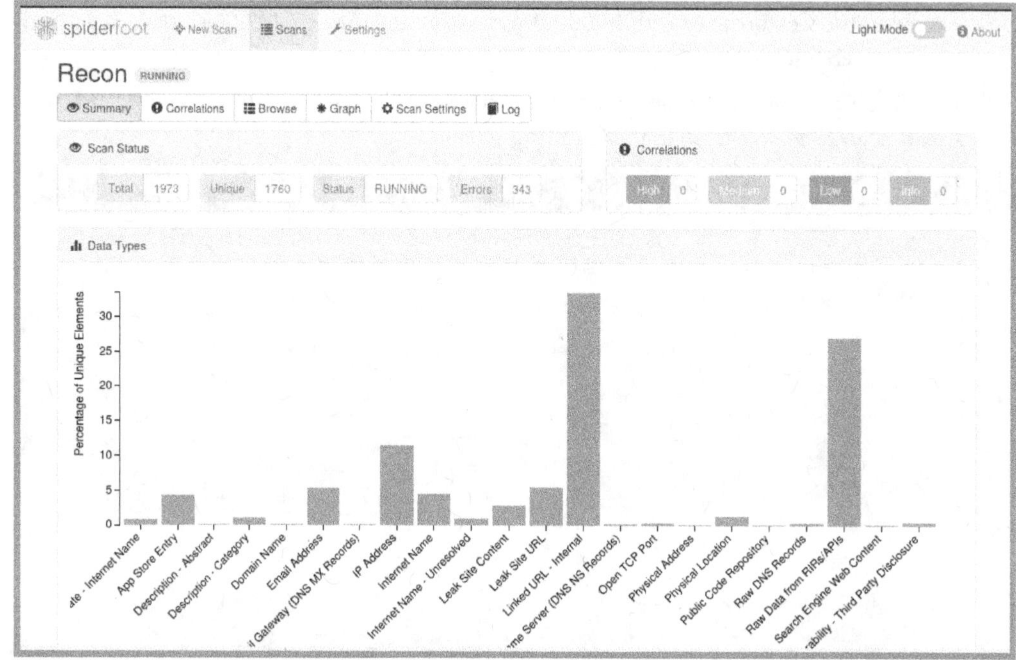

Figure 4.54 – Summary data

6. Next, selecting the **Browse** sub-tab within the current scan will display the number of collected data elements for each type of artifact:

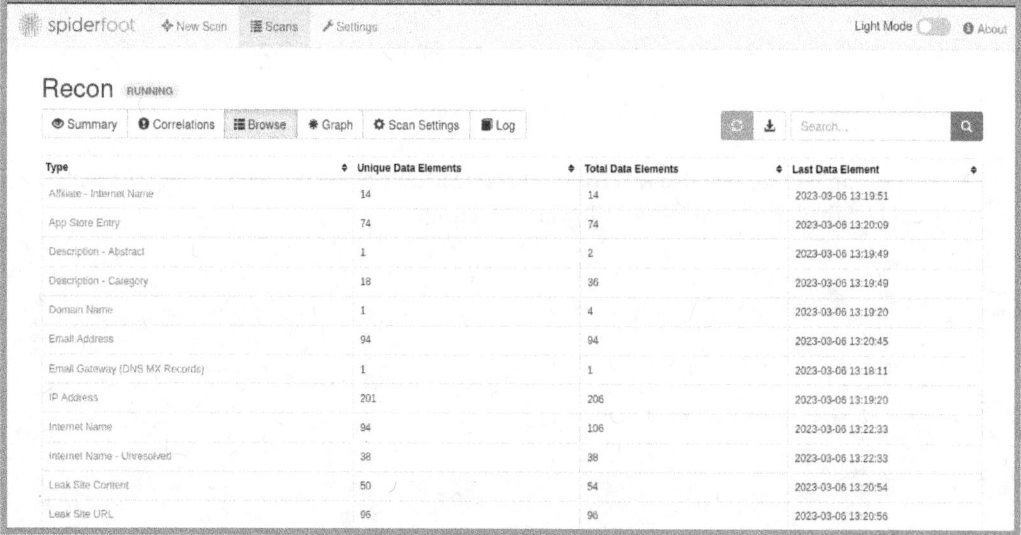

Type	Unique Data Elements	Total Data Elements	Last Data Element
Affiliate - Internet Name	14	14	2023-03-06 13:19:51
App Store Entry	74	74	2023-03-06 13:20:09
Description - Abstract	1	2	2023-03-06 13:19:49
Description - Category	18	36	2023-03-06 13:19:49
Domain Name	1	4	2023-03-06 13:19:20
Email Address	94	94	2023-03-06 13:20:45
Email Gateway (DNS MX Records)	1	1	2023-03-06 13:18:11
IP Address	201	206	2023-03-06 13:19:20
Internet Name	94	106	2023-03-06 13:22:33
Internet Name - Unresolved	38	38	2023-03-06 13:22:33
Leak Site Content	50	54	2023-03-06 13:20:54
Leak Site URL	96	96	2023-03-06 13:20:56

Figure 4.55 – Viewing artifact data

7. Next, select the **Internet Name** artifact to view a list of sub-domains for the targeted domain name:

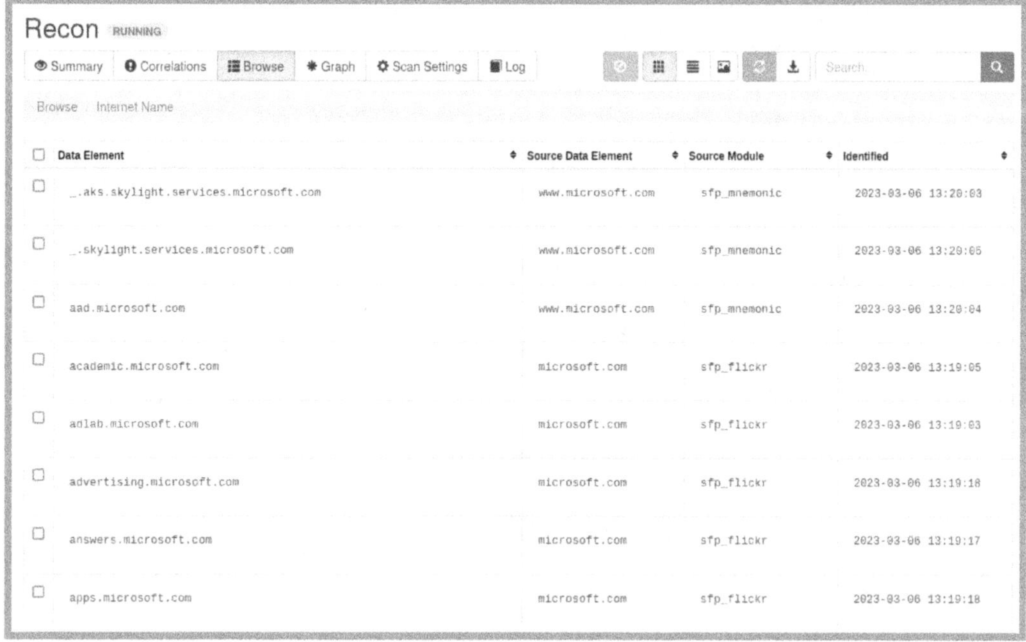

Figure 4.56 – Viewing sub-domains

8. Next, click on **Browse** > **Physical Location** to view a list of the target's geo-location data:

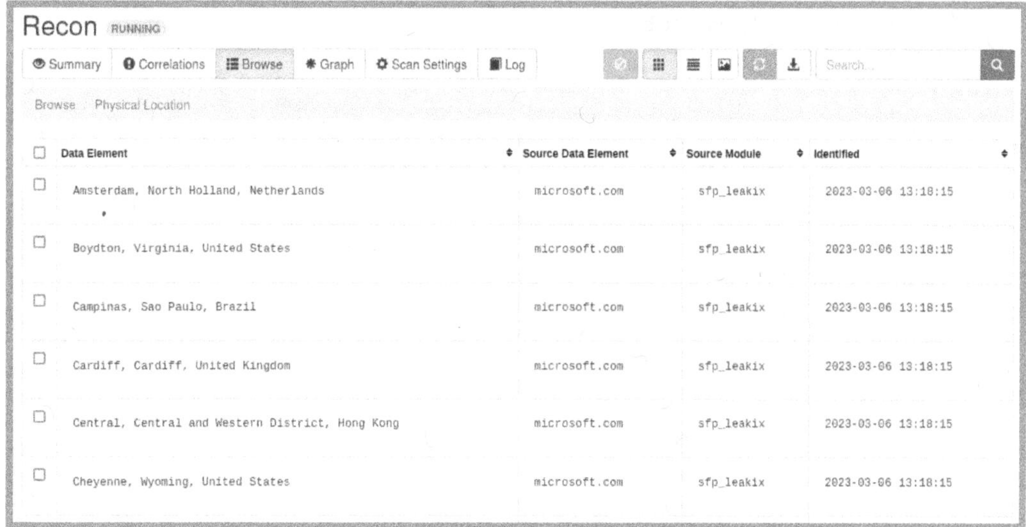

Figure 4.57 – Geo-locations

After the entire scan is completed, ensure you visit all the artifact categories to view all collected OSINT about the target to improve the profile of the target and identify any security vulnerabilities that can be exploited to gain a foothold.

Having completed this section, you have learned how to use various tools and techniques to efficiently discover sensitive information about a target such as its hostname, IP address, and even geo-locations.

Summary

In this chapter, you have learned how to use leverage internet search engines to discover sensitive information about organizations. Additionally, you have gained hands-on skills to perform Google hacking techniques to find interesting domain names and sub-domains for a company. Furthermore, you have explored how to use various techniques and tools to discover and collect OSINT on domain names, sub-domains, and DNS information.

I hope this chapter has been informative for you and helpful in your journey in the cybersecurity industry. In the next chapter, *Organizational Infrastructure Intelligence*, you will gain the practical skills needed to identify the network and organizational infrastructure of a targeted organization.

Further reading

- DNS servers: `https://www.cloudflare.com/learning/dns/what-is-a-dns-server/`
- DNS records: `https://www.cloudflare.com/learning/dns/dns-records/`
- SpiderFoot: `https://github.com/smicallef/spiderfoot`

5

Organizational Infrastructure Intelligence

As more systems and networks are connecting to the internet, there's a lot more data and information publicly available to everyone around the world. Sometimes, organizations leak data about their assets and network infrastructure without realizing it, and this enables threat actors to strategically collect and analyze those data leaks to plan their future attacks on a target. As an aspiring ethical hacker, it's important to understand the **Tactics, Techniques, and Procedures** (**TTPs**) that are commonly used by threat actors, and how such knowledge can be used to help organizations safeguard their infrastructure and reduce their attack surface to prevent a real hacker from compromising their systems and networks.

During the course of this chapter, you will learn how to use various online and offline tools to harvest data from the internet using **Open Source Intelligence** (**OSINT**) techniques to identify the infrastructure of an organization. Additionally, you will discover how ethical hackers and threat actors are able to discover vulnerable and exposed systems and devices on the internet and gain a better understanding of how attackers are able to gain a foothold into a target's network.

In this chapter, we will cover the following topics:

- Harvesting data from the internet
- Discovering exposed systems
- Collecting social media OSINT

Let's dive in!

Technical requirements

To follow along with the exercises in this chapter, please ensure that you have met the following hardware and software requirements:

- Kali Linux: `https://www.kali.org/get-kali/`
- Trace Labs OSINT VM: `https://www.tracelabs.org/initiatives/osint-vm`
- Sherlock: `https://github.com/sherlock-project/sherlock`

Harvesting data from the internet

The internet contains lots of websites, open databases, and servers that store data about people, networks, and organizations. Before an adversary launches a cyber-attack on their target, the attacker spends sufficient time researching the target to better understand their infrastructure and identify any security vulnerabilities that can be exploited to gain a foothold. Attackers gather and analyze OSINT from multiple data sources to create a profile of their target, which helps to identify targeted hosts, servers, operating systems, network block information, IP addresses, geo-location, subdomains, and so on. Such information is very useful in planning a cyber-attack as it enables the attacker to determine the attack surface of the targeted organization.

As an ethical hacker with a good moral compass and intentions, using the same TTPs as adversaries can help organizations identify how they are intentionally or unintentionally leaking sensitive data about their infrastructure, and how an attacker can leverage the same information collected from data leaks to compromise their systems and networks.

The following are popular websites and open databases for gathering information:

- `www.netcraft.com`: Identify servers, their host operating systems, and web application
- `www.shodan.io`: Identify servers and **Internet of Things** (**IoT**) devices
- `censys.io`: Identify attack surface on servers and other connected systems
- `hunter.io`: Identify email addresses of employees
- `urlscan.io`: Identify web applications and web servers
- `intelx.io`: Gather OSINT from multiple data sources
- `www.wigle.net`: Identify wireless networks around the world
- `fullhunt.io`: Collect attack surface data
- `vulners.com`: A search engine for security intelligence and vulnerabilities
- `viz.greynoise.io`: Identify threat intelligence
- `builtwith.com`: Identify the web infrastructure of a domain

The preceding list of search engines will be valuable within your arsenal of tools and resources when performing passive information gathering on your target as an ethical hacker. Ensure you take the time to visit each of these websites to gain a better understanding of how they work, and the type of data they provide about a target.

Over the next few subsections, you will learn how to use both Netcraft and Maltego to collect and analyze OSINT about an organization.

Netcraft

Netcraft provides various internet security services that collect and analyze data on various systems and networks found on the internet. One of the most popular services that is leveraged by cybersecurity professionals such as ethical hackers and penetration testers is the **Web Server Survey** tool, sometimes referred to as the **Site Report** tool on the Netcraft website. This tool enables anyone to identify the technologies and usage of web servers, their web applications, host operating systems, software version, hosting providers, and network blocks.

Netcraft provides the following data about a targeted domain and its web servers:

- **Background**: Provides information about the website title, rank, description, and when it was first seen on the internet
- **Network**: Provides information about the IPv4 and IPv6 addresses, network block owner, hosting providers and country, name servers, and domain registrar details
- **IP delegation**: Contains information about the allocation of IP addresses to the domain
- **IP Geolocation**: Provide geo-location details about the domain name and its associated IP address, such as city and country
- **SSL/TLS**: Provides information about the digital certificate for the domain name
- **Site Technology**: Provides information about the web technologies on the web server
- **Hosting History**: Provides information on the network block owner, name of the web application and host operating systems, and IP address of the web server

While the data provided by Netcraft is intended to help both IT and cybersecurity professionals to improve their organization's security posture, threat actors can also leverage such information for malicious activities and planning future attacks on a target. However, as an ethical hacker, you can also leverage the information from Netcraft to help you better understand what type of data is being leaked about your organization's web servers and its infrastructure, and whether there are any known security vulnerabilities on them.

To get started using Netcraft to profile a web server, please use the following instructions:

1. Open your web browser and go to `https://sitereport.netcraft.com/`, then insert a target domain name such as `https://microsoft.com` within the URL input field, and click on **Look up**, as shown in the following screenshot:

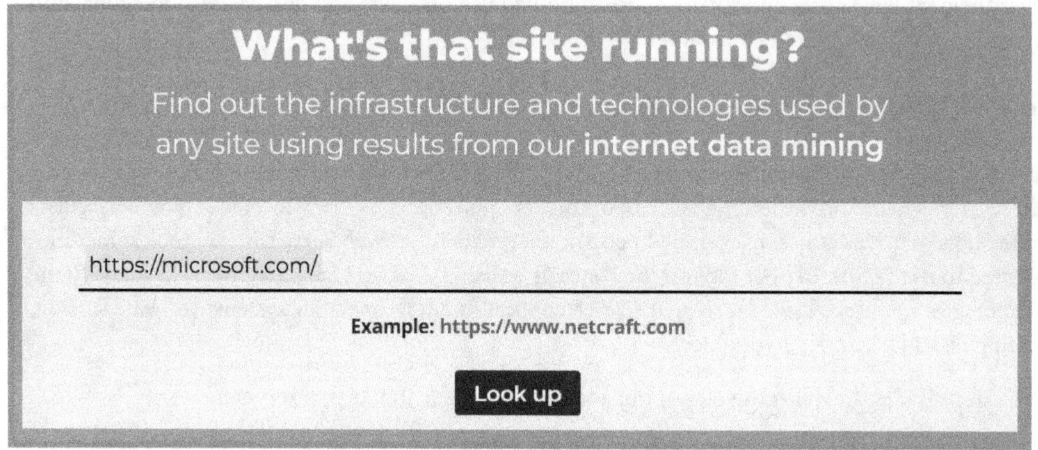

Figure 5.1 – Netcraft lookup

2. Netcraft will take a few seconds to analyze the targeted domain and its technologies and then present the information, as shown in the following screenshot:

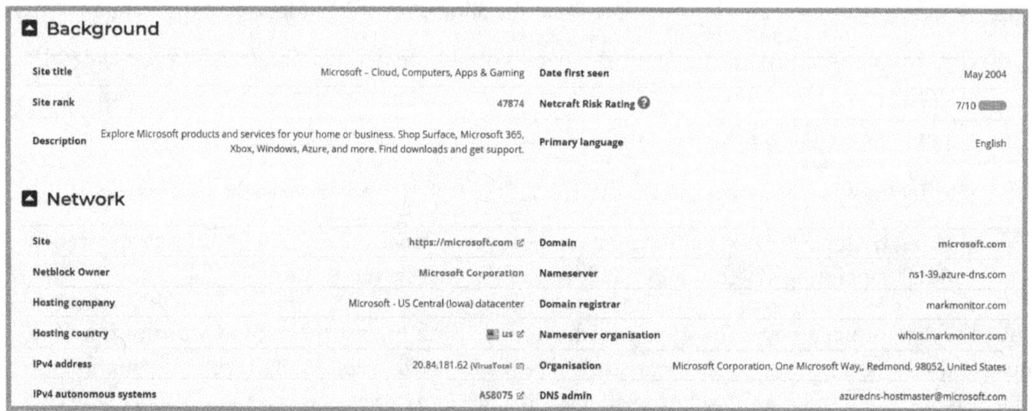

Figure 5.2 – Netcraft data

3. Next, scroll down to the **SSL/TLS** section to identify the geo-location of the company and subdomains from the digital certificate:

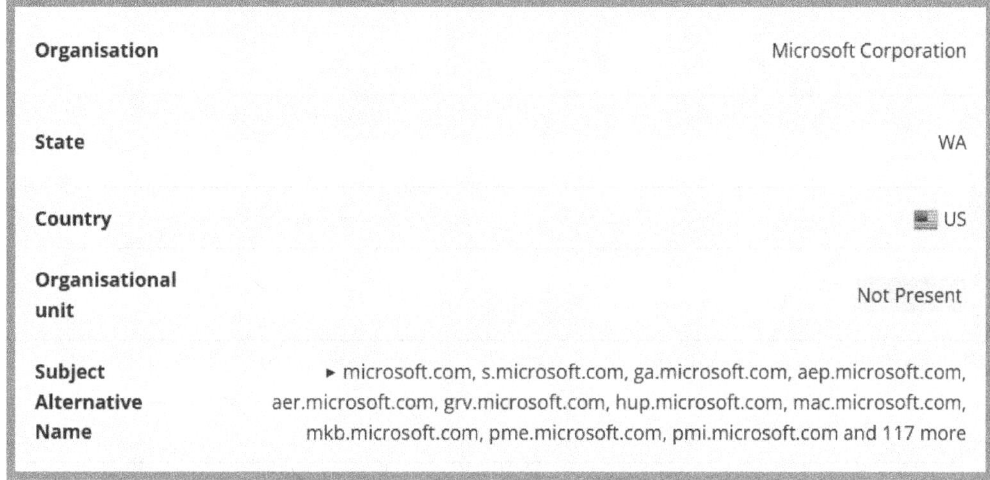

Organisation	Microsoft Corporation
State	WA
Country	US
Organisational unit	Not Present
Subject Alternative Name	▸ microsoft.com, s.microsoft.com, ga.microsoft.com, aep.microsoft.com, aer.microsoft.com, grv.microsoft.com, hup.microsoft.com, mac.microsoft.com, mkb.microsoft.com, pme.microsoft.com, pmi.microsoft.com and 117 more

Figure 5.3 – Identifying interesting data

As shown in the preceding snippet, the digital certificate indicates the country and state of the organization. Additionally, the **Subject Alternative Name** field indicates the various subdomains that are associated with this digital certificate.

4. Next, scroll down to the **Hosting History** section to identify the host operating system, running web application, and IP addresses of the target's web server on the internet:

Hosting History

Netblock owner	IP address	OS	Web server	Last seen
Microsoft Corporation One Microsoft Way Redmond WA US 98052	20.112.52.29	Linux	Kestrel	15-Mar-2023
Microsoft Corporation One Microsoft Way Redmond WA US 98052	20.84.181.62	Linux	Kestrel	7-Mar-2023
Microsoft Corporation One Microsoft Way Redmond WA US 98052	20.81.111.85	Linux	Kestrel	26-Feb-2023
Microsoft Corporation One Microsoft Way Redmond WA US 98052	20.112.52.29	Linux	Kestrel	17-Feb-2023
Microsoft Corporation One Microsoft Way Redmond WA US 98052	20.81.111.85	Linux	Kestrel	9-Feb-2023
Microsoft Corporation One Microsoft Way Redmond WA US 98052	20.112.52.29	Linux	Kestrel	26-Jan-2023

Figure 5.4 – Hosting History

Determining the host operating system and running web applications helps an ethical hacker to improve their planning and identify known security vulnerabilities that can be exploited to gain access to the target's network infrastructure.

5. Lastly, Netcraft can be used to provide a list of subdomains of a parent domain; go to `https://searchdns.netcraft.com/` and enter `.microsoft.com` within the search field and click on **Search**, as shown in the following screenshot:

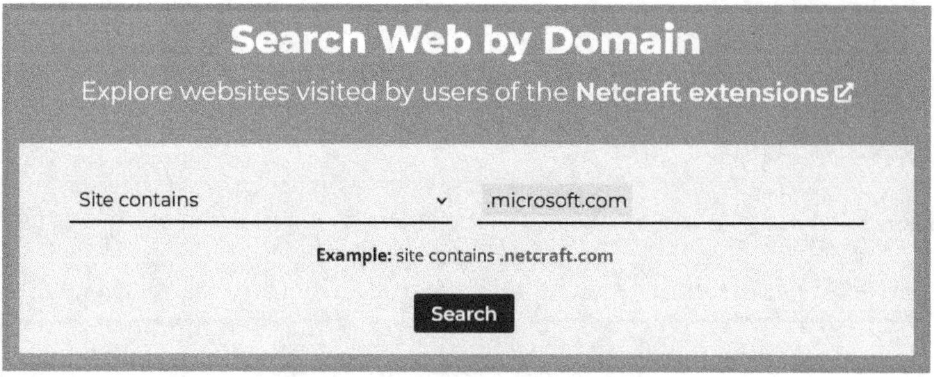

Search Web by Domain

Explore websites visited by users of the Netcraft extensions

Site contains .microsoft.com

Example: site contains .netcraft.com

Search

Figure 5.5 – Subdomain lookup

After a few seconds, Netcraft will provide the results showing a list of subdomains, their network blocks, host operating systems, and the date first seen, as shown in the following screenshot:

Rank	Site	First seen	Netblock	OS	Site Report
32	teams.microsoft.com	November 2016	Microsoft Corporation	Windows Server 2008	
40	learn.microsoft.com	July 2015	Akamai International, BV	unknown	
66	support.microsoft.com	October 1997	Akamai Technologies	unknown	
92	www.microsoft.com	August 1995	Akamai Technologies, Inc.	Linux	
169	admin.microsoft.com	November 2017	Microsoft Corporation	Windows Server 2008	
180	security.microsoft.com	December 2006	Microsoft Corporation	Windows Server 2008	
202	answers.microsoft.com	August 2009	Akamai International, BV	Linux	

Figure 5.6 – Finding subdomains

As shown in the preceding snippet, Netcraft can be used to collect and identify subdomains of a target organization.

Having completed this exercise, you have learned how to use Netcraft to identify the public infrastructure of an organization. Next, you will learn how to use Maltego to collect and analyze OSINT from multiple data sources to improve the profile of a target.

Maltego

Maltego is a powerful data collection, visualization, and analysis tool for performing passive reconnaissance on people, networks, and organizations. This tool was created to assist cybersecurity investigators and analysts in efficiently collecting and analyzing large amounts of OSINT from various data sources, such as online databases, websites, and even social media platforms.

Maltego helps cybersecurity professionals such as ethical hackers and penetration testers to better understand the relationships between complex datasets by showing patterns and associations between different pieces of collected data, which can help identify interesting patterns. This tool is commonly used by ethical hackers during their reconnaissance phase when collecting and analyzing data about their targets.

To get started using Maltego for reconnaissance, please use the following instructions.

Part 1 – setting up Maltego

To set up Maltego, follow these steps:

1. Firstly, open your web browser and go to `https://www.maltego.com/ce-registration/` to register for a Maltego CE account on the official website.
2. Next, open the **VirtualBox Manager** application, power on the Trace Labs virtual machine (TL OSINT VM 2022.1), and log in using `osint/osint` as the username and password.
3. Once you're logged in to the Trace Labs virtual machine, click on the **Kali Linux** icon in the top-left corner of the desktop and select **Open Source Intelligence** | **Frameworks** | **maltego**, as shown in the following screenshot:

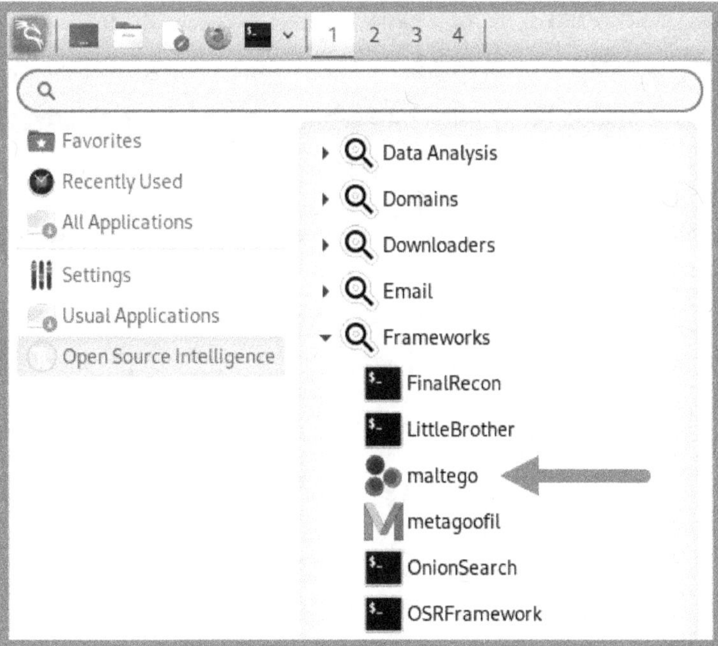

Figure 5.7 – OSINT Frameworks

4. Next, the Maltego application will load on the desktop and display the **Product Selection** window; select **Maltego CE (Free)** and click **Run**, as shown in the following screenshot:

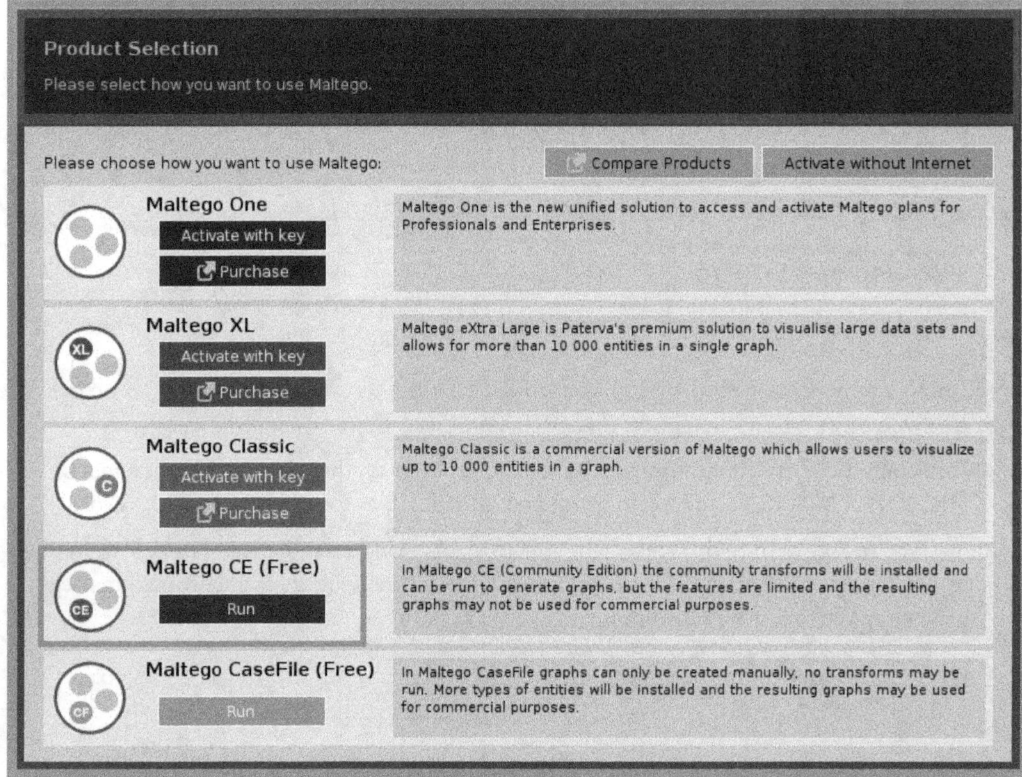

Figure 5.8 – Production selection

5. Next, the **Configure Maltego** setup menu will appear; ensure you check **Accept** and click on **Next** > to continue:

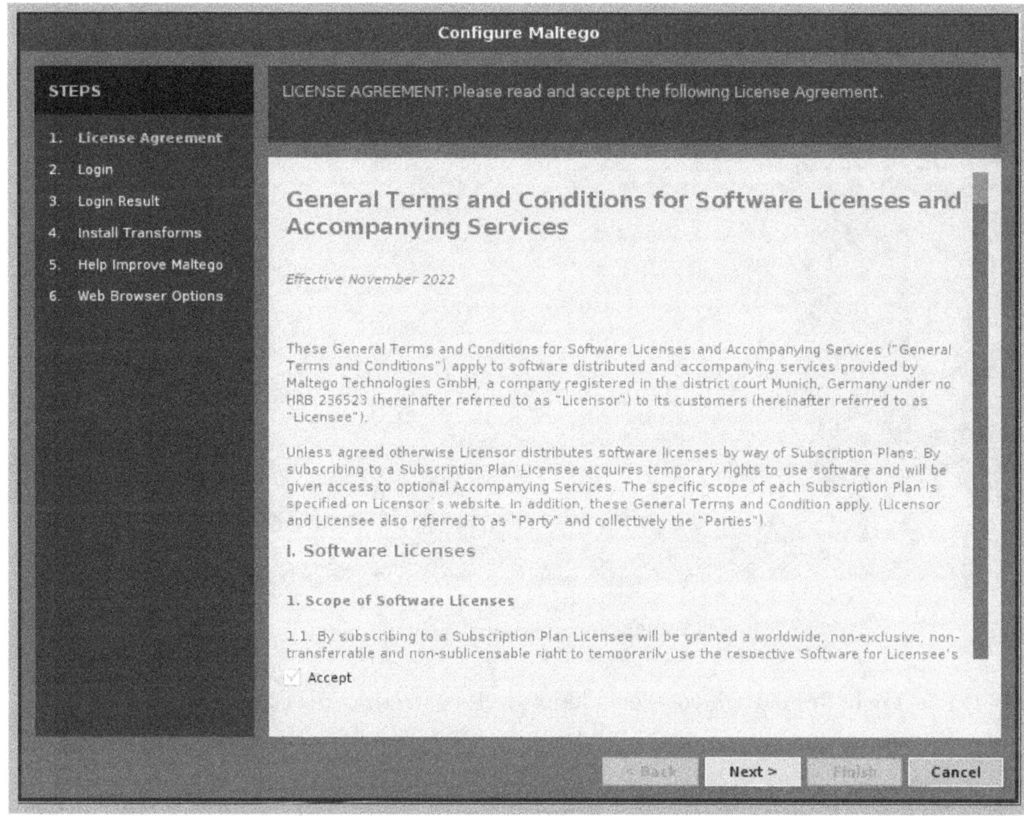

Figure 5.9 – License Agreement

6. Next, the **Login** window will appear; ensure you use the credentials that were created during *Step 1* and click on **Next >** to continue:

Figure 5.10 – Logging in to Maltego

7. On the **Login Result** page, your login results will appear; simply click on **Next** >.
8. On the **Install Transforms** page, the transforms set will install automatically; then click on **Next** >.
9. On the **Web Browser Options** page, select **Firefox** as the preferred web browser and click on **Finish**.

Part 2 – working with Maltego

To work with Maltego, follow these steps:

1. Once the Maltego CE application loads, click on the **Maltego** icon in the top-left corner of the application window, then select **New** to create a new case file (project), as shown in the following screenshot:

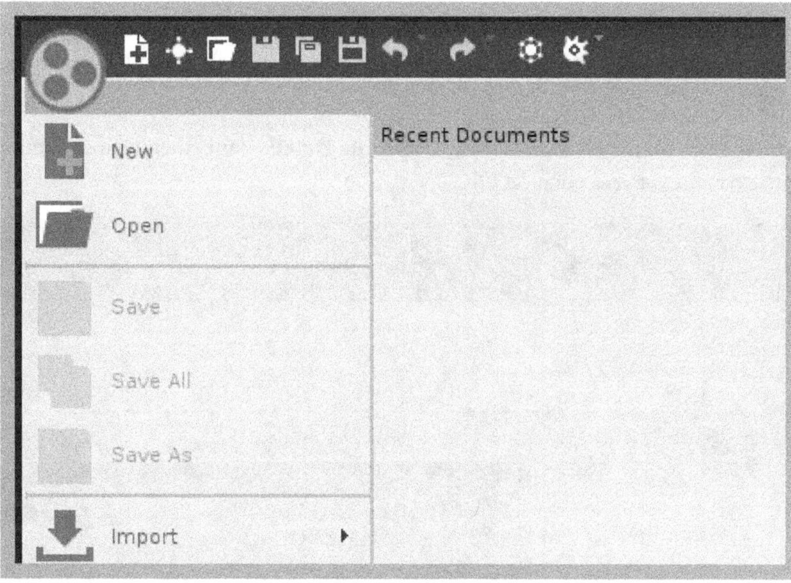

Figure 5.11 – Creating a new case file

2. Next, to get started with identifying the infrastructure of a targeted organization, search for the **Domain** entity from **Entity Palette**, then drag and drop the **Domain** entity anywhere onto the graph pane, as shown in the following screenshot:

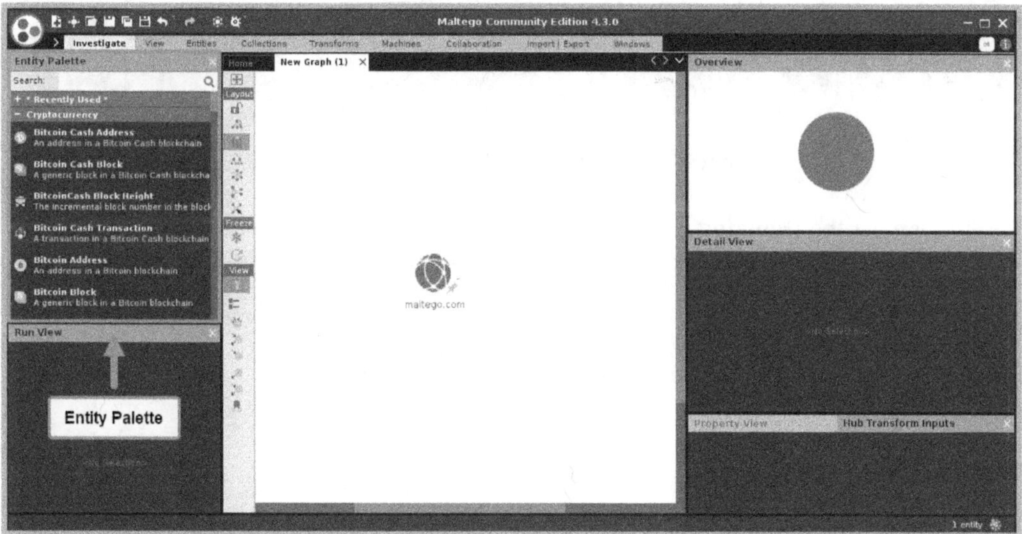

Figure 5.12 – Domain entity

As shown in the preceding snippet, **Entity Palette** contains various objects that enable us to collect and analyze data on a target based on their identity, social media, organization, and even network.

3. Next, double-click on the **Domain** entity to open its **Details** pane, then change **Domain Name** to `microsoft.com` and click on **OK**:

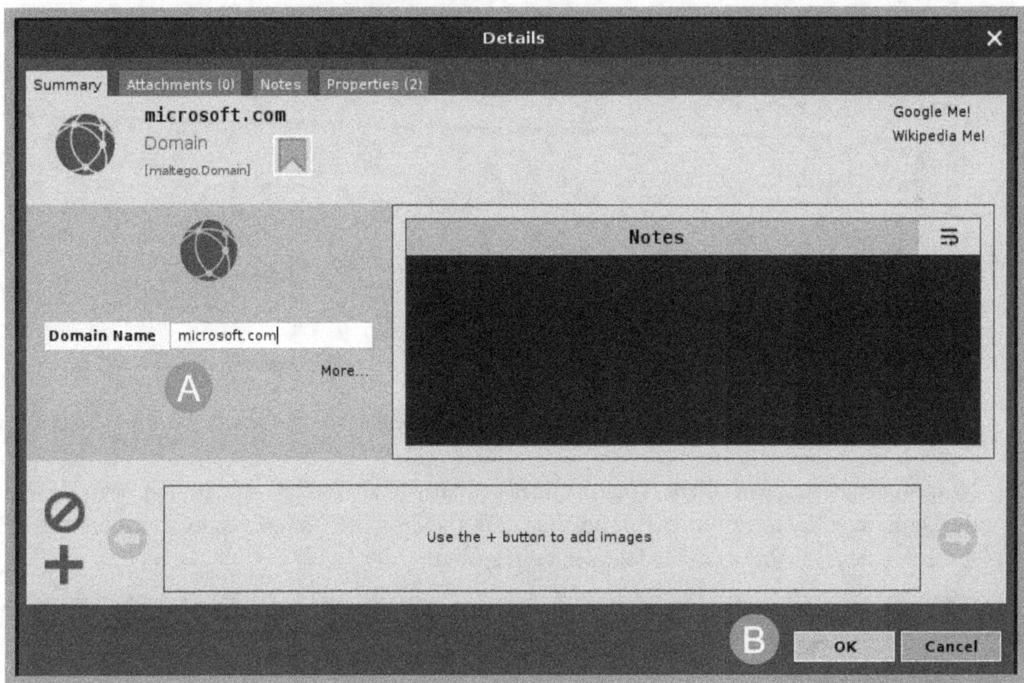

Figure 5.13 – Setting a target domain name

4. To discover the **Domain Name System** (**DNS**) records of the domain, right-click on the **Domain** entity | **All Transforms** | **To DNS Name – NS** (**name server**), the result is shown in the following screenshot:

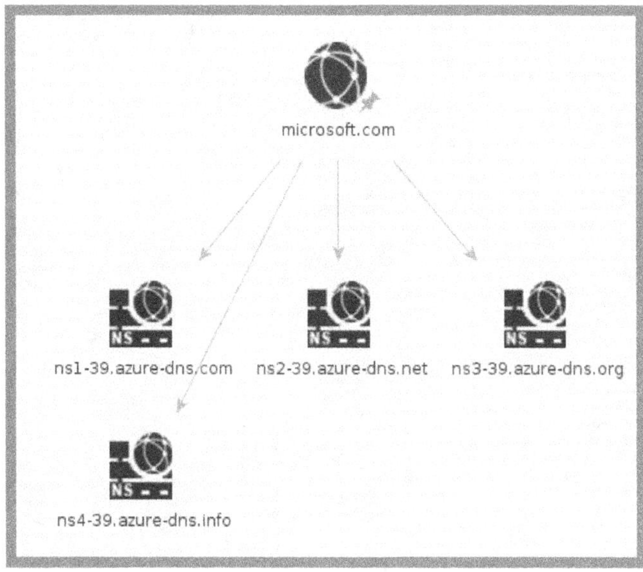

Figure 5.14 – Identifying name servers

As shown in the preceding snippet, Maltego was able to retrieve the name servers for the target domain. The name servers are used to map the hostnames of a domain to IP addresses.

5. To discover the email exchange servers of a domain, right-click on the **Domain** entity | **All Transforms** | **To DNS Name - MX (mail server)**, the result is shown in the following screenshot:

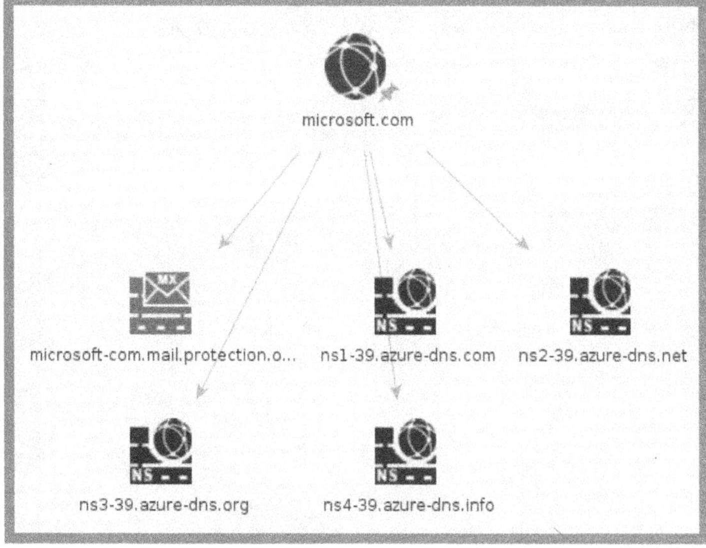

Figure 5.15 – Identifying email servers

6. To identify the IP address of the email server, right-click on the **Email Server** entity | **All Transforms** | **To IP Address** [**DNS**], the result is shown in *Figure 5.16*:

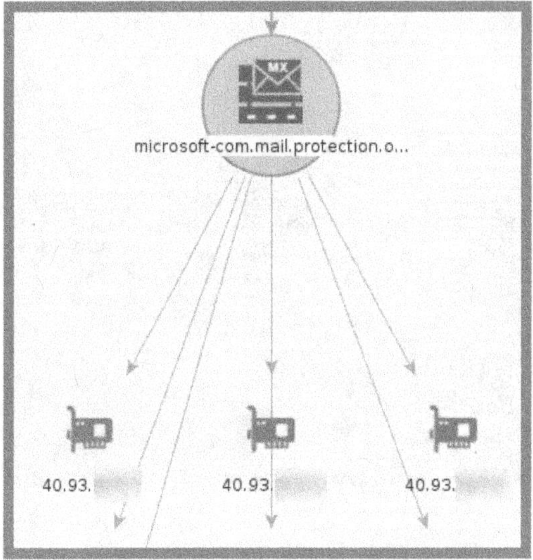

Figure 5.16 – Identifying IP addresses

7. To find the website that is associated with the domain name, right-click on the **Domain** entity | **All Transforms** | **To Website** [**Quick lookup**].
8. Next, to identify the IP address(es) of the website, right-click on the **Website** entity | **To IP Address** [**DNS**], the result is shown in the following screenshot:

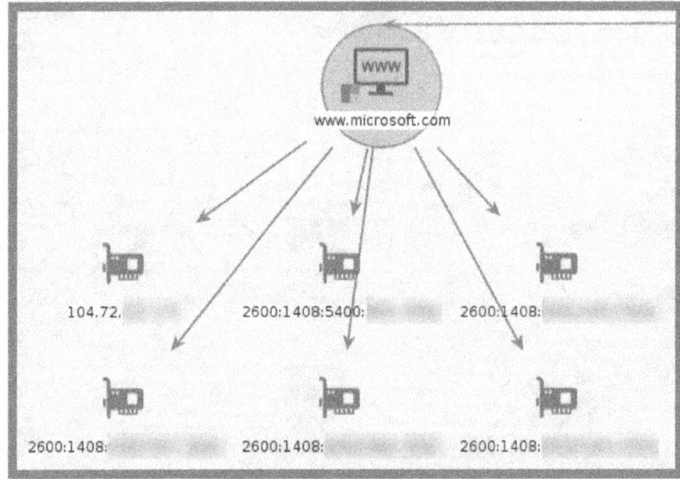

Figure 5.17 – IP addresses associated with the website

As shown in the preceding snippet, Maltego was able to retrieve the IPv4 and IPv6 addresses of the target website.

9. To retrieve email addresses that are associated with the domain, right-click on the **Domain** entity | **Email Addresses from Domain** | **To Email Addresses** [PGP], the result is shown in the following screenshot:

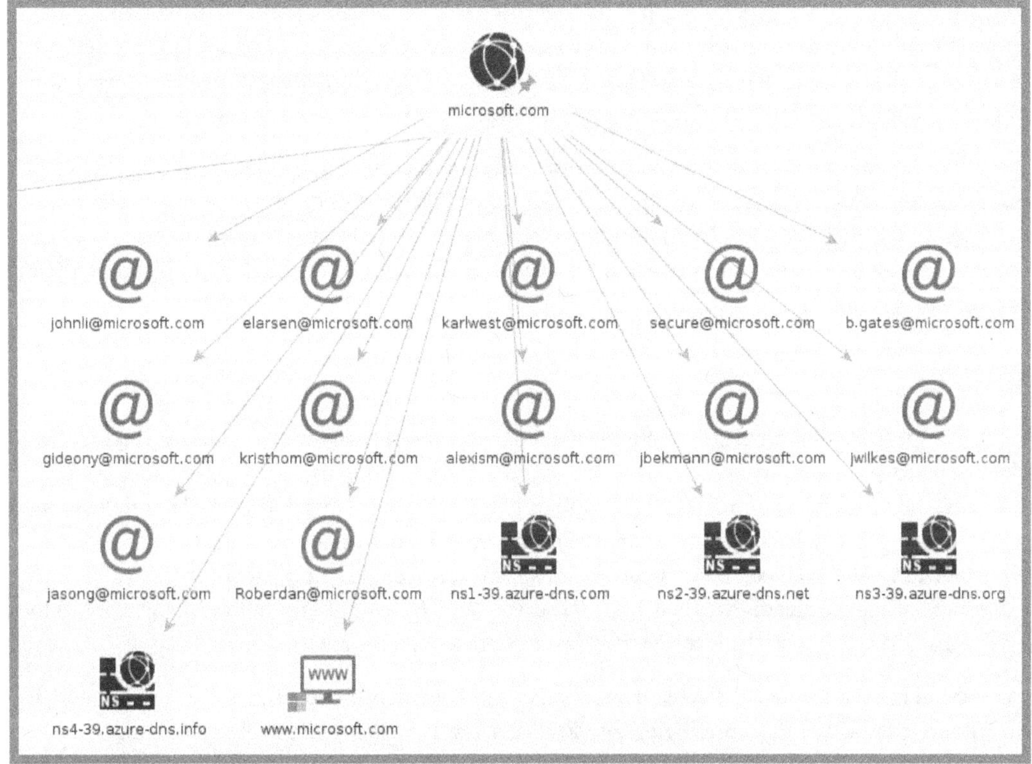

Figure 5.18 – Harvesting email addresses

As shown in the preceding snippet, Maltego was able to harvest the email addresses of users that are associated with the domain. Email harvesting can be leveraged for social engineering attacks as an attack vector for gaining access to the target's systems and networks.

Having completed this section, you have learned how to harvest OSINT from the internet and identify the infrastructure of an organization using Netcraft and Maltego. In the next section, you will learn how to discover unintentionally exposed systems and networks on the internet.

Discovering exposed systems

During the reconnaissance phase, it's essential for ethical hackers and penetration testers to identify the internal and external network infrastructure of their targets, as such information is useful for identifying the attack surface and attack vectors and developing exploits for future operations. Organizations often connect their systems and networks to the internet without performing a reconnaissance or OSINT penetration test on their own infrastructure to determine whether any of their assets are unintentionally exposed on the internet. Ethical hackers and penetration testers are hired by organizations to identify how their systems and network infrastructures are exposed and how their attack surface can be reduced to prevent future cyber-attacks and threats.

During this section, you will learn how threat actors and cybersecurity professionals can collect OSINT from specialized search engines to identify an organization's infrastructure and exposed service ports on servers. In addition, you will learn how recruiters often leak sensitive data on job boards about their organization's systems and network infrastructure, which enables attackers to improve their planning and operations.

Shodan

Shodan is a specialized internet search engine that enables cybersecurity professionals such as ethical hackers to discover exposed systems such as servers, IoT devices, modems, video surveillance systems, and **Industrial Control Systems** (**ICSes**). Shodan helps cybersecurity professionals to determine whether their organizations' systems are exposed on the internet and what type of data can be collected by an adversary. For instance, many organizations have connected their network infrastructure to the internet without realizing that their internal servers and systems are unintentionally exposed to everyone and everything on the internet, such as hackers and malware.

For an ethical hacker, the data collected from Shodan can help identify exposed servers, open service ports, running services and technologies, and known security vulnerabilities. Such information can be used by real adversaries when planning their attack on an organization. However, ethical hackers use the same techniques and information to simulate real-world cyber-attacks to identify hidden security vulnerabilities, while providing insights on how to reduce the attack surface. It's important to remember that not all organizations have the same level of security awareness training, a dedicated security team, and the know-how to safeguard their assets. Hence, ethical hackers and penetration testers are white-hat hackers who use their skills to help companies improve their cyber defenses and security posture to prevent a real cyber-attack in the future.

To get started using Shodan, please use the following instructions:

1. On your computer, open your web browser, go to `https://www.shodan.io/` and register for a free user account.
2. Once your user account is created, log in to the Shodan website, as shown in *Figure 5.19*:

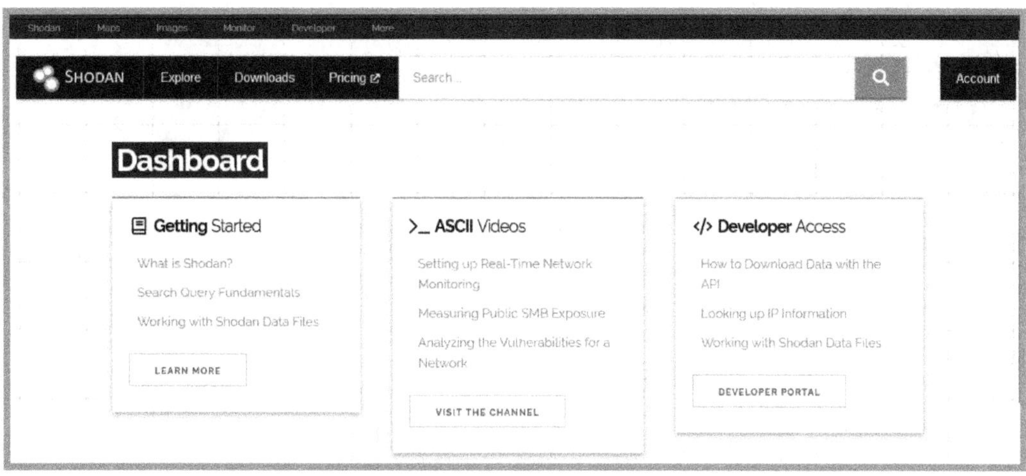

Figure 5.19 – Shodan website

3. Using the Shodan search bar, perform a search for `windows server 2008` to find all systems on the internet that are running Microsoft Windows Server 2008, as shown in the following screenshot:

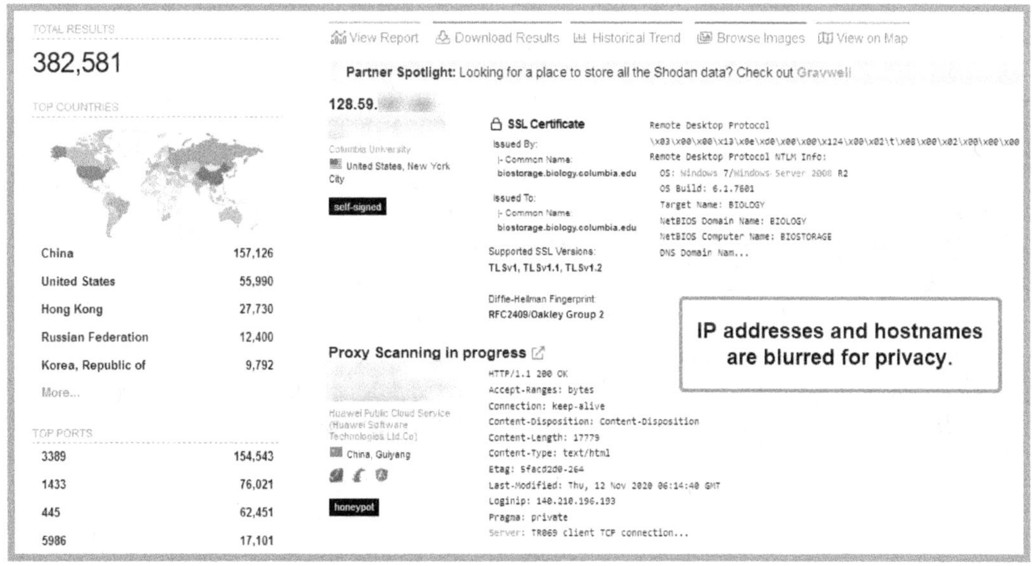

Figure 5.20 – Identifying Windows Server machines

4. Next, to filter the search results based on a specific country, append the `country:"country code"` syntax to the search, such as `windows server 2008 country:"US"`, as shown in the following screenshot:

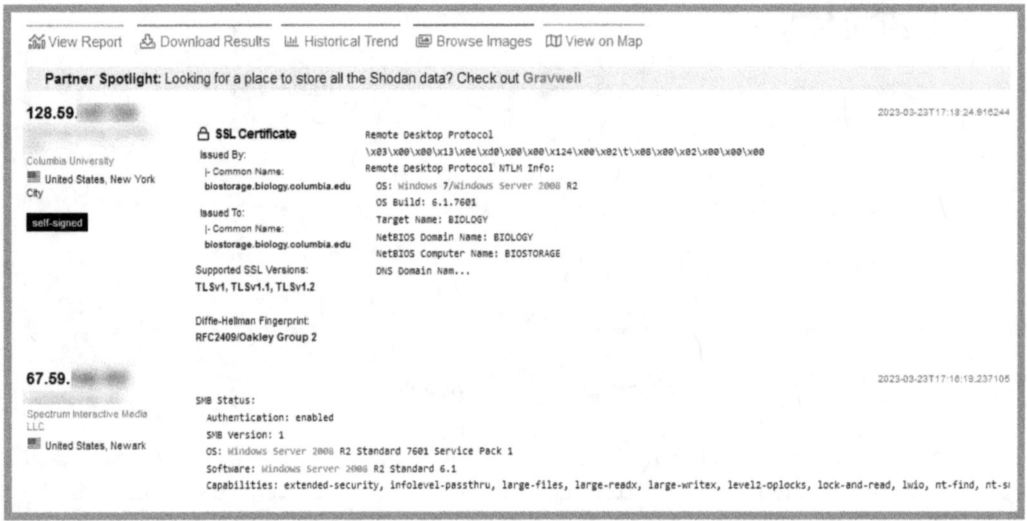

Figure 5.21 – Identifying devices based on their location

As shown in the preceding snippet, the country syntax enables us to filter results for a specific country or geo-location. This syntax is useful for identifying the servers of a company within a specific location.

> **Tip**
>
> To learn more about Shodan filters, please see `https://www.shodan.io/search/filters`.

5. To identify servers and devices that are running a specific service or open port, add the `port:port_number` syntax to the search bar, such as `windows server 2008 country:"US" port:3389`, as shown in the following screenshot:

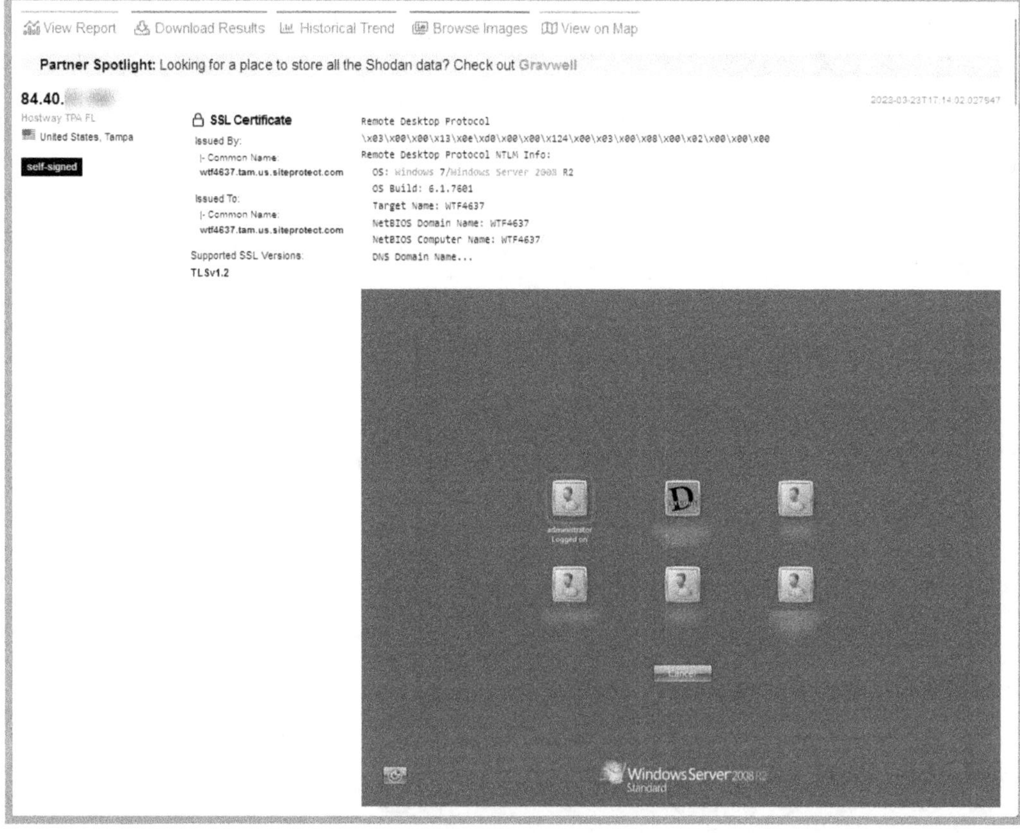

Figure 5.22 – Identifying devices with specific running services

As shown in the preceding snippet, Shodan was able to identify a Microsoft Windows Server 2008 system within the United States, which has service port `3389` and is running **Remote Desktop Protocol** (**RDP**). Additionally, Shodan was able to perform a screen capture on the login window and display users' accounts.

6. Next, click on any system on the search results page to view specific details about the server such as open service ports, running services, and whether there are any known security vulnerabilities:

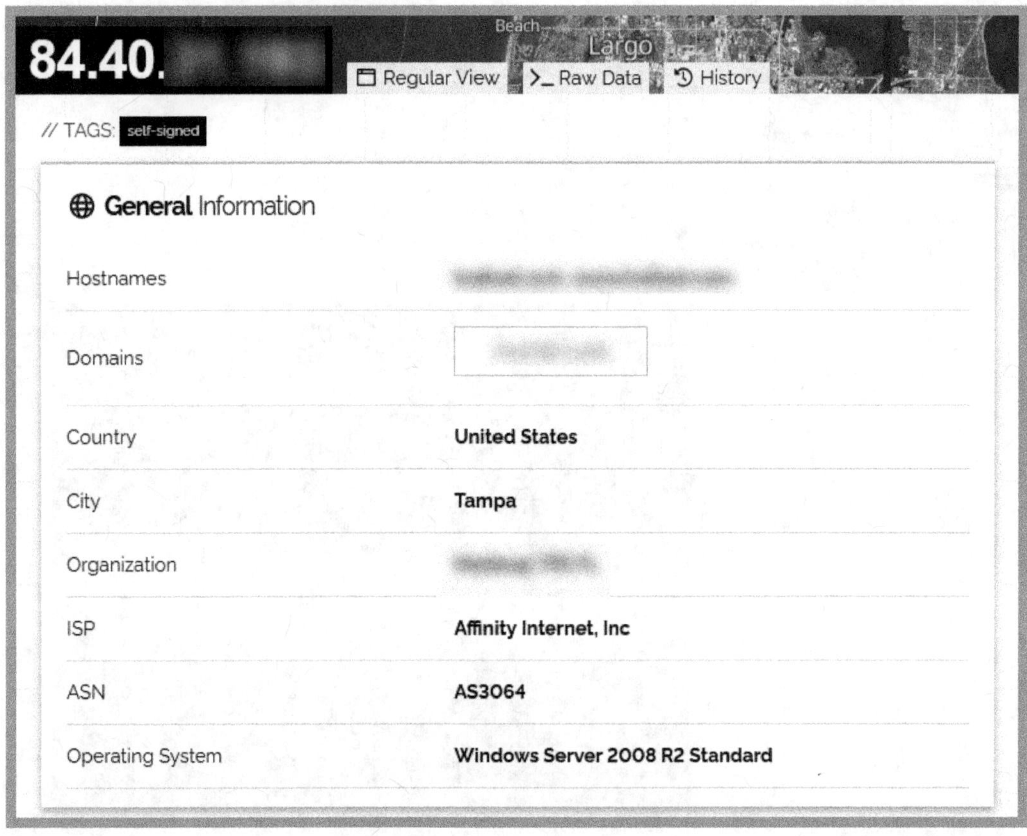

Figure 5.23 – Geo-location data

As shown in the preceding snippet, Shodan collects geo-location details, which can be used by an ethical hacker to identify the physical location of their target during a penetration testing assessment.

The following snippet shows the open service port numbers on the server and running services:

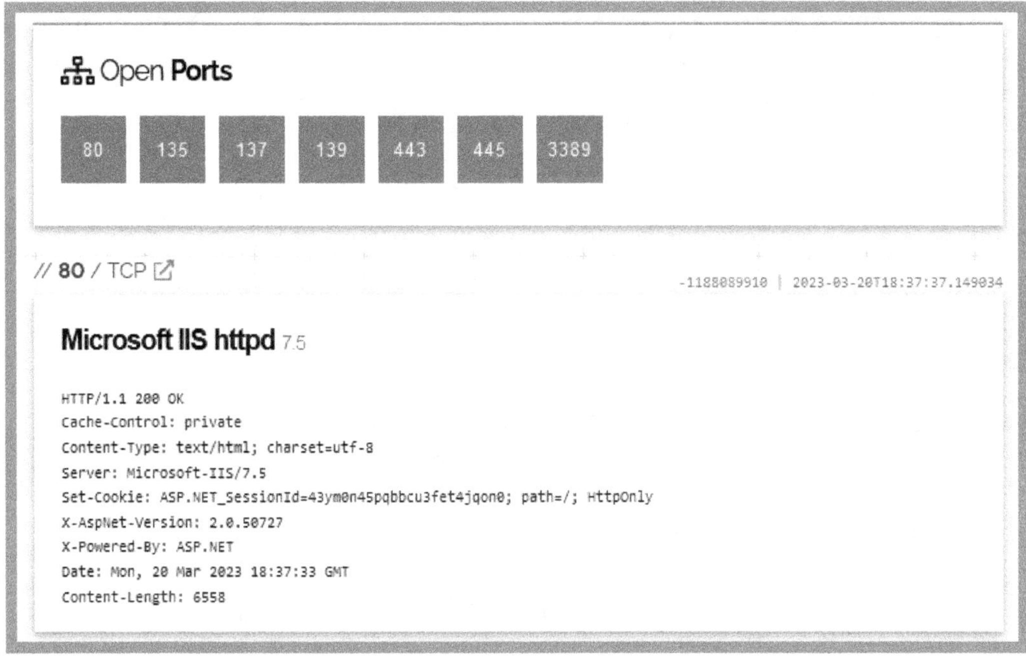

Figure 5.24 – Identifying open ports and services

Whenever a service is running on a server, a service port is open to listen for incoming messages. Ethical hackers can profile a service and determine the number of services that are running by identifying open service ports, such as **Hypertext Transfer Protocol** (**HTTP**) – port `80`, **Remote Procedure Call** (**RPC**) – port `135`, **NetBIOS** – ports `137` and `139`, **Hypertext Transfer Protocol Secure** (**HTTPS**) – port `443`, **Server Message Block** (**SMB**) – port `445`, and RDP – port `3389`. Then, the ethical hacker can identify security vulnerabilities within each running service that can be exploited to gain access to the target system.

Important note

To learn more about each service port number, please see the service name and port number registry at `https://www.iana.org/assignments/service-names-port-numbers/service-names-port-numbers.xhtml`.

Lastly, Shodan is able to identify known security vulnerabilities on servers and devices, listing their **Common Vulnerabilities and Exposures** (**CVE**) identifier and description, as shown in the following screenshot:

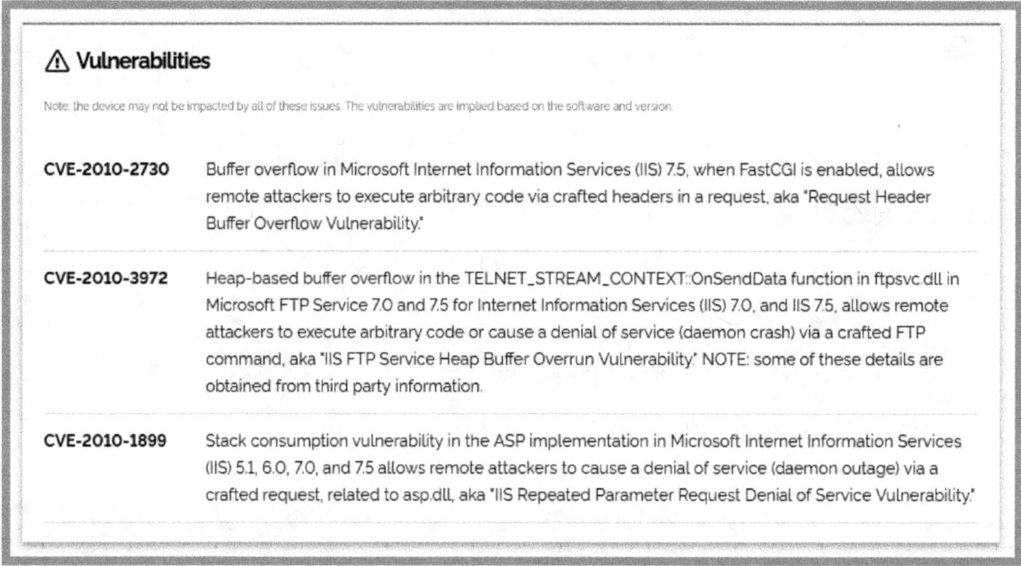

Figure 5.25 – Vulnerabilities on the server

As shown in the preceding steps, ethical hackers can collect the data found on Shodan to create a profile of their target's system and network infrastructure, identify host operating systems and running services, open service port numbers, vulnerabilities, and even usernames for remote services.

Important note

To gain more information about a reported vulnerability, visit `https://cve.mitre.org/`, which is an open database that enables security professionals to record and track security vulnerabilities. In addition, each CVE record provides details about the vulnerability such as how it can be exploited and recommendations on resolving the issue.

Having completed this exercise, you have learned how to use Shodan to discover vulnerable and exposed systems on the internet. Next, you will learn how to leverage the data from Censys to identify systems owned by an organization.

Censys

Censys is a search engine that enables ethical hackers to collect information about servers and identify their attack surfaces. It operates by indexing data from domain names, IP addresses, digital certificates, and other elements found on the internet. The data is collected, analyzed, and presented in a user-friendly format, thus enabling ethical hackers to easily find specific data about their target during the reconnaissance phase.

To get started with using Censys to identify the attack surface of an organization, please use the following instructions:

1. On your computer, open the web browser and go to `https://search.censys.io/`. Register for an account and log in:

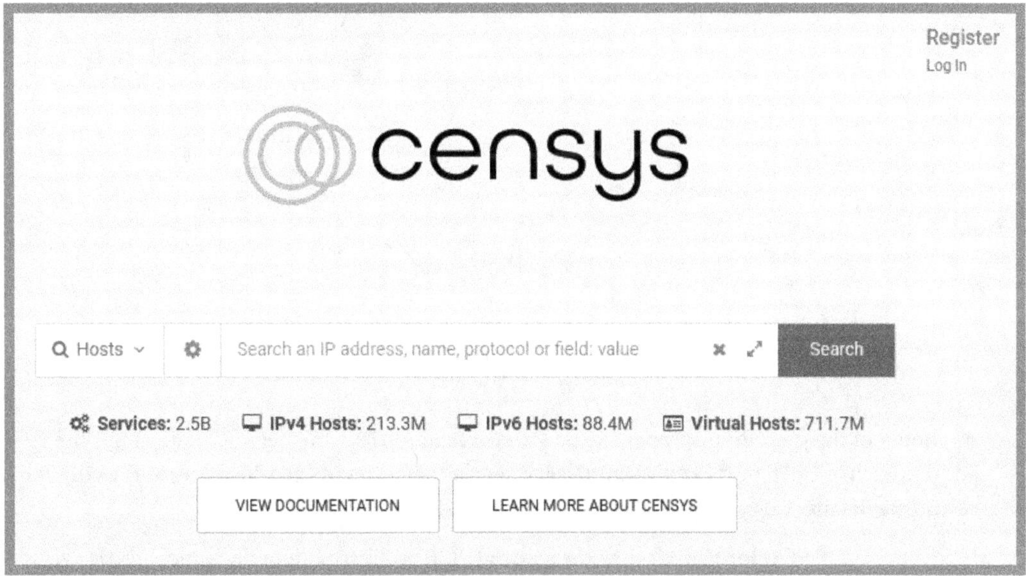

Figure 5.26 – Censys search

2. Next, in the **Search** field, enter the IP address or domain name of a target and click on **Search** to perform a lookup, as shown in *Figure 5.27*:

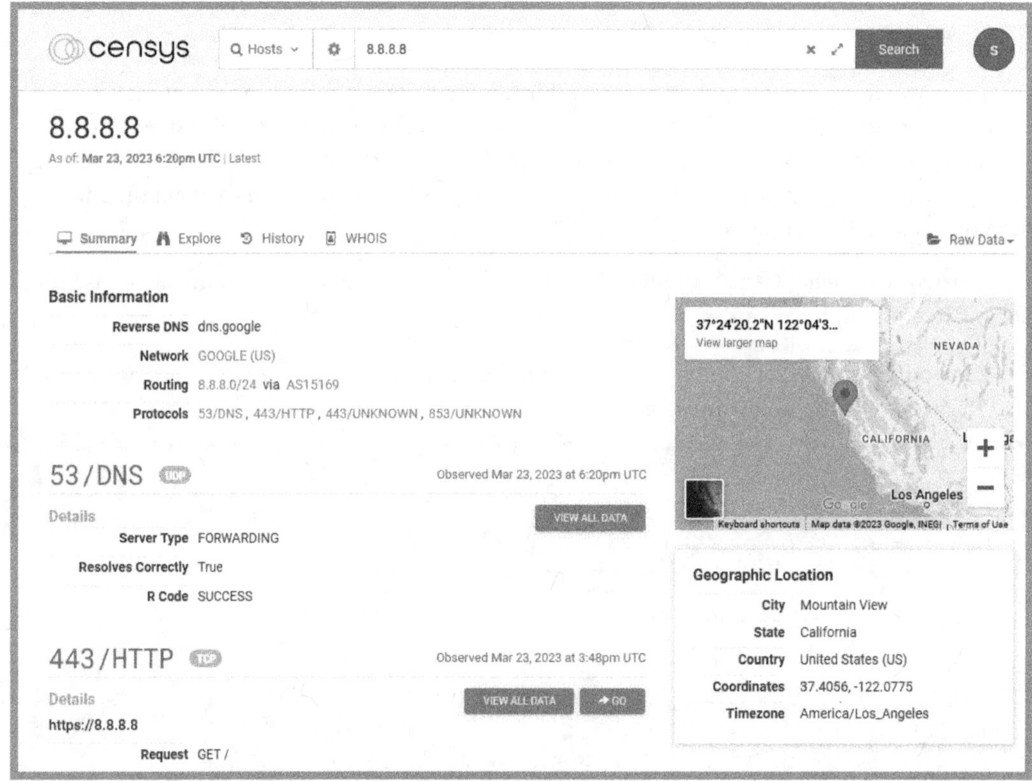

Figure 5.27 – IP lookup

As shown in the preceding snippet, Censys was able to retrieve the network data about the IP address such as open ports and running services on the server, its geo-location, and its digital certificate details.

3. Next, click on the **Explore** tab to view the associated IP addresses, domain names, and servers:

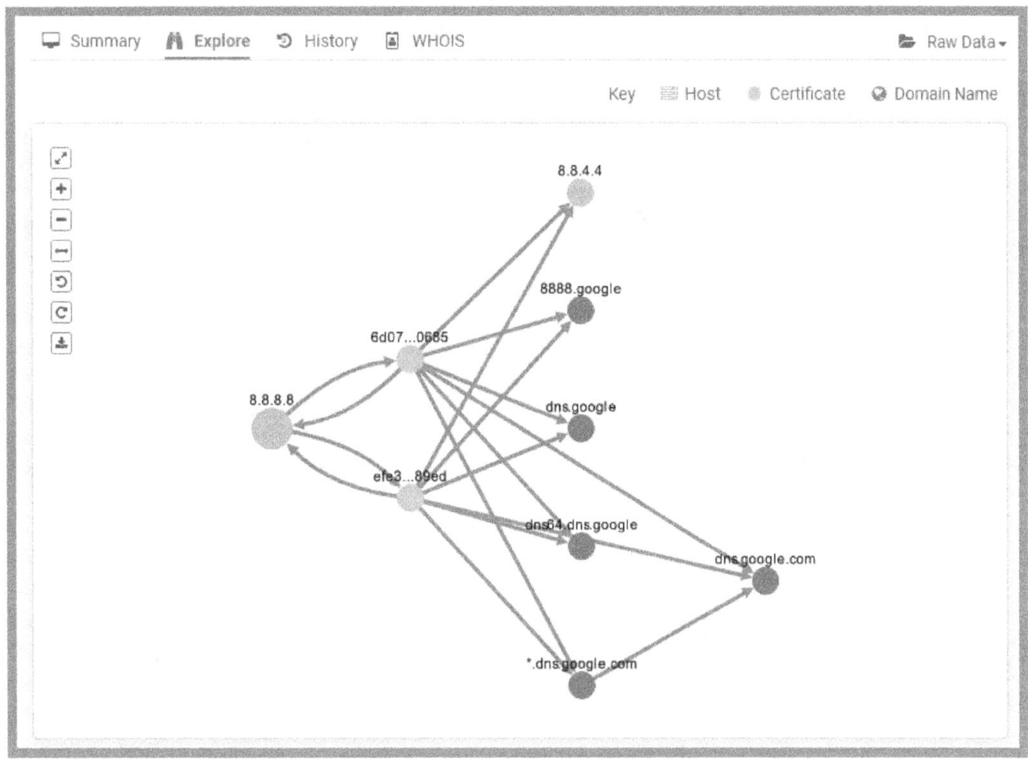

Figure 5.28 – Viewing associated systems

If you right-click on any system shown on the **Explore** tab, you can further gather intelligence on that system such as its hostname and certificate details. This is useful when performing external network penetration testing and identifying additional assets owned by the target.

4. Click the **History** tab to view a list of changes that are observed on the target system. History data is valuable to ethical hackers as it indicates software or configuration changes.
5. Lastly, click on the **WHOIS** tab to gather the network, hosting, and registrar details of the target, as shown in the following screenshot:

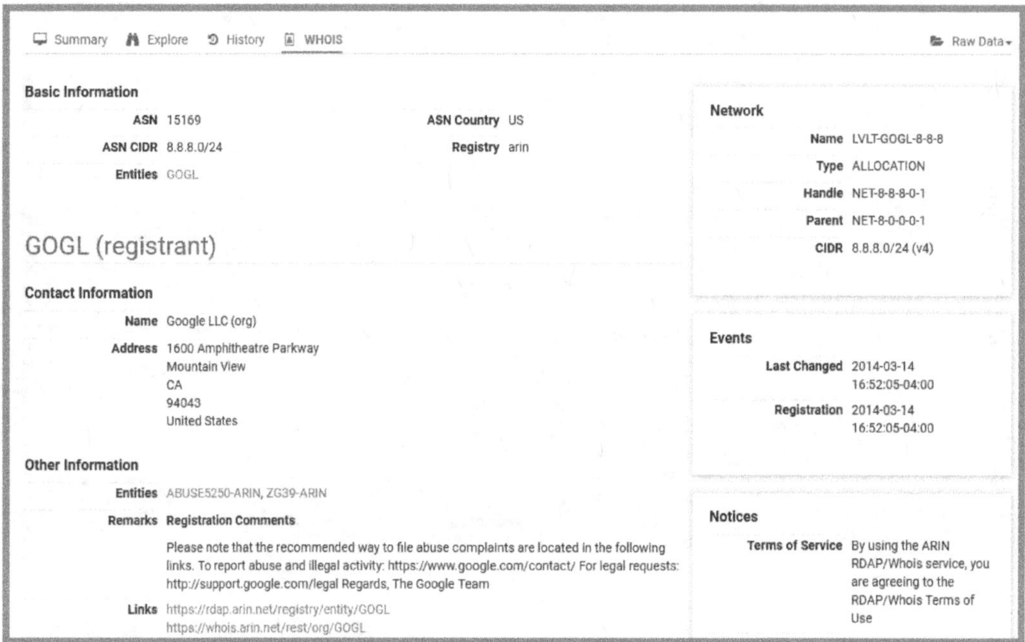

Figure 5.29 – WHOIS data

The information provided by Censys can be used to improve the profile of a target, such as identifying additional assets, open ports, and running services on servers. Such information is useful for determining the attack surface and security vulnerabilities on the systems and networks of an organization. In addition, ethical hackers can provide insights on how their organization is leaking technical data about their assets on the internet and how it can be leveraged by adversaries.

Job boards

In the workforce around the world, people commonly move between different companies and jobs according to their preference, and companies expand, creating new opportunities to hire additional staff. As a result, whenever an employee leaves their current job role either for a promotion or to exit the organization, the **Human Resources (HR)** department starts the off-boarding process and starts looking for a new candidate to fill the existing role. In doing so, the HR recruiter will post details about the vacant position on one or more recruiting websites, enabling anyone to view the description and requirements for a preferred candidate.

However, organizations sometimes leak too much sensitive data about their systems and network infrastructure when posting a job vacancy for a technical role within their company. From a HR perspective, they are listing all the necessary details for a potential candidate to be aware of before applying for the position. For instance, the job description contains the summary, role, responsibilities (duties), and required qualifications to help the candidate better understand whether the position is suitable for them.

From an adversary's perspective, the technical details found on a job post can be leveraged as OSINT as they identify the technologies, operating systems, vendor devices, systems, and network infrastructure of the company's internal network. An attacker can use this information to research security vulnerabilities in the IT infrastructure, create exploits and malware, and determine which attack vectors are suitable for delivering the payload to the target. As an ethical hacker or penetration tester, recruiting websites are valuable data sources when collecting OSINT on a targeted organization, as the recruiter or HR personnel may leak a lot of sensitive data without realizing it.

To gain a better understanding of how adversaries and ethical hackers can leverage the data found on a job post, let's analyze the following snippet:

JOB REQUIREMENTS

Knowledge/Experience:

- Bachelor's degree in Computer Science, Information Technology or related field.
- Minimum three (3) years' experience in a supervisory or managerial role.
- Be thoroughly familiar with all aspects of the technology environment.
- Extensive knowledge of and experience with (but not limited to):

1. Cloud-based infrastructure
2. Microsoft 365 proficient including Azure AD
3. Wireless networks
4. Routing, Switching, Firewalls, VPNs
5. Windows Server Environments
6. Backup and disaster recovery systems
7. Network/workstation peripherals; print servers; firewalls, ticketing software; project and task management software and computer hardware.

Figure 5.30 – Data leakage on job post

As shown in the preceding snippet, the recruiter listed the main qualifications and experience/knowledge required for an ideal candidate, such as Microsoft 365, Azure **Active Directory** (**AD**), wireless networking, routing and switching, firewalls, **Virtual Private Networks** (**VPNs**), Windows Servers, disaster recovery systems, and so on.

The following can be derived from the technical list found in the preceding snippet:

- The organization is using a cloud computing provider to host some of its services and resources. Therefore, if their IT team has misconfigured a cloud service, the ethical hacker may gain a foothold and access to the cloud resources owned by the target.
- The organization is using Microsoft 365 applications and services, perhaps SharePoint and cloud-based email services. This means the ethical hacker can use social engineering techniques to trick employees into revealing their user credentials to gain access to users' accounts.
- There's a wireless network infrastructure available at the physical location(s) of the company. Therefore, an ethical hacker can identify the physical location of the company and compromise the wireless network to gain a foothold in the internal network.
- There is an enterprise wired network infrastructure with routers and switches. It's important to have a solid foundation in networking and understand how major networking vendor devices are configured and how they operate.
- The organization has one or more firewall appliances for filtering traffic between networks.
- The company is using a site-to-site VPN between branch offices and/or a remote access VPN for various employees. Therefore, compromising a branch office with weak security enables the ethical hacker to pivot the attack from one branch office to another. Additionally, compromising the VPN user's account enables the ethical hacker to remotely connect to the targeted network infrastructure while pretending to be an authorized user.
- There are on-premises Windows Servers and the organization is using Microsoft Azure AD services for replication and redundancy. Therefore, the ethical hacker can create or obtain exploits for Microsoft Windows Server operating systems and research vulnerabilities within Microsoft AD roles and services.

Furthermore, when an organization posts a job vacancy on a recruiting website, it's recommended to visit the company's social media page and official website to identify any additional data leaks that can be used to improve the profiling of the target. Sometimes, the target may create a social media post indicating a new technology was recently implemented within their networks. However, an ethical hacker can use the information to research security vulnerabilities in the system to better plan their attack methodology.

Having completed this section, you have discovered various tactics and techniques that are commonly used by threat actors and ethical hackers to collect and analyze OSINT to create a profile of their target.

Collecting social media OSINT

Social networking platforms provide a medium that enables people to digitally connect and share experiences and life moments with their friends and families beyond traditional boundaries. With social media platforms, a person can create an account and update their profile as their life changes, share photographs and videos of precise moments with others, and join discussion groups for mutual topics. With social media platforms, people no longer need to be physically present at a specific location to participate in a forum; they can simply join an online group that enables them to post their opinions and comment on other discussions.

People use social media for various purposes ranging from connecting with friends and family to starting and advertising an online business from their homes. Social media has changed the way that humans connect with each other and share information. For instance, news companies use social media to post about events in real time to share awareness with their online followers, hence enabling the company to expand its reachability to a wider audience. A lot of organizations around the world use social media platforms to market their products and services to new and existing customers each day and gain feedback from consumers on how they can improve their services.

To put it simply, social media platforms help many people and organizations to share information and connect with others around the world. However, threat actors also use social platforms to collect information about their targets, such as people and organizations. Organizations commonly post about new products, services, and job vacancies on their social media pages. A hacker can use the list of followers from the company's social media page to identify the employees of the targeted organization. Such information can be used for social engineering attacks on the employees, with the intention of tricking them into revealing their user credentials or installing malware onto the company's systems. Sometimes, an organization may advertise a new job vacancy and leak a lot of technical details about its internal infrastructure to the public, which can be leveraged by attackers.

As an ethical hacker who is gathering information on a target, social media platforms contain an abundance of data about people and organizations. For instance, the following types of data can be found and collected from a person's social media profile:

- **Contact details**: Email addresses and telephone numbers can be used for future social engineering attacks.
- **Photos and videos**: These media types may contain pictures of their employee badges, workspace, computers, office location, and sensitive objects in the background. For instance, a picture of an employee badge enables attackers to create a forged badge to gain physical access to a building. Pictures and videos may reveal passwords written on paper, applications running on a computer, and the location of where the photo was taken.
- **Their favorite place for lunch**: Such information can be used to determine the likelihood of this person visiting the particular restaurant during lunchtime. This creates the idea of setting up a *watering hole attack* at the restaurant location. **Watering hole attacks** enables the threat actor to target a specific group of people by infecting commonly visited websites with malware or even compromising the wireless network of the restaurant in the hope that the employee connects their mobile device to the wireless network and becomes infected with malware from the threat actor.
- **View likes and dislikes**: Such information can be useful to determine the person's preferences and whether they are a disgruntled employee.

- **Friends and connections**: The friends and connections list may contain additional employees of the targeted organization, and their profiles might be leaking sensitive data about the company.
- **Usernames**: Social media platforms enable users to list their additional social media handles on their profiles. The handles are the usernames for their online accounts.

Various online tools are commonly used to help ethical hackers quickly find the social media accounts of people and organizations. For instance, **Social Searcher** (`www.social-searcher.com`) enables you to find social media accounts based on mentions (hashtags), users, and trends.

The following snippet shows the results when I search my name on Social Searcher:

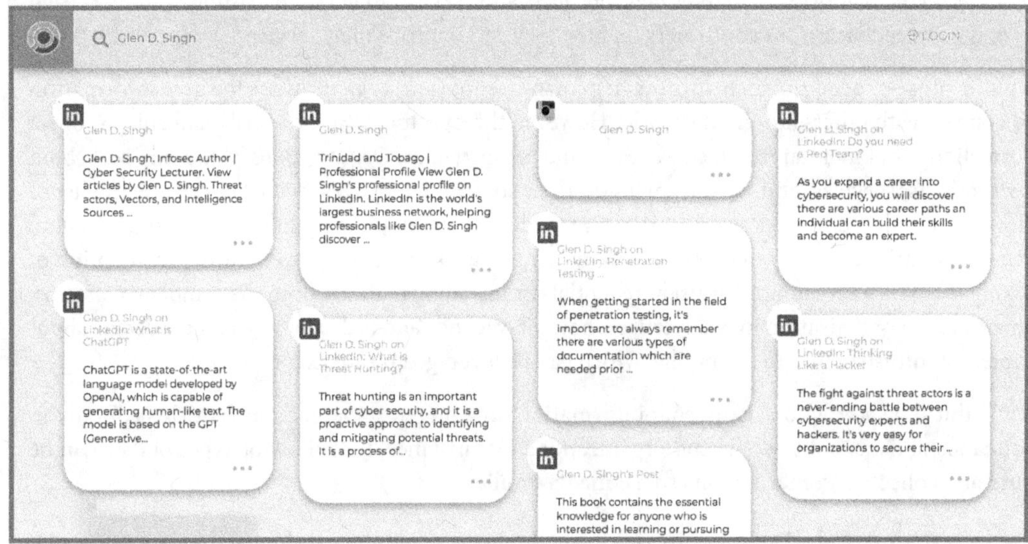

Figure 5.31 – Social Searcher

The preceding snippet is a proof-of-concept in which I've performed a lookup using my name, and the results show various social media profiles and posts I've made in the past. Similarly, an ethical hacker can use this online tool to quickly discover the social media profiles and accounts of a target, whether the target is a person or an organization.

Over the next few subsections, you will learn how to collect data on social media accounts using various tools and techniques.

Sherlock

Sherlock is a popular OSINT tool that enables ethical hackers to efficiently perform username lookups on multiple social media websites to identify the accounts of their targets. This tool quickly searches over 200 social media platforms for the target's username, automates the checking process on each website, and provides a report upon completion.

To get started using Sherlock to discover where a targeted username is found on the internet, please use the following instructions:

1. Firstly, power on your Kali Linux virtual machine.
2. Next, open the Terminal and execute the following commands to update the package repository source list and download Sherlock from its official GitHub repository:

```
kali@kali:~$ sudo apt update
kali@kali:~$ git clone https://github.com/sherlock-project/
sherlock
```

3. Next, use the following commands to change the working directory to the Sherlock folder and install its requirements and dependencies:

```
kali@kali:~$ cd sherlock
kali@kali:~/sherlock$ python3 -m pip install -r requirements.txt
```

4. Once the installation is complete, use the `python3 sherlock <username>` syntax to perform a username lookup on various social media platforms, as shown here:

```
kali@kali:~/sherlock$ python3 sherlock microsoft --timeout 5
```

The following snippet shows the execution of the preceding commands with `--timeout` used to instruct Sherlock to spend no more than 5 seconds on any of the social media websites:

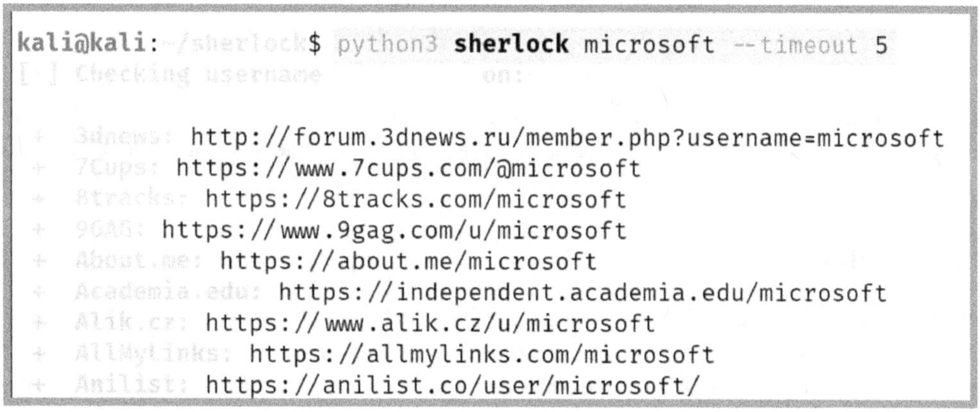

```
kali@kali:~/sherlock$ python3 sherlock microsoft --timeout 5
[*] Checking username            on:

[+] 3dnews: http://forum.3dnews.ru/member.php?username=microsoft
[+] 7Cups: https://www.7cups.com/@microsoft
[+] 8tracks: https://8tracks.com/microsoft
[+] 9GAG: https://www.9gag.com/u/microsoft
[+] About.me: https://about.me/microsoft
[+] Academia.edu: https://independent.academia.edu/microsoft
[+] Alik.cz: https://www.alik.cz/u/microsoft
[+] AllMyLinks: https://allmylinks.com/microsoft
[+] Anilist: https://anilist.co/user/microsoft/
```

Figure 5.32 – Sherlock

5. After Sherlock completes its lookup, it automatically creates a text file within its current directory with a list of social media websites where the username was found:

```
kali@kali:~/sherlock$ ls
CODE_OF_CONDUCT.md  docker-compose.yml  images   microsoft.txt
CONTRIBUTING.md     Dockerfile          LICENSE  README.md

kali@kali:~/sherlock$ cat microsoft.txt
http://forum.3dnews.ru/member.php?username=microsoft
https://www.7cups.com/@microsoft
https://8tracks.com/microsoft
https://www.9gag.com/u/microsoft
https://about.me/microsoft
https://independent.academia.edu/microsoft
https://www.alik.cz/u/microsoft
https://allmylinks.com/microsoft
https://anilist.co/user/microsoft/
https://developer.apple.com/forums/profile/microsoft
```

Figure 5.33 – Displaying the collected data

As shown in the preceding snippet, Sherlock has extracted the results and inserted them into a text file. This enables ethical hackers to semi-automate their data collection process when performing multiple username lookups at the same time. However, ensure you manually check each URL within the results to ensure it's valid.

> **Tip**
> To learn more about Sherlock and its functionality, check out its official GitHub repository at `https://github.com/sherlock-project/sherlock`.

As an ethical hacker, the data collected from Sherlock can be further used to discover usernames and data leakage of sensitive information and find employees and contact details, which can be used to plan future operations such as social engineering attacks.

Facebook IDs

Facebook is a popular online social networking platform that enables people and organizations to digitally connect and share information with each other. When a user registers for an account, a user **profile** is automatically created and enables the person to update their profile details, share media, and connect with others. Once the account is created, the user can join or create groups and pages on Facebook. **Groups** are simply large bodies of people with similar interests or discussions, and **pages** are like digital places and enable organizations, professionals, and artists to connect with their customers and supporters.

When an account (profile), page, or group is created, Facebook automatically assigns a unique identifier to it. If there are two or more accounts with the same name, the identifier is a unique number that can be used to distinguish one account from the other. As an ethical hacker, you can obtain these identifiers when you're collecting information on a target who uses Facebook as their social media platform. Such information can be used to identify a targeted account and filter their posts to find sensitive details about the account holder.

When you're logged in to Facebook as a user, you can use the search bar to find people, groups, and pages. If you're looking for common places that were visited by the target, then you'll need to tweak the Facebook search filters until you're able to find what you're looking for. However, we can leverage the **Intelligence X** platform to help us find specific types of data about an account, group, or page on Facebook. Intelligence X is a search engine and data archival platform that helps ethical hackers to find specific data about a target from multiple data sources.

To get started using **Intelligence X** to gather social media intelligence, please use the following instructions:

1. Firstly, ensure you're logged in to Facebook at `www.facebook.com` using your **sock puppet** account.
2. Next, find your target's account or page by using the Facebook search field. For instance, we can use a public figure page for this exercise:

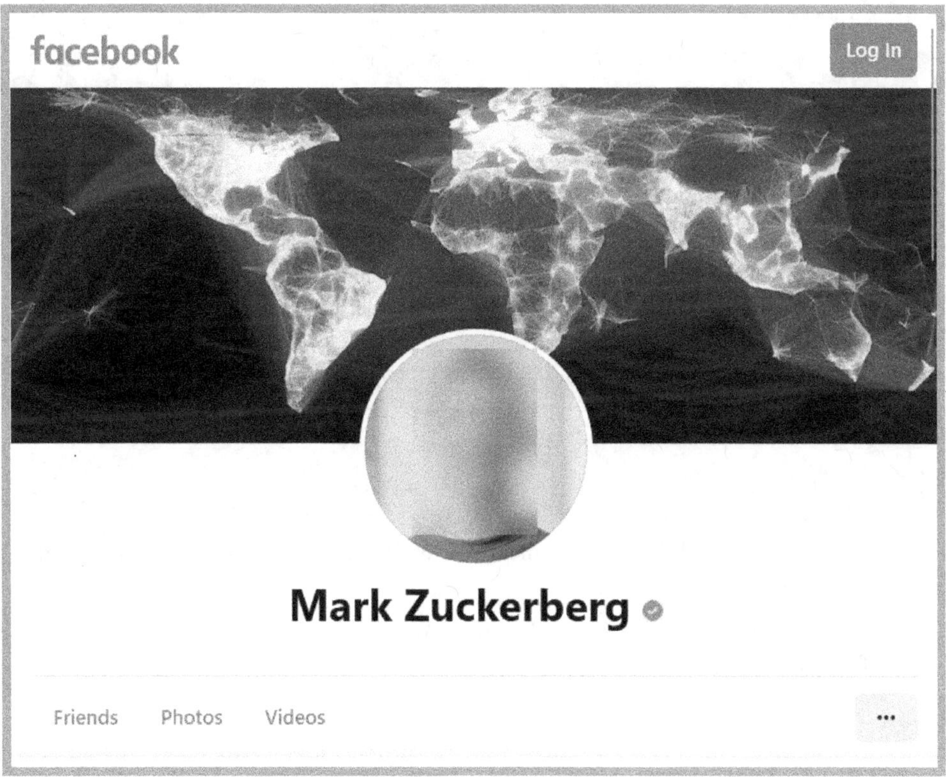

Figure 5.34 – Facebook page

On the account or page, you can see a lot of information such as recently visited places, photos with sensitive details, videos, and so on.

3. When you're on a target's profile or page, right-click anywhere on the page and select **View Page Source** to view the web coding.
4. Next, use the **Find** feature on your web browser and search for `userID`, as shown in the following screenshot:

```
{"__dr":"ProfileCometRoot.react"}],"resource":
{"__dr":"ProfileCometLoggedOutRoot.react"},"props":
{"collectionToken":"YXBwX2NvbGxlY3Rpb246NDoyMzI3MTU4MjI3OjIwMg==","userID":"4","userVanity
":"zuck","viewerID":"0","eligibleForProfilePlusEntityMenu":false,"cometLoginUpsellType":nu
ll},"entryPoint":
{"__dr":"ProfileCometLoggedOutRouteRoot.entrypoint"}},"tracePolicy":"comet.profile.logged_
out","meta":{"title":"Mark
Zuckerberg","accessory":null,"favicon":null},"prefetchable":true,"timeSpentConfig":
{"has_profile_session_id":true},"entityKeyConfig":{"entity_type":
{"source":"constant","value":"profile"},"entity_id":
{"source":"prop","value":"userID"},"section":
{"source":"constant","value":"timeline"}},"hostableView":{"allResources":
[{"__dr":"ProfileCometLoggedOutRoot.react"},
{"__dr":"ProfileCometLoggedOutRouteRoot.entrypoint"},
{"__dr":"ProfileCometRoot.react"}],"resource":
{"__dr":"ProfileCometLoggedOutRoot.react"},"props":
{"collectionToken":"YXBwX2NvbGxlY3Rpb246NDoyMzI3MTU4MjI3OjIwMg==","userID":"4","userVanity
":"zuck","viewerID":"0","eligibleForProfilePlusEntityMenu":false,"cometLoginUpsellType":nu
ll},"entryPoint":
```

Figure 5.35 – Finding the userID value

As shown in the preceding snippet, the `userID` value for the profile or page is `4`. Ensure you record this value as it will be needed in the later steps.

Important note

Facebook accounts (profiles), groups, and pages contain a unique identifier that can be found in their source code. These can be found by searching for `userID`, `entity_id`, `page_id`, and `group_id` within the source code.

5. Next, open a new tab on the same web browser and go to the **Intelligence X** tools page at `https://intelx.io/tools`. In the **Social Media Tools** category, click on **Facebook** to open **Facebook Graph Searcher**.
6. Next, enter the `userID` value (`4`) in the **ID of the user** field and a keyword in the **Talking about** field to filter the search results and click on **Search**, as shown in the following screenshot:

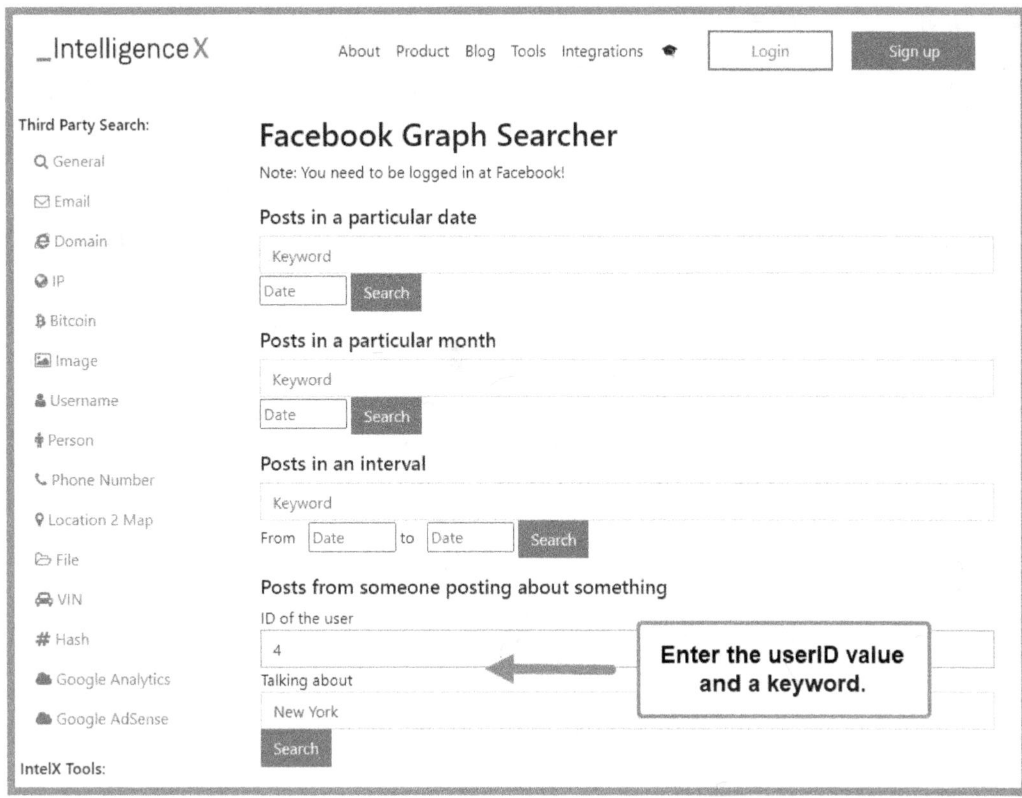

Figure 5.36 – Facebook Graph Searcher

7. Next, a new tab will automatically open and will load the search results for the targeted profile or page using the keyword to filter the results, as shown in the screenshot:

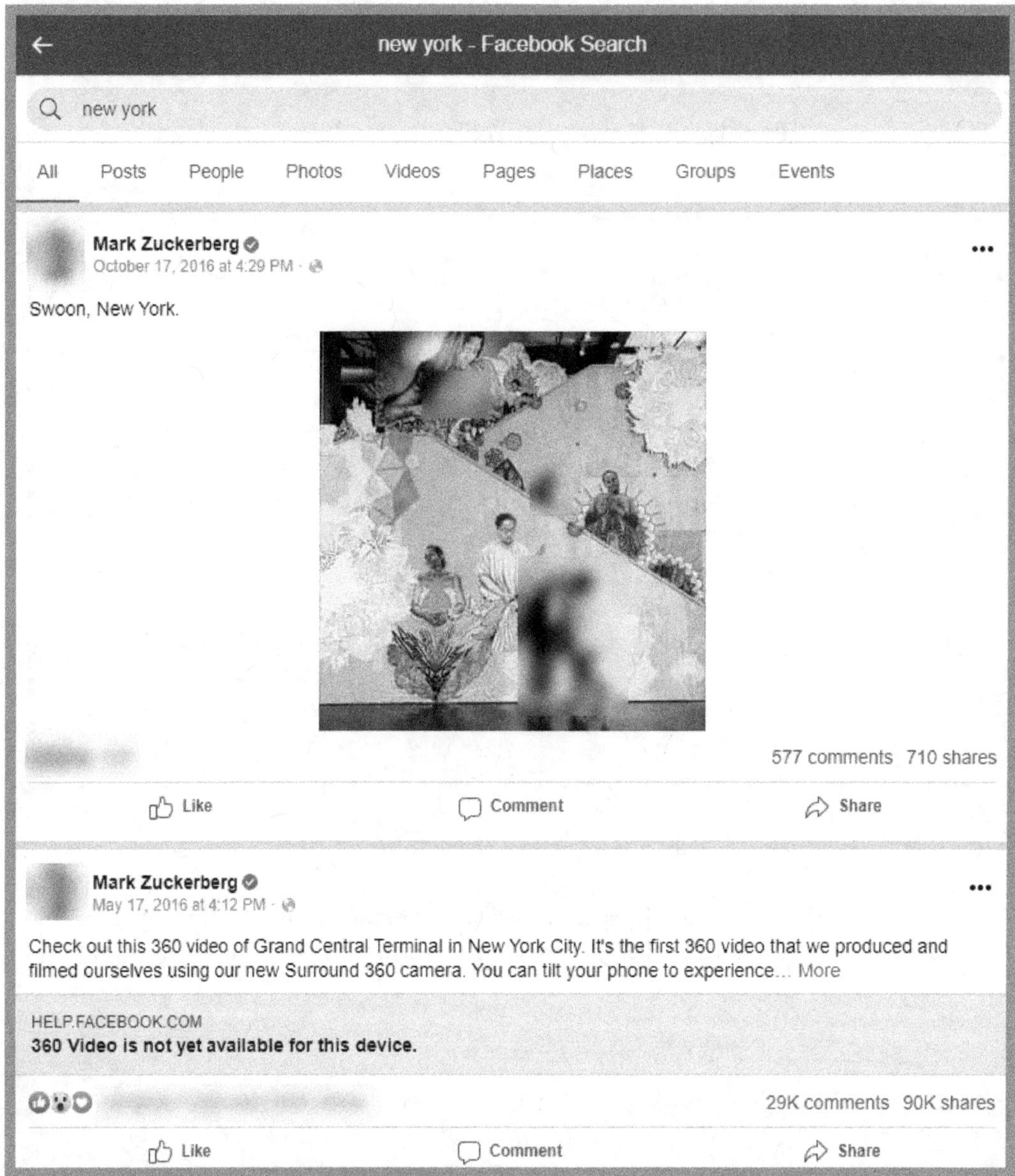

Figure 5.37 – Filtering social media posts

As shown in the preceding snippet, the results were filtered using the keyword `New York`.

As an ethical hacker, Intelligence X provides the capabilities of quickly filtering search results for posts, pictures, and videos that are made by a target. In addition, the keyword helps to filter the search results, enabling us to find specific types of data.

Instagram

Instagram is a social media platform that allows people to share photos and videos with their followers. As an ethical hacker, you will commonly notice people sharing photos of themselves and other objects without considering what's in the background of these pictures or videos.

The following are key details to consider when collecting information from Instagram:

- Who are the people in the photos? Are these people all connected? If so, how?
- What are the objects in the background?
- Can you identify the geo-location of the photo based on the background objects?
- When and where was the photo taken?

Sometimes, people or employees of an organization upload photos of themselves but reveal their desktop icons or files on their computer's monitor. A **threat actor** who is targeting an organization can find photos that are uploaded by employees and look for anything within each picture that can help improve the exploitation phase of the cyber-attack. For instance, desktop icons on a computer can reveal the type of applications and software that are commonly used by employees, and the attacker can identify the host operating system based on the desktop interface and any user credentials that are written on paper on the desk.

Collecting data on social media platforms has become more challenging over the past years as improved security and privacy features are implemented to protect users. However, by visiting an Instagram user's profile or page, you can collect the following data:

- Usernames
- Email addresses
- Telephone numbers
- Addresses
- Hashtags

Each profile or page displays the account's username as their handle; therefore, ethical hackers can perform username harvesting of targeted accounts. Many businesses list their contact details and addresses on their Instagram profile or page to help their potential customers communicate with them. However, the contact details and addresses can be used for planning social engineering attacks and determining the geo-location of the company. Sometimes, people and organizations use hashtags to create a trend. Therefore, monitoring hashtags enables you to track and get notifications whenever the target or anyone uses the hashtag in their posts.

As a proof-of-concept, the following snippet shows my personal Instagram page with some intelligence data:

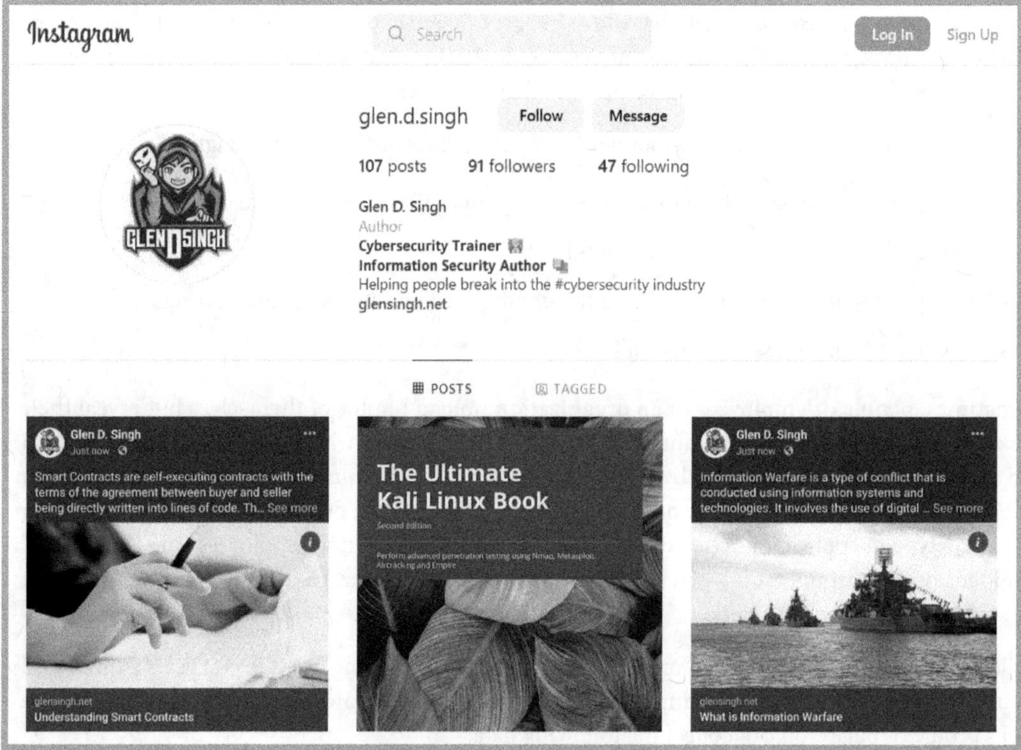

Figure 5.38 – Instagram page

As shown in the preceding snippet, the following data is revealed about the account holder:

- Username
- Identity
- Profession
- The purpose of the profile
- Websites
- Pictures

Visiting a more popular Instagram account will contain hundreds of pictures, videos, and stories (short clips), and analyzing the large quantity of data on the live website can become challenging. **Instaloader** enables ethical hackers to efficiently collect various types of data such as photos, videos, locations, stories, and more from a public profile.

To get started using Instaloader to collect data on a person or organization's Instagram account, please use the following instructions:

1. Firstly, power on the Trace Labs OSINT virtual machine and log in using `osint/osint` as the username and password.
2. Open the web browser and find a target and record the username on the profile.
3. Next, open the Terminal and execute the `instaloader --help` command to view various commands and their usage.
4. Next, use the `instaloader profile <username>` syntax to download everything that's publicly available on the account, as shown here:

   ```
   osint@osint:~$ instaloader profile zuck
   ```

   The preceding command will download images, videos, locations, stories, and feed data and save it within the `/home/osint/` directory, as shown in the following screenshot:

   ```
   osint@osint: $ pwd
   /home/osint

   osint@osint: $ ls /home/osint/zuck
   2022-10-17_13-57-43_UTC.json.xz              2022-12-02_15-01-03_UTC.mp4
   2022-10-17_13-57-43_UTC.txt                  2022-12-02_15-01-03_UTC.txt
   2022-10-17_13-57-43_UTC.webp                 2022-12-13_15-59-27_UTC.json.xz
   2022-10-18_14-34-04_UTC_profile_pic.jpg      2022-12-13_15-59-27_UTC.txt
   2022-10-19_15-08-54_UTC.jpg                  2022-12-13_15-59-27_UTC.webp
   ```

   Figure 5.39 – Download files

   As shown in the preceding snippet, Instaloader downloaded pictures, videos, and geo-location and feed data offline for post-analysis.

As an ethical hacker, ensure you check each download file during your analysis phase to determine whether it contributes toward building a profile of your target and future attack operations.

LinkedIn

LinkedIn is a social networking site that's designed for academics, industry professionals, students, and researchers to connect, collaborate, and share ideas with each other. As an ethical hacker, you can collect a lot of valuable information about your target from LinkedIn, whether your target is a person or an organization. The following are common types of data that can be gathered from a person's LinkedIn profile:

- Name
- Location
- Current and past jobs

- Educational history
- Pictures of employee badges and workplace
- Past projects

A LinkedIn profile is a like a resume and curriculum vitae in the form of a web page for each user, enabling others to view and connect based on mutual interests. While LinkedIn has a lot of privacy features that enable users to restrict access to their profiles, threat actors can create a fake account and trick their target into connecting with them, allowing the attacker to view the details on their LinkedIn profile.

Let's imagine that every time a person changes their job, they update their work experience to reflect the change. This may include adding a new job title at a new company and including a description of their role and responsibilities. While this information is useful for HR professionals and recruiters, an attacker can look for any technical details within a job description to identify the organization's network infrastructure and system. Furthermore, if the user works on a technical project, the employee may include the technical details on their LinkedIn profile, which provides better insights into the organization's infrastructure, technologies, and whether any security vulnerabilities exist.

Another technique is to map the list of IT certifications that are obtained by an employee to their work history from their LinkedIn profile. Many employees obtain IT certifications to become proficient in their jobs and better understand the technologies used within the organization. For instance, if a LinkedIn user has been employed at the targeted organization for the past 4 years and obtained specific Microsoft certifications for the duration, this is an indication the company may be using (or has implemented) those specific technologies. Hence, such information can be harvested from LinkedIn to collect data from users' online resumes to profile an organization. As an ethical hacker, these techniques will enable you to better profile and gain insights into the technologies used by your target.

As an ethical hacker, having a sock puppet on LinkedIn can help you examine the connections between both your target and their employees. Using the following Google Dork syntax enables you to find LinkedIn profiles by specifying a person's name:

```
site:linkedin.com NAME intitle:professional
```

The following snippet shows a proof-of-concept using my first name:

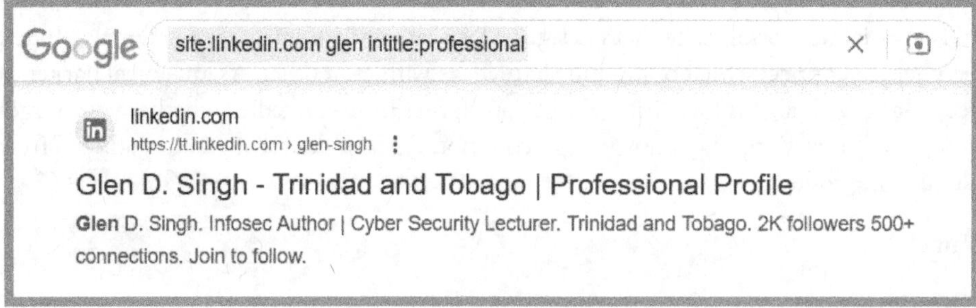

Figure 5.40 – Google Dorking

As you may realize already, people share a lot of information on LinkedIn and such information can be harvested and analyzed. Hence, sharing too much information can be used for both good and bad actions.

Twitter

Twitter is another popular social networking platform that enables people to post short messages, pictures, and video clips and allows them to follow other users and get notifications when they post something new. Over the past few years, people have posted almost everything about their thoughts and daily activities on the platform. As an ethical hacker, you can use a sock puppet to harvest the following types of data from a target's Twitter account:

- **Profile handle**: Can be used to identify other users who tagged your target within their posts
- **Profile description**: Can be used to better understand your target
- **Location**: Can be used to identify the physical location of the target
- **Website**: Enables you to identify their website and collect data from it for other types of attacks
- **Joined date**: Indicates the age of the profile, which helps to determine how often the target makes a new post and identifies the types of devices that were used to post on the platform
- **Media**: Enables you to view any photos and videos that may contain sensitive details and geo-location data
- **Likes**: Helps you determine your target's preferences and interests
- **Tweets**: View the actual content made by your target
- **Followers**: Helps to identify employees of the company

Having completed this section, you have learned how ethical hackers are able to harvest sensitive data from social media platforms to better profile their targets.

Summary

During the course of this chapter, you learned how to collect and analyze OSINT from various data sources to identify the network infrastructure of organizations. Additionally, you discovered how to use various online websites such as Shodan, Censys, social media platforms, and even job boards to find data leaks and gain insights into how a threat actor can leverage the data to profile an organization and determine their attack surface.

I hope this chapter has been informative for you and is helpful in your journey in the cybersecurity industry. In the next chapter, *Imagery, People, and Signals Intelligence*, you will learn how to analyze images and use mapping applications to identify the physical location of a target, collect user credentials from online databases, and identify wireless devices within an area.

Further reading

- *DNS records*: `https://www.cloudflare.com/learning/dns/dns-records/`
- *Maltego Foundations 1*: `https://courses.maltego.com/courses/maltego-foundations-1-NEW`
- *MITRE Reconnaissance*: `https://attack.mitre.org/tactics/TA0043/`

6
Imagery, People, and Signals Intelligence

Photographs can tell you a lot and threat actors collect imagery intelligence on their targets to identify their geo-locations and any sensitive data in the background of pictures that can be used to improve their operations. People around the world are always posting pictures of themselves, their family members, and their locations on social media platforms without fully understanding how adversaries can track their whereabouts.

In this chapter, you will learn how to analyze imagery and maps to identify the physical location of a target, and collect and analyze data from public databases to identify valid usernames and passwords of people. Additionally, you will learn how to perform wireless signals intelligence to identify wireless devices within an area and profile a targeted wireless network infrastructure.

In this chapter, we will cover the following topics:

- Image and metadata analysis
- People and user intelligence
- Wireless signals intelligence

Let's dive in!

Technical requirements

To follow along with the exercises in this chapter, please ensure that you have met the following hardware and software requirements:

- Kali Linux: `https://www.kali.org/get-kali/`
- Trace Labs OSINT VM: `https://www.tracelabs.org/initiatives/osint-vm`
- Kali Linux ARM: `https://www.kali.org/get-kali/`

- Google Earth Pro: `https://www.google.com/earth/versions/`
- Raspberry Pi Imager: `https://www.raspberrypi.com/software/`
- Rufus: `https://rufus.ie/`
- Raspberry Pi 3 B+
- A 32-GB Samsung EVO+ microSD card
- A microSD card reader
- A micro-USB cable
- An Alfa AWUS036NHA - Wireless B/G/N USB adapter
- A VK-162 G-Mouse USB GPS Dongle Navigation Module
- A portable power bank

Image and metadata analysis

Exchangeable Image File (**EXIF**) is a standard that specifies the formatting of sounds and images that are commonly used in image handling devices such as scanners and digital cameras. Put simply, EXIF is the metadata embedded into photo files taken by a digital camera and includes camera data such as the geo-location, time and date, manufacturer, resolution, and so on. Images with EXIF data can be leveraged by ethical hackers for reconnaissance and social engineering.

As an ethical hacker, you can collect publicly available images of your target and analyze them for EXIF data. The collected EXIF data may reveal the target's geo-location, the type of device used for capturing the photo, and the time the picture was taken. For instance, if a picture was taken with a digital camera within a server room of an organization, an ethical hacker who is hired to simulate real-world cyberattacks can identify whether the geo-location data is available and use it determine the physical location of the server room and the targeted company. Additionally, the EXIF data can be used by ethical hackers to improve their social engineering attacks by identifying the geo-location of the photo and sending spear phishing email messages that appear to originate from the same physical location as the target, therefore attempting to trick the target into thinking the email originates from a trusted source.

However, many smartphone and mobile operating system vendors have improved the security of their devices and implemented security mechanisms to prevent sensitive EXIF data from being attached to photos and videos captured using their devices. Imagine if all the photos uploaded on the internet contained EXIF data; then cyber criminals would be able to track their potential victims a lot easier than before by using a smartphone or computer. Therefore, the EXIF location data has been made a bit more secure over the years and reduces the risk of someone collecting the **Global Positioning System** (**GPS**) data from a photo, and some social media platforms have implemented security mechanisms to remove the EXIF data from photos before they are published on the online platform.

> **Important note**
>
> Digital forensic investigators can use EXIF data from a photo to determine when the picture was taken and where. Such information is useful during investigations as it helps identify a potential suspect and focus on the timeline of an attack.

EXIF provides a lot of data, which can be leveraged the right way by an ethical hacker to determine the whereabouts of a target, the device used for capturing the picture, the direction to target was facing, the device settings, and much more.

EXIF data analysis

In this exercise, you will learn how to identify and analyze EXIF data from a photo and determine the geo-location of where it was captured. To get started with this exercise, please use the following instructions:

1. Power on the **Trace Labs OSINT virtual machine** and log in using `osint/osint` as the username and password.
2. Once you're logged in to the Trace Labs OSINT virtual machine, open the **web browser** and go to `https://osint.tools` for a practice file. In the menu on the left column, you will see a list of resources – click on **Exif Example 1** as shown:

Figure 6.1 – EXIF file

3. Next, a picture will load on the web page. Right-click on it and select **Save Image As...** to download and save the image to the **Trace Labs virtual machine** as shown:

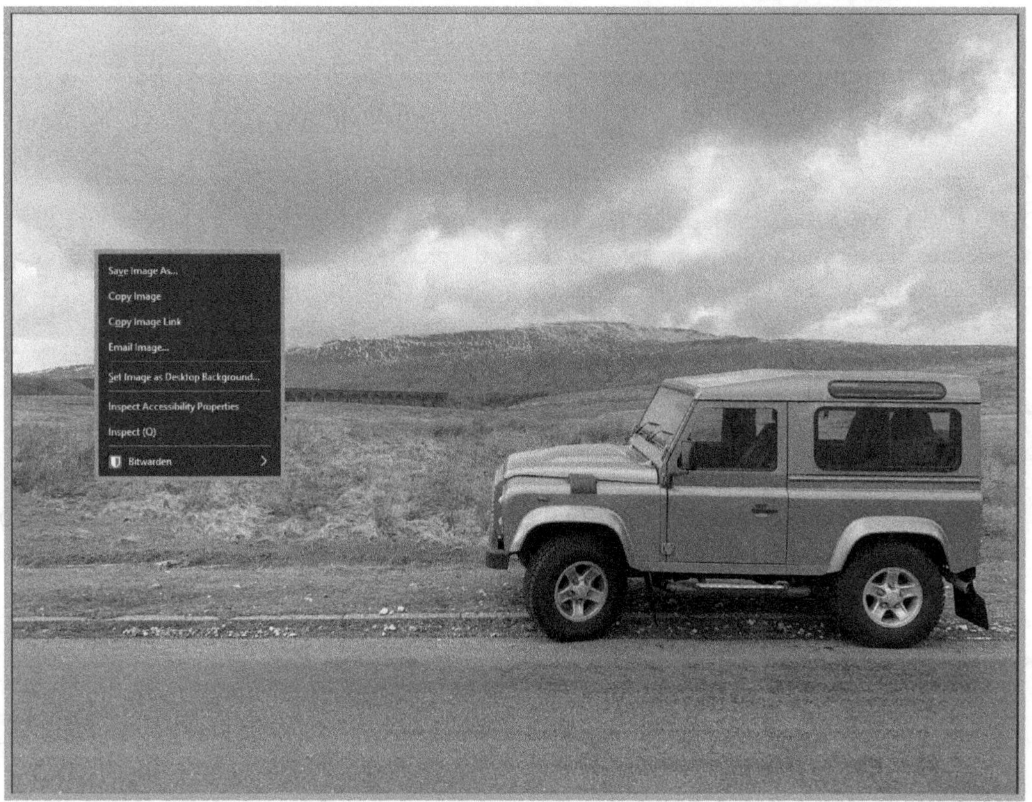

Figure 6.2 – EXIF picture

By default, all downloads are automatically saved in the `/home/osint/Downloads` directory on the Trace Labs virtual machine.

4. Next, use `exiftool` to extract the EXIF data from the image file. Open the **Terminal** and execute the following command:

```
osint@osint:~$ exiftool Downloads/exif1.jpg
```

The following screenshot shows the output from executing the preceding command:

```
osint@osint: $ exiftool Downloads/exif1.jpg
ExifTool Version Number         : 12.40
File Name                       : exif1.jpg
Directory                       : Downloads
File Size                       : 5.9 MiB
File Modification Date/Time     : 2023:03:27 11:23:06-04:00
File Access Date/Time           : 2023:03:27 11:23:06-04:00
File Inode Change Date/Time     : 2023:03:27 11:23:06-04:00
File Permissions                : -rw-r--r--
File Type                       : JPEG
File Type Extension             : jpg
MIME Type                       : image/jpeg
Exif Byte Order                 : Big-endian (Motorola, MM)
Make                            : Apple
Camera Model Name               : iPhone XS Max
Orientation                     : Horizontal (normal)
X Resolution                    : 72
Y Resolution                    : 72
```

Figure 6.3 – EXIF data

As shown in the preceding screenshot, we are able to view the EXIF data from the image. The time and date, geo-location, device type and model, and resolution can be collected and analyzed.

5. Next, let's use the `exifprobe` tool to extract EXIF data from the same picture:

```
osint@osint:~$ exifprobe Downloads/exif1.jpg
```

As shown in the following screenshot, the exifprobe tool extracted the EXIF and geo-location data from the picture:

```
@0x0000792=1938   :  <GPS IFD> (in IFD 0) 13 entries starting at file offset 0x794=1940
@0x0000794=1940   :    <0x0001=    1> LatitudeRef               [2 =ASCII       2]  = 'N'
@0x00007a0=1952   :    <0x0002=    2> Latitude                  [5 =RATIONAL    3]  = @0x834=2100
@0x00007ac=1964   :    <0x0003=    3> LongitudeRef              [2 =ASCII       2]  = 'W\000'
@0x00007b8=1976   :    <0x0004=    4> Longitude                 [5 =RATIONAL    3]  = @0x84c=2124
@0x00007c4=1988   :    <0x0005=    5> AltitudeRef               [1 =BYTE        1]  = 0
@0x00007d0=2000   :    <0x0006=    6> Altitude                  [5 =RATIONAL    1]  = @0x864=2148
@0x00007dc=2012   :    <0x000c=   12> SpeedRef                  [2 =ASCII       2]  = 'K\000'
@0x00007e8=2024   :    <0x000d=   13> Speed                     [5 =RATIONAL    1]  = @0x86c=2156
@0x00007f4=2036   :    <0x0010=   16> DirectionRef              [2 =ASCII       2]  = 'T\000'
@0x0000800=2048   :    <0x0011=   17> Direction                 [5 =RATIONAL    1]  = @0x874=2164
@0x000080c=2060   :    <0x0017=   23> BearingRef                [2 =ASCII       2]  = 'T\000'
@0x0000818=2072   :    <0x0018=   24> Bearing                   [5 =RATIONAL    1]  = @0x87c=2172
@0x0000824=2084   :    <0x001f=   31> GPS_0x001f                [5 =RATIONAL    1]  = @0x884=2180
@0x0000830=2096   :    **** next IFD offset 0
@0x0000834=2100   :    ============ VALUES, GPS IFD ============
@0x0000834=2100   :    Latitude                    = 54,12,30.67
@0x000084c=2124   :    Longitude                   = 2,21,41.53
@0x0000864=2148   :    Altitude                    = 280.202
@0x000086c=2156   :    Speed                       = 0
@0x0000874=2164   :    Direction                   = 311.281
@0x000087c=2172   :    Bearing                     = 311.281
@0x0000884=2180   :    GPS_0x001f                  = 4.57993
-0x000088b=2187   :  </GPS IFD>
```

Figure 6.4 – The exifprobe tool

6. Next, open the web browser again and go to `https://fotoforensics.com/`, where you can upload a picture to extract the EXIF data. Simply click on **Browse...** to attach the file and select **Upload** to perform the extraction and analysis as shown:

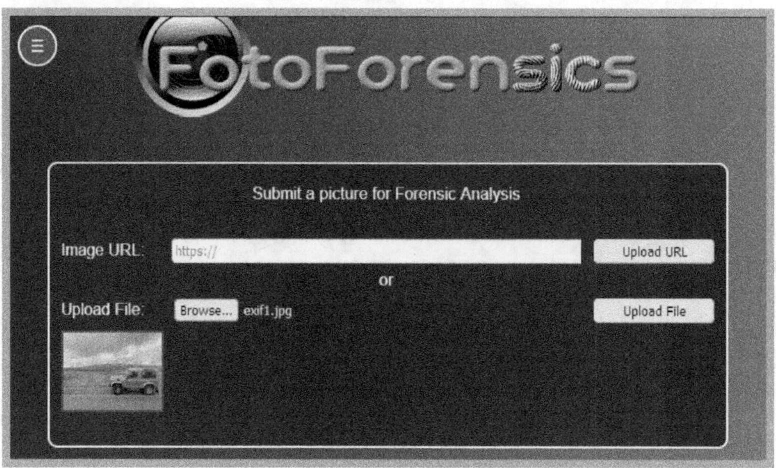

Figure 6.5 – Fotoforensics website

After a few seconds or minutes, the website will display all the EXIF data and provide various specific categories of data to view.

7. Next, select **Analysis** | **Metadata** to view the EXIF data:

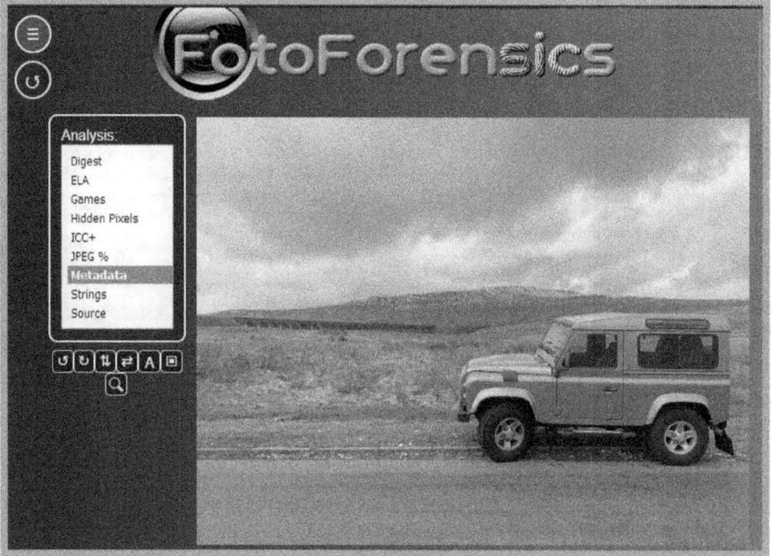

Figure 6.6 – Viewing metadata

The following screenshot shows all the EXIF data that was found in the file:

File	
File Type	JPEG
File Type Extension	jpg
MIME Type	image/jpeg
Exif Byte Order	Big-endian (Motorola, MM)
Image Width	4032
Image Height	3024
Encoding Process	Baseline DCT, Huffman coding
Bits Per Sample	8
Color Components	3
Y Cb Cr Sub Sampling	YCbCr4:2:0 (2 2)
EXIF	
Make	Apple
Camera Model Name	iPhone XS Max
Orientation	Horizontal (normal)
X Resolution	72
Y Resolution	72
Resolution Unit	inches
Software	13.3.1
Modify Date	2020:02:29 14:37:44
Y Cb Cr Positioning	Centered
Exposure Time	1/1032
F Number	1.8
Exposure Program	Program AE
ISO	25
Exif Version	0231
Date/Time Original	2020:02:29 14:37:44
Create Date	2020:02:29 14:37:44

Figure 6.7 – EXIF data from the uploaded image

8. Next, scroll down the end of the EXIF results page to view the geo-location of the image:

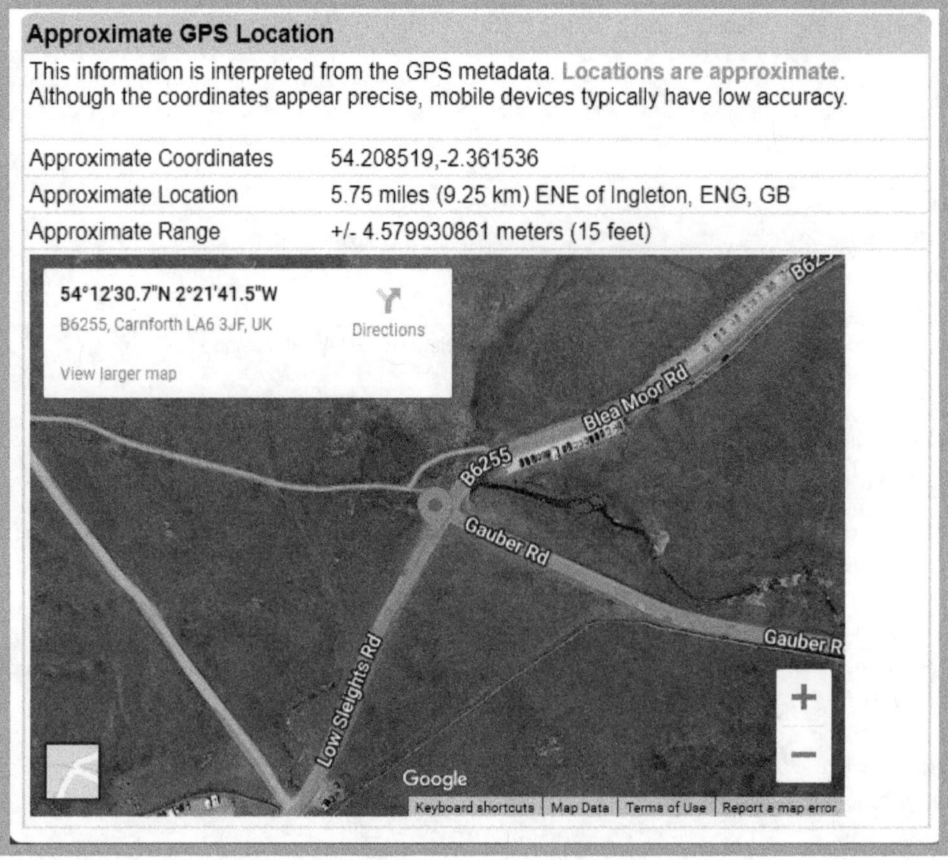

Figure 6.8 – Geo-location data

As shown in the preceding screenshot, the GPS data was found and applied to a mapping system to identify the approximate location of where the photo was taken.

9. Lastly, you can copy the GPS coordinates from the previous step and enter them into `https://maps.google.com` (Google Maps) to get a second opinion on the location as shown:

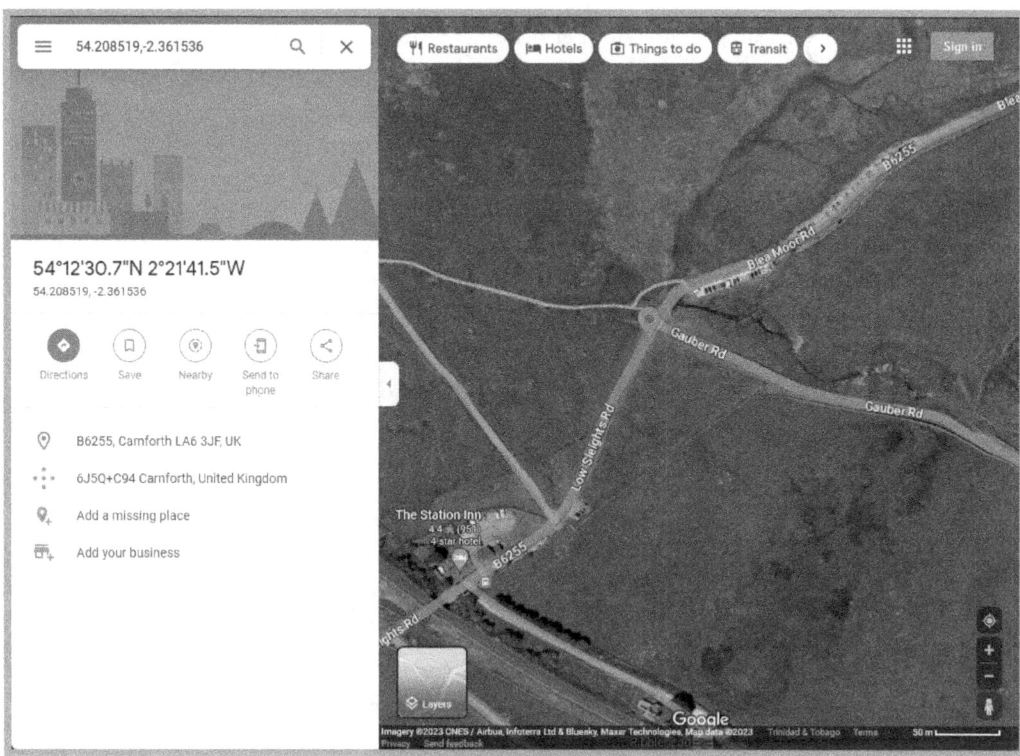

Figure 6.9 – Google Maps

As shown in the preceding screenshot, GPS coordinates can be used on Google Maps to gain better insights into the vicinity and street view of a location.

Reverse image search

Reverse image search allows anyone to perform an internet search for pictures by either uploading a photo or providing a URL that points to an image. Just as most people are familiar with using internet search engines to find websites and domain names based on keywords, reverse image search enables us to find similar pictures of various sizes amd resolutions, different versions of the same picture, and even pictures that contain the same objects.

Reverse image search is very useful for ethical hackers as it enables them to identify the source and location of the picture, and determine where the picture has been used on the internet. If you're trying to identify the location of where a picture was taken by your target, this technology provides you with the capabilities to find the location.

The following are popular search engines for performing reverse image searches:

- Google: `https://images.google.com`
- Bing: `https://www.bing.com`
- Yandex: `https://yandex.com`
- TinEye: `https://tineye.com`

To gain a better understanding of how reverse image search helps ethical hackers to identify the geo-location of a target, let's consider the following scenario:

1. You're given a photo and need to identify the location of where the picture was taken. In this exercise, we are using a publicly available photo taken by photographer Mauro Lima from `https://unsplash.com` as shown:

Figure 6.10 – An example photograph

As shown in the preceding figure, it's a lovely picture taken by a photographer showing a religious place of worship. Sometimes, it's recommended to remove any objects or subjects from a photo before performing a reverse image search to ensure the search engine is better able to identify similar pictures on the internet.

> **Tip**
>
> Being able to identify specific architecture styles, signage, flora, and fauna can help you determine the specific region of the world a picture was taken. For instance, specific types of trees can only be found in certain countries.

2. We can use **Cleanup.pictures** (`https://cleanup.pictures`) to remove specific objects from the picture, such as the person and vehicles. The following screenshot shows the Cleanup.pictures website with the upload field, upload your image here:

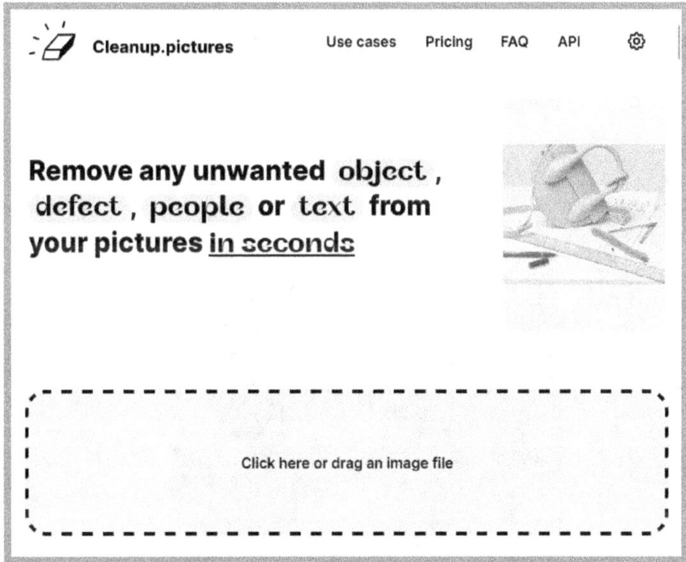

Figure 6.11 – Cleanup.pictures website

3. Next, use the eraser brush to select the objects and people you want to remove and enable the **Artificial Intelligence** (**AI**) to automatically fill in the highlighted area:

Figure 6.12 – Objects and people

4. Next, ensure you download and save the AI-modified image:

Figure 6.13 – AI modified image

> **Tip**
>
> You can use **Remote Background** at `https://www.remove.bg` to remove backgrounds from pictures.

5. Next, to perform a reverse image search, go to **Google Images** at `https://images.google.com` and upload the modified image to view the results as shown:

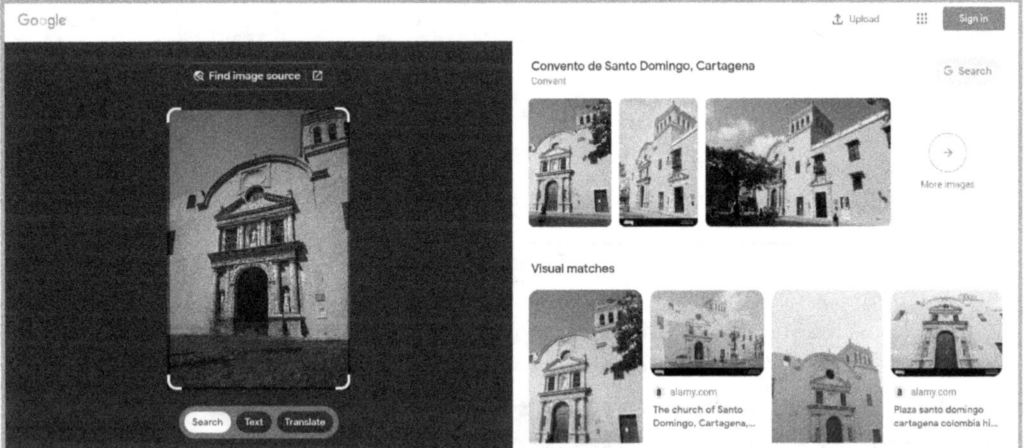

Figure 6.14 – Reverse image search results

As shown in the preceding screenshot, the location and similar images were found.

> **Important note**
>
> **Google Images** has integrated **Google Lens** to improve its image search. This technology enables you to select a specific area of the image, or the entire thing.

6. Next, enter the location `Convento de Santo Domingo, Cartagena` within **Google Maps** at `https://maps.google.com` as shown:

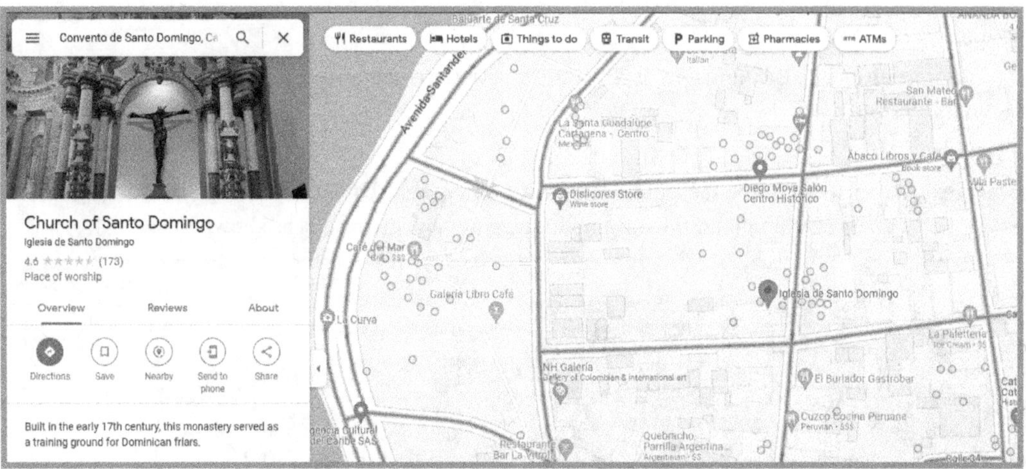

Figure 6.15 – Google Maps

7. Lastly, using the **Street View** feature on Google Maps, you can virtually navigate around the location to collect additional intelligence:

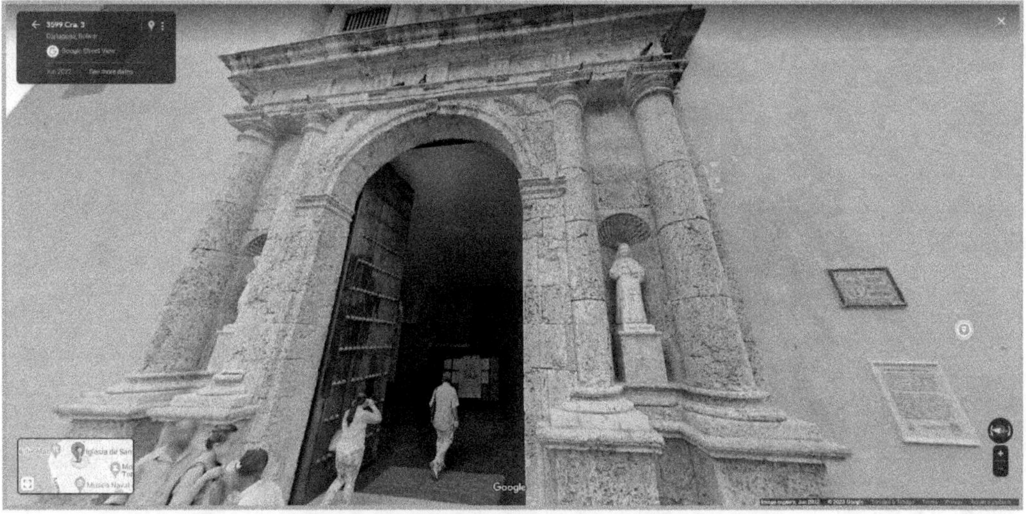

Figure 6.16 – Google Maps Street View

As shown in the preceding screenshot, we have found the location where the photo was taken by the photographer.

As you have now learned, reverse image searching helps ethical hackers to identify the geo-location and whereabouts of a target, along with using global mapping systems to view the surroundings of the location. This knowledge is also useful when performing a physical penetration testing assessment to identify the points of entry into a building or compound with the least resistance.

Geo-location analysis

Mapping systems are excellent data sources for ethical hackers to identify the physical location of a targeted organization and its vicinity. Imagine you need to perform a wireless penetration test on a targeted organization during a black-box assessment. You can perform Google searches on the organization's name to identify its websites and check whether its address is listed. Additionally, you can use publicly available online maps such as **Google Maps** to identify the physical location of an organization.

For instance, we can use Google Maps to find the physical location of Twitter HQ (red pin) and gain a better understanding of the surroundings as shown:

Figure 6.17 – Locating a company

As shown in the preceding screenshot, Google Maps was able to provide us with a picture of the building to better identify the target, its street address, its hours of operations, and its contact details, which can be used during future operations such as social engineering attacks.

Changing to Satellite view on Google Maps helps us to gain a better visual of the layout and surrounding areas of the company, such as identifying a place to set up wireless attacks as an ethical hacker:

Figure 6.18 – Satellite view

Furthermore, we can observe there are nearby restaurants that employees are likely to visit frequently and may connect their mobile devices to the restaurants' wireless networks. As an ethical hacker, you could set up a rogue wireless network to trick an employee of your target into connecting and redirect their traffic to phishing websites and perform **Domain Name System** (**DNS**) poisoning attacks.

Google Maps provides the Street View feature, which enables ethical hackers to interact with panoramic views of streets, allowing you to view the vicinity of the target's building and look for car parks, points of entry, and security checkpoints.

The following screenshot shows you a street view of the Twitter headquarters without you leaving your computer:

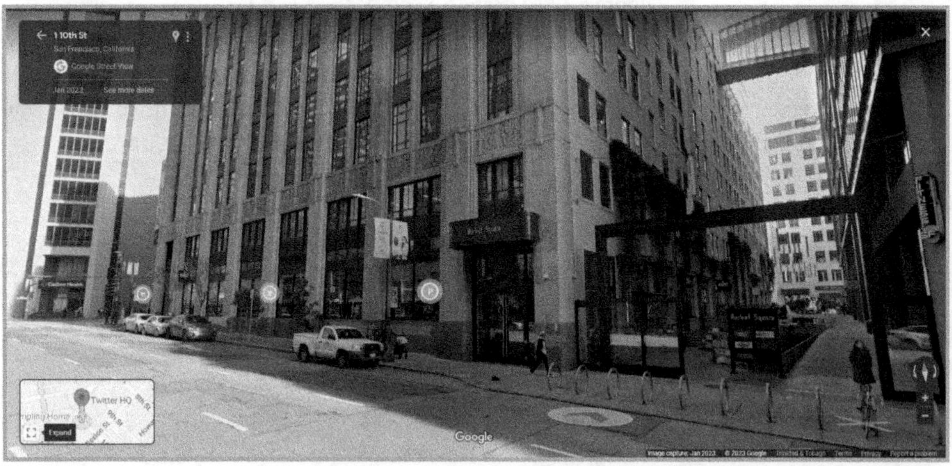

Figure 6.19 – Street View

As an ethical hacker, you can leverage Street View to look at the sides and other pathways that may be convenient and have less security presence for physical penetration testing and wireless network assessments.

The following screenshot shows a side view of the Twitter building with a small restaurant:

Figure 6.20 – Side view

While ethical hackers use these tactics and techniques to help organizations, threat actors are also using them for malicious purposes. Hence, it's important that organizations ensure they perform OSINT penetration testing to ensure their cyber defenses are improved and to reduce their attack surface.

During this section, you gained the hands-on skills to collect and analyze imagery and geo-location intelligence and gain further insights into how threat actors are able to use publicly available information to profile their targets. In the next section, you will learn how to collect intelligence on people and users from organizations.

People and user intelligence

People OSINT focuses on collecting information about a person's names, address, telephone numbers, usernames, and even their social media accounts. Such information is useful when identifying the employees of a targeted organization and improving social engineering attacks to gain a foothold within the target's network. People often create social media accounts and add their company, job title, and contact details, which helps threat actors to easily harvest such data when planning their attacks. For instance, employees sometimes include their company's email address in their contact information on social media platforms such as LinkedIn and Facebook.

Organizations use various formats for their employees' email addresses, such as the following:

- `firstname @ domain . com`
- `firstnameinitial + lastname @ domain . com`
- `firstname.lastnameinitial @ domain . com`
- `firstname + lastnameinitial @ domain . com`
- `fullname @ domain . com`

Therefore, if an attacker is collecting OSINT on a targeted organization, collecting email addresses from current employees found on social media allows them to determine the email format used within the organization. Some companies even publicly post their organizational charts on their websites outlining various high-profile employees and their departments. Such information can be used to identify specific persons and determine their email address format to plan spear-phishing attacks.

Furthermore, there are many online forums where students, professionals, and researchers can get help with technical issues such as **Stack Overflow** (`https://stackoverflow.com`). Attackers can use the information on Stack Overflow to identify the technologies and security vulnerabilities that exist within a targeted organization. For instance, imagine the technical team is experiencing some issues with an application and decides to create an account on Stack Overflow using their real name, job title, and organizational name on their profiles. They then proceed to make a post containing the technical details of their infrastructure, host operating systems, application name, and service version. While this information is useful for anyone providing assistance with good intentions, such information reveals a lot to a threat actor about their target and makes it easier to identify the attack surface.

As an ethical hacker, it's important to fully understand the tactics and techniques that are commonly used by threat actors to collect and analyze data leaks from the internet. Sometimes, finding people can be a challenging task, whether you're helping law enforcement or identifying the employees of a targeted organization. There are many people-based search engines that can be used for collecting names, addresses, telephone numbers, social media accounts, and usernames of individuals. The following is a list of both free and commercial people-based search engines:

- `www.peekyou.com`
- `thatsthem.com`
- `radaris.com`
- `www.beenverified.com`
- `www.skopenow.com`

During a penetration test, ethical hackers commonly perform a technique known as **password spraying**, whereby a common password is used with multiple usernames on the same application or system. The purpose of using password spraying is to identify users that have configured their accounts with weak passwords or used the same one as others within the organization. Sometimes, the name (user) portion of an email address is an employee's username for accessing a company-owned system or application. In some cases, the entire email address is used as the employee's username.

When harvesting email addresses from various data sources on the internet, it's essential to identify whether an address is valid or blacklisted and whether there's an online social media account associated with it. This can be done by performing a search on social media websites using the target's email address to identify their social media account on the platform if it exists.

The following websites can be used to determine the reputation and validity of an email address:

- MXToolBox: `https://mxtoolbox.com/emailhealth`
- Simply Email Reputation: `https://emailrep.io`
- Cisco Talos: `https://www.talosintelligence.com/reputation_center/email_rep`

In addition, **Hunter** (`https://hunter.io`) enables ethical hackers to discover employees, their job titles, and any email addresses associated with a targeted domain. Hunter allows you to perform a lookup using the domain name of your target, then searches its database and multiple data sources on the internet for individuals' contact details related to this domain name.

The following screenshot shows a lookup that was performed on the `microsoft.com` domain:

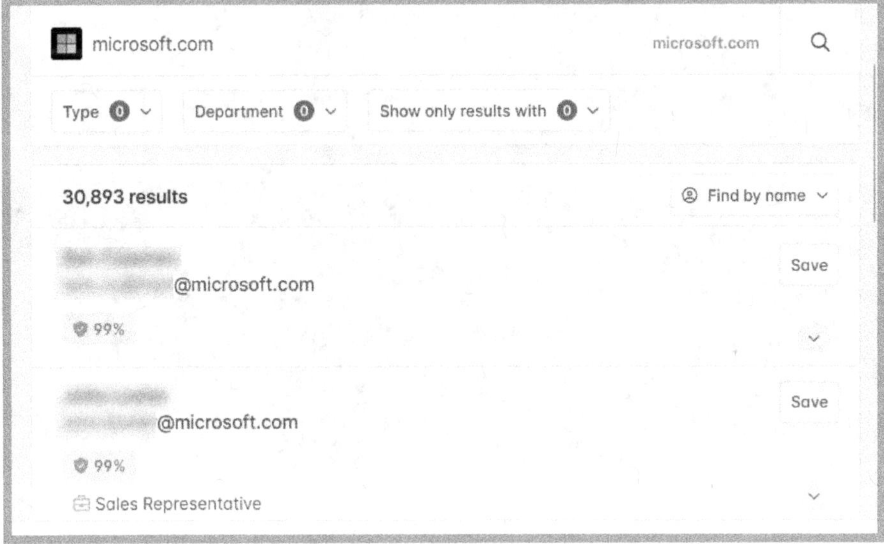

Figure 6.21 – Hunter domain lookup

As shown in the preceding screenshot, Hunter is a useful tool for ethical hackers as it helps you to harvest organizational and people intelligence from the internet. The data collected from Hunter can be used to plan social engineering attacks on a targeted organization.

People and geolocation

There are many platforms on the internet that enable people to upload and share pictures of themselves and places they've visited. Social media and photo-sharing platforms are valuable repositories for ethical hackers to identify targets and their locations. For instance, imagine you're given the task of identifying the high-profile employees of a targeted organization, determining their whereabouts, and identifying any places of interest associated with your target.

You can use social media platforms such as LinkedIn, Facebook, and Twitter to perform a lookup/search using the organization's name and filter the results as needed. On an employee's profile, they may include the target organization and their location. Some people often use the check-in feature on social media, which provides useful data for threat actors including insights into how often a person visits a particular location, travels for work, and when and where the person goes on vacation.

To get a better understanding of how data can be collected and leveraged by adversaries, let's observe the following Twitter profile of a famous person of our times:

Figure 6.22 – Twitter profile

The preceding screenshot shows their public profile with lots of data. For instance, if you're looking for a photo of a target, their social media profile will most likely have an up-to-date picture of themselves. In addition, the public profile picture can then be used with reverse image lookup to identify additional data sources that may contain the same or similar pictures on the internet. The preceding screenshot also shows the location of the individual, their website, and the age of their profile. As previously mentioned, the location helps you identify the target's general location or city. The website on the profile can be used to find additional intelligence. Lastly, you can view all the posts and media uploaded by the target; much sensitive information can be found within these to help you identify the infrastructure of a given target organization.

Social media apps on smartphones enable a user to include their geo-location data on their posts to allow their followers, connections, and friends to know their location at the time the post was made. Imagine you want to identify all the employees of a target organization but all you have is the address of the company. As an ethical hacker, you can use **Google Maps** to find the specific location of the organization and obtain the GPS coordinates.

To get a better understanding of how to geolocate people based on their GPS data, please use the following instructions:

1. Open the web browser and go to **Google Maps** at `https://maps.google.com`.
2. Next, left-click anywhere on the map to place a point as shown:

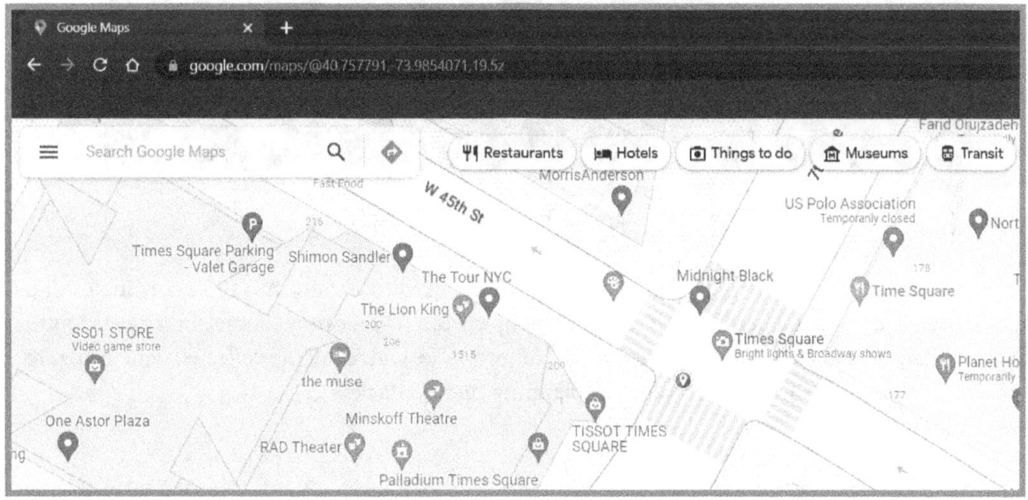

Figure 6.23 – Collecting GPS data

As shown in the preceding screenshot, Times Square was used as an example. Once you've clicked on a specific spot on Google Maps, a pin is inserted and the URL is updated to include the GPS data coordinates: `40.757791,-73.9854071`.

3. Next, to find all the people who've posted from this location, go to `https://twitter.com` and enter the following syntax within the **Search** field:

```
geocode:40.757791,-73.9854071,1km
```

The preceding syntax will show the results for all profiles that posted within 1 km of the GPS coordinates as shown:

Figure 6.24 – Filtering posts based on GPS data

The preceding exercise is a proof-of-concept that can be used to obtain the GPS coordinates of a targeted organization and filter the posts made by people from that location. Some of the posts found will most likely be made by employees; hence, it's important to analyze all the collected data to ensure it's useful for your OSINT operations and for planning future attacks.

When profiling a person or employee, it's important to ensure you're not collecting data on a fake profile or performing a reverse image lookup on a fake profile picture. A website such as **WhatsMyName** (`https://whatsmyname.app`) is useful for ethical hackers and penetration testers to search usernames across multiple data sources and websites on the internet to identify where a specific username is registered. The results help you to determine additional areas on the internet where a target may be posting more useful information.

The following screenshot shows the WhatsMyName website with the lookup field:

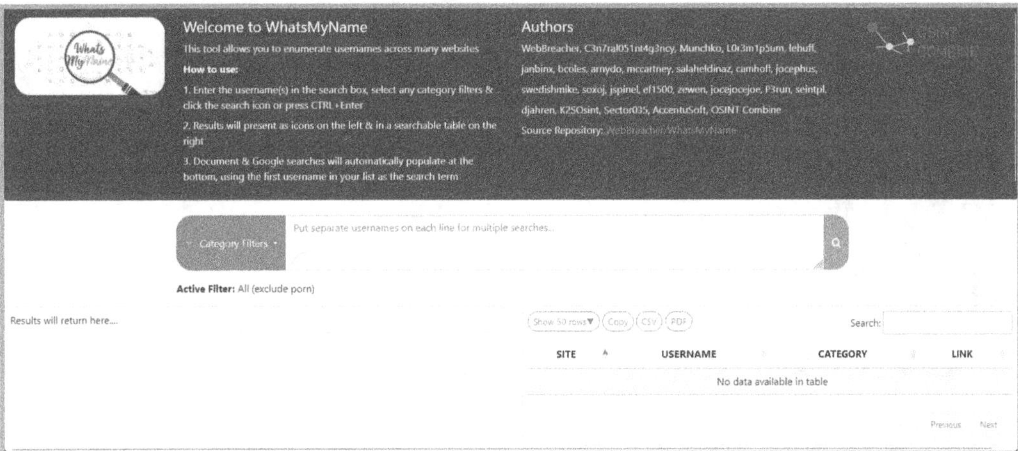

Figure 6.25 – WhatsMyName lookup

Lastly, you can use websites such as **TinEye** (`https://tineye.com`) and **PimEyes** (`https://pimeyes.com`) to perform a reverse image search on a person's picture to identify where on the internet similar images or the same images are found. These tools are useful for law enforcement when trying to find a missing person and track their movement.

The following screenshot shows the search results for a fake picture of a non-existent person on TinEye:

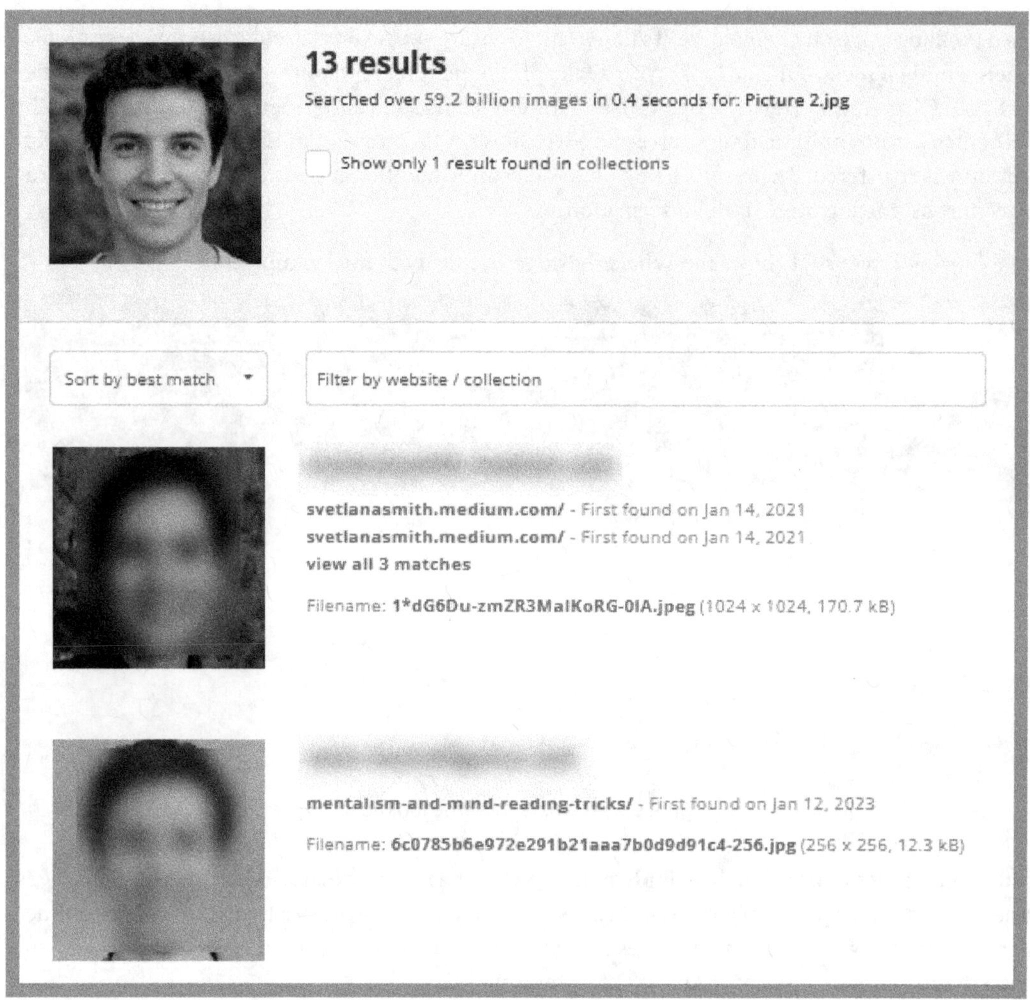

Figure 6.26 – TinEye results

As shown in the preceding screenshot, TinEye was able to identify similar pictures on the internet with their data sources, URLs, filenames, and first-seen data.

PimEyes uses face recognition to better identify the person within a picture. The following screenshot shows the PimEyes results of a fake picture:

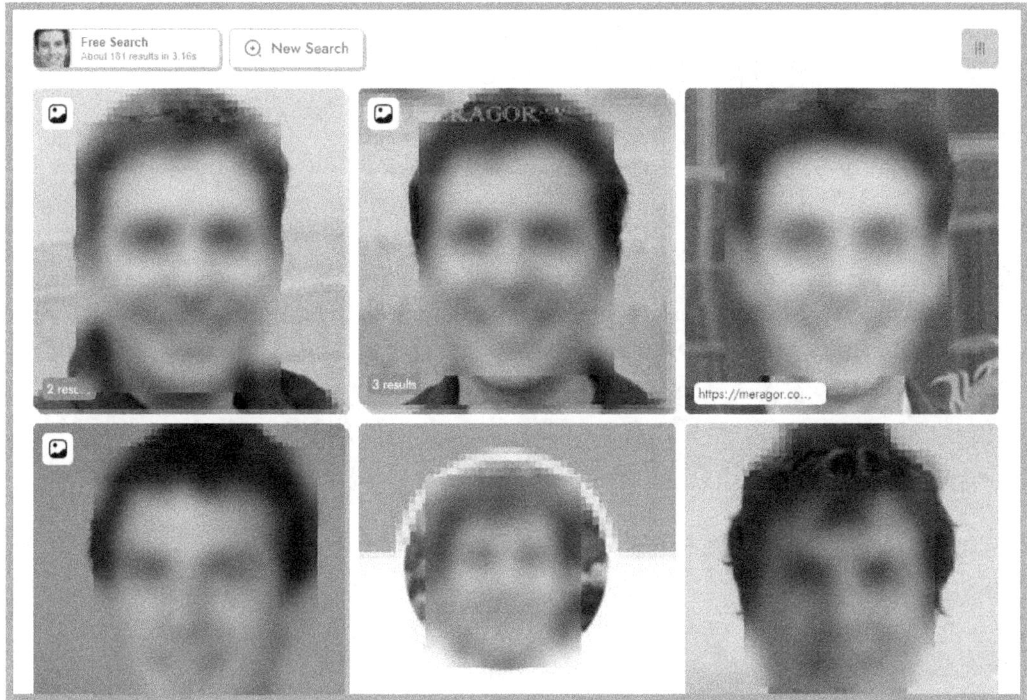

Figure 6.27 – PimEyes results

As shown in the preceding screenshot, PimEyes was able to identify people with very similar facial structures and attributes, while providing the source of the data and more.

User credential OSINT

Many people commonly configure weak passwords or have experienced their accounts being compromised by a hacker. Often, when accounts are compromised, the breached data is used by adversaries to expand their operations and for future attacks, and can also be sold on the dark web for profit. As an ethical hacker, access to breached data is beneficial as it provides us with user credentials for a targeted organization.

For instance, imagine you're performing a penetration test on a network and attempting to gain a foothold into Windows-based systems that are connected to the domain. If the company was breached prior to the penetration test, there's a chance the breached data is available on the dark web or online databases. Within the breached data, you may find users' credentials for their domain accounts. Some users may change their passwords after a breach, while others may not. For users who haven't changed their password after their accounts were compromised, you can leverage their user credentials from the breached data to gain a foothold into the network and perform various types of password-based attacks to identify additional security vulnerabilities.

The following are free and commercial repositories and databases that help ethical hackers and penetration testers to identify the user credentials of their target:

- Have I Been Pwned?: `https://haveibeenpwned.com`
- Intelligence X: `https://intelx.io`
- Dehashed: `https://www.dehashed.com`

Ethical hackers can use **Have I Been Pwned?** to identify whether email addresses and telephone numbers were found in batches of leaked data. For instance, an ethical hacker can perform a lookup on their target's email address as shown here:

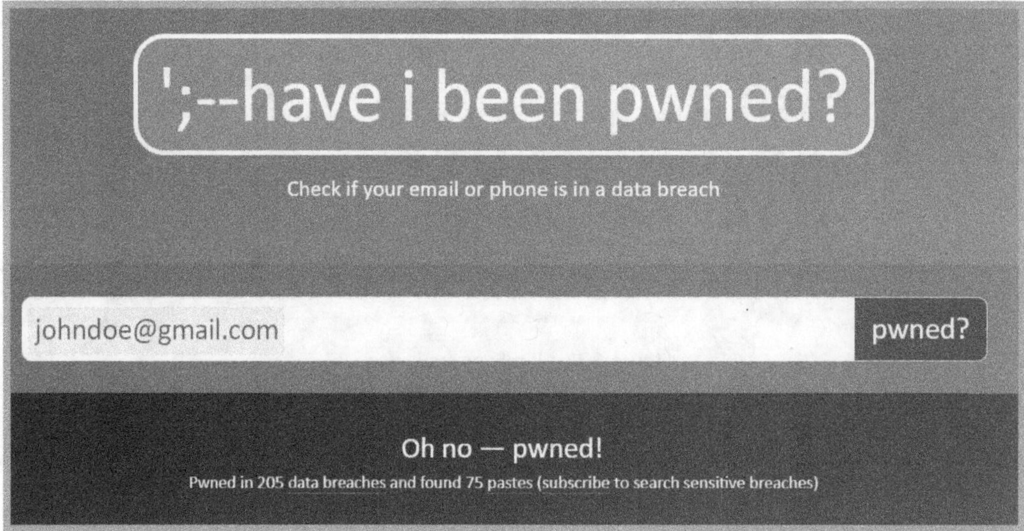

Figure 6.28 – Have I Been Pwned?

Have I Been Pwned? provides results showing which data breaches contained the email address as shown here:

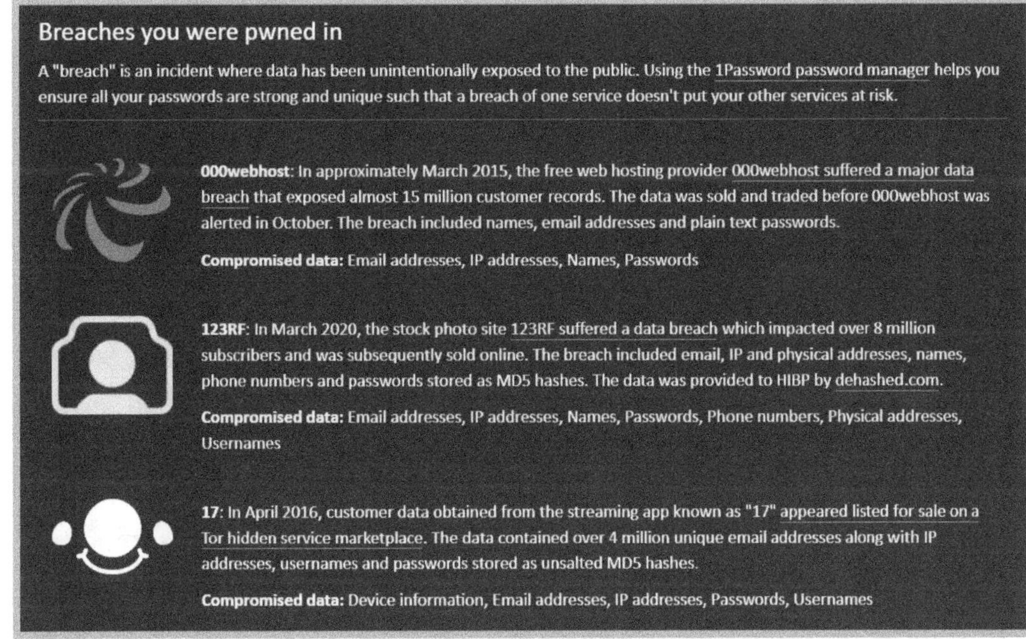

Figure 6.29 – Search results

The preceding data can be analyzed by ethical hackers to further determine the validity of the email address and its holder, and the websites on which the target used the email address to register for an account. In addition, hackers usually expose compromised account data such as usernames and passwords on the internet. Without some additional research, you may be able to find the password for the email account. Sometimes, users are not aware of data breaches and don't always change their passwords often. You may be lucky enough to find valid user credentials belonging to your target and use them to gain a foothold into their accounts, systems, or networks.

Furthermore, if you scroll all the way down to the end of the results page on *Have I Been Pwned?*, you will get the pastes (sources) for all data breaches containing the given email address, as shown in the following screenshot:

Pastes you were found in

A paste is information that has been published to a publicly facing website designed to share content and is often an early indicator of a data breach. Pastes are automatically imported and often removed shortly after having been posted. Using the 1Password password manager helps you ensure all your passwords are strong and unique such that a breach of one service doesn't put your other services at risk.

Paste title	Date	Emails
siph0n.in	Unknown	7,842
siph0n.in	Unknown	59,437
balockae.online	Unknown	178,410
pred.me	Unknown	4,788,657
pxahb.xyz	Unknown	4,161
www.pemiblanc.com	Unknown	2,909,066
xn--e1alhsoq4c.xn--p1ai	Unknown	4,788,657
is-bad-at.tech	Unknown	1,878,845

Figure 6.30 – Breached data sources

Furthermore, **Intelligence X** helps ethical hackers to locate user credentials from breached data. For instance, you can perform an email lookup to identify all relevant data associated with the email address, as follows:

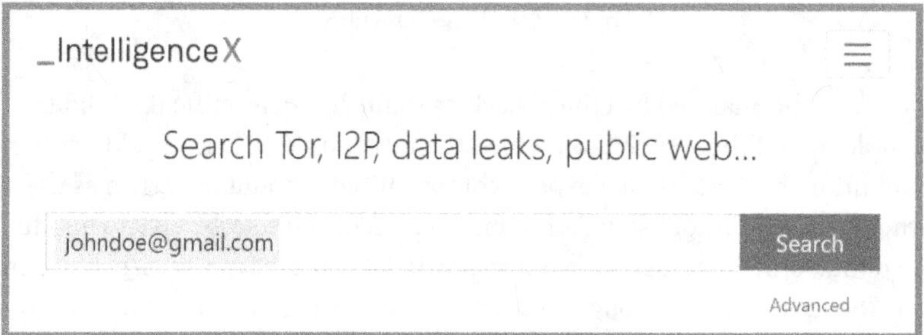

Figure 6.31 – Intelligence X email lookup

The following screenshot shows the data provided by Intelligence X:

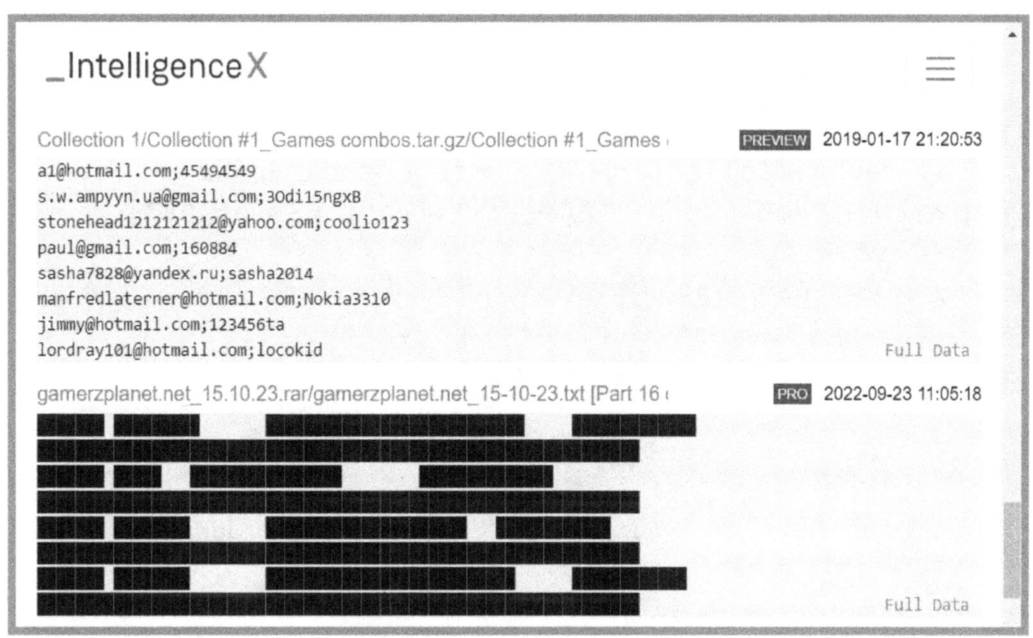

Figure 6.32 – Intelligence X results

As shown in the preceding screenshot, Intelligence X provides some results freely, while other results contain redacted data. A paid account/subscription is required to view the redacted data on Intelligence X.

As an ethical hacker, you can click on a data collection to view all its contents, as shown in the following screenshot:

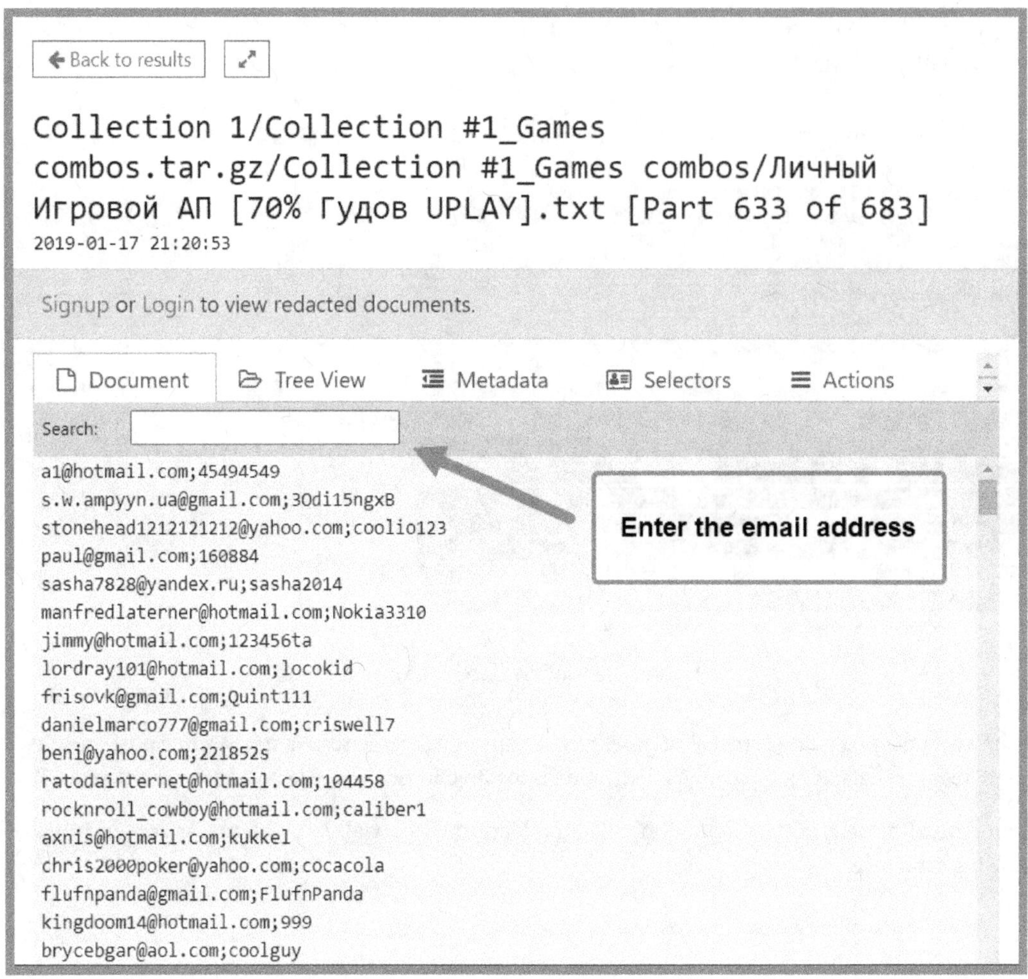

Figure 6.33 – Data collection

As shown in the preceding screenshot, some data collections contain a lot of entries, so using the search field will help you find specific results within a data collection.

Lastly, **Dehashed** is an online database that allows ethical hackers to purchase user credentials from data breaches. This website is a valuable tool for ethical hackers and penetration testers who are performing password-based attacks on accounts and systems of a targeted organization. It allows you to perform lookups on email addresses, usernames, IP addresses, names, addresses, telephone numbers, and domain names.

Having completed this section, you have learned how to collect and analyze OSINT on people and their geo-locations, and locate data breaches that contain user credentials. In the next section, you will learn how to collect and analyze data on wireless signals.

Wireless signals intelligence

Wireless **Signals Intelligence** (**SIGINT**) is a common technique used by ethical hackers, penetration testers, and red teamers to intercept and analyze wireless traffic to discover and exploit security vulnerabilities found in both radio and wireless transmissions. SIGINT is typically used in military operations by governments to monitor the communication channels of other countries and identify potential threats. However, using various tools and techniques, cybersecurity professionals can employ SIGINT to identify their target's wireless network infrastructure (such as access points and associated clients), intercept and monitor **Wireless Local Area Network** (**WLAN**) frames, which can then be used to retrieve passphrases for accessing the network, and locate devices within a vicinity.

Organizations obtain the following benefits from performing SIGINT techniques on their wireless infrastructure:

- **Identifying security vulnerabilities**: Ethical hackers can capture and analyze the WLAN traffic on their target's network to identify any security vulnerabilities such as weak wireless security standards, weak encryption algorithms, and insecure authentication protocols
- **Reconnaissance**: Ethical hackers can perform wireless SIGINT to gain insights into things such as the targeted wireless network infrastructure, the type of wireless devices and their usage, and what security countermeasures can be implemented to reduce the attack surface
- **Testing security controls**: After implementing cyber defenses and security controls, its good practice to test the effectiveness of those countermeasures to ensure they are working as expected to reduce the risk of a potential cyberattack or threat
- **Improving incident response and handling**: If an attacker was able to gain a foothold into an organization's network via their wireless infrastructure, cybersecurity professionals could perform wireless SIGINT to identify the tools, techniques, and source of the attack to improve their **Cyber Threat Intelligence** (**CTI**)

Generally, wireless SIGINT helps organizations to better understand the attack surface of their wireless network infrastructure from an adversary's perspective and the techniques that can be used to potentially compromise the network. The data collected during SIGINT helps ethical hackers to provide recommendations to their customers and organizations for improving their security posture, situational awareness, and incident response and handling. However, keep in mind that ethical hackers must obtain legal permission from the relevant authorities before performing wireless SIGINT.

Next, you will learn how to build a portable/mobile infrastructure for performing wireless SIGINT.

Building a SIGINT infrastructure

Many ethical hackers and penetration testers use a laptop with an external wireless network adapter that supports monitoring mode and packet injection to collect and analyze wireless signals within a vicinity. However, if you want to create a more compact setup that can be placed in a backpack for war-walking or mounted on an aerial drone for war-flying, then a laptop isn't the best choice due to its size and weight. We can leverage the power of microcomputers such as a **Raspberry Pi** running **Kali Linux ARM** to perform reconnaissance on a target's wireless network.

The following is a list of materials that will be needed to set up our portal wireless signals infrastructure:

- Raspberry Pi 3 B+
- A 32-GB Samsung EVO+ microSD card
- A micro-USB cable
- An Alfa AWUS036NHA Wireless B/G/N USB adapter
- A VK-162 G-Mouse USB GPS Dongle Navigation Module
- A portable power bank

Raspberry Pi is a microcomputer with sufficient computing power to run **Kismet**, a wireless signals intelligence tool for monitoring IEEE 802.11, Bluetooth, and RF signals and detecting wireless network intrusions. The microSD card will be used to store the Kali Linux ARM operating system and all the data collected. The Alfa AWUS036NHA wireless network adapter supports monitoring mode and packet injection and connects to Raspberry Pi. The VK-162 G-Mouse USB GPS Dongle is used to collect the GPS coordinates of wireless stations (clients) and access points when they are found. The portable power bank will supply power to Raspberry Pi while war-walking, war-driving, or war-flying.

The following diagram shows the signals intelligence infrastructure when assembled:

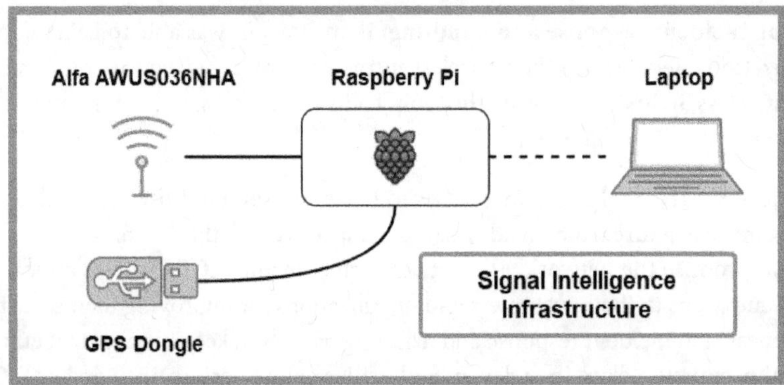

Figure 6.34 – Signals intelligence infrastructure

In addition, we'll configure Raspberry Pi as an access point with a **Dynamic Host Configuration Protocol** (**DHCP**) server to provide IP addresses and subnet masks to any device that's connected to its `wlan0` interface. This enables us to easily connect our laptop and use **Secure Shell** (**SSH**) to securely access the Kali Linux ARM operating system that's running on Raspberry Pi during our wireless penetration testing assessments.

To get started with setting up Raspberry Pi for signals intelligence, please use the following instructions:

Part 1 – Installing Kali Linux ARM

1. First, download and install **Raspberry Pi Imager** from `https://www.raspberrypi.com/software/`.
2. Next, download the official **Kali Linux ARM** operating system from `https://www.kali.org/get-kali/` for your version of Raspberry Pi:

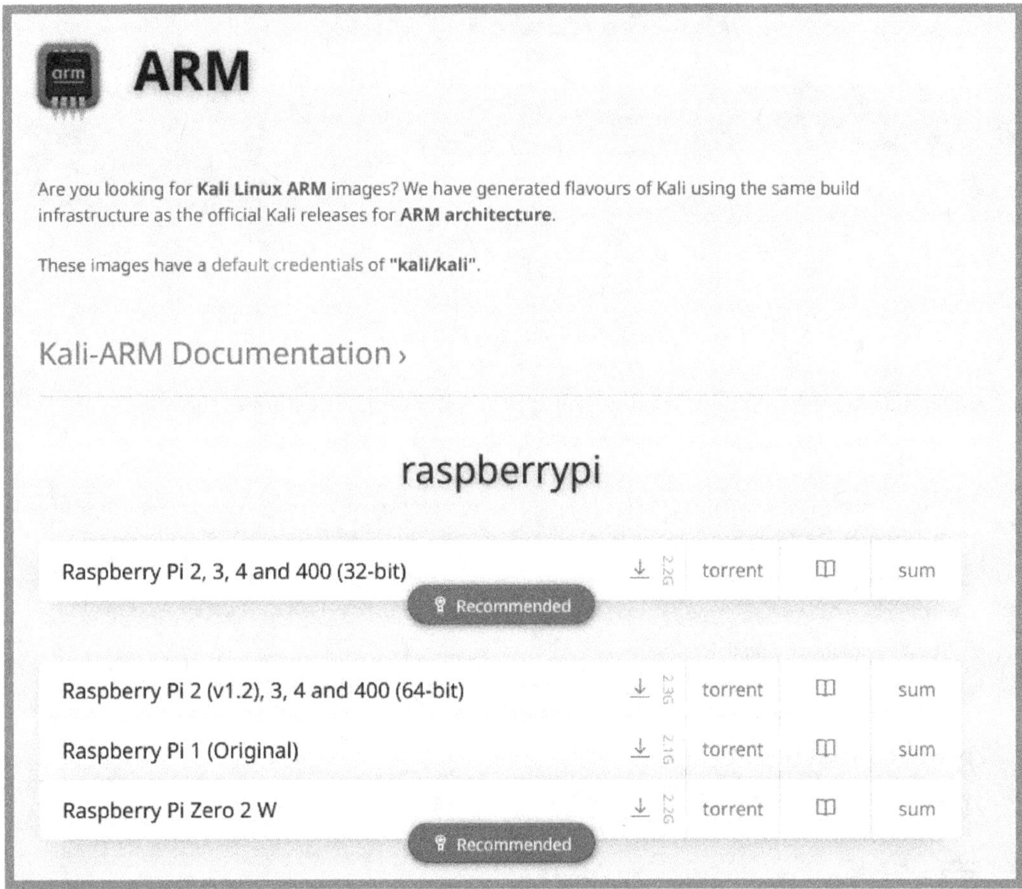

Figure 6.35 – Kali Linux ARM

3. Next, connect the microSD card to your host computer using an SD card reader.
4. To load the Kali Linux ARM operating system onto the microSD card, launch **Raspberry Pi Imager** and click on **CHOOSE OS** as shown:

Figure 6.36 – Raspberry Pi Imager

5. Next, the **Operating System** window will appear. Click on **Use custom** as shown:

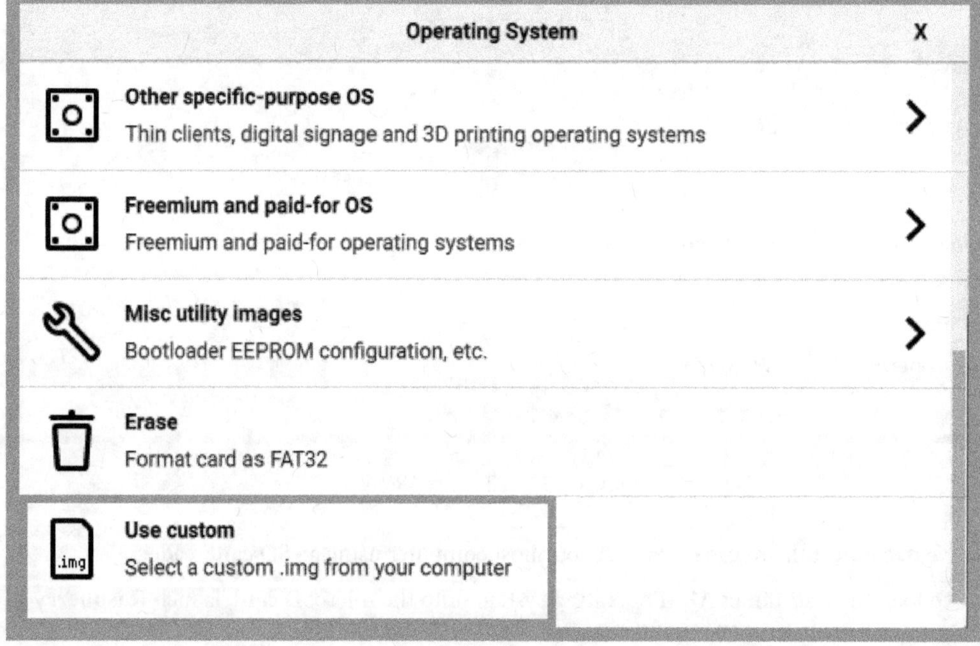

Figure 6.37 – Operating System menu

6. Next, select the **Kali Linux ARM** file and click on **Open** to load it onto **Raspberry Pi Imager**:

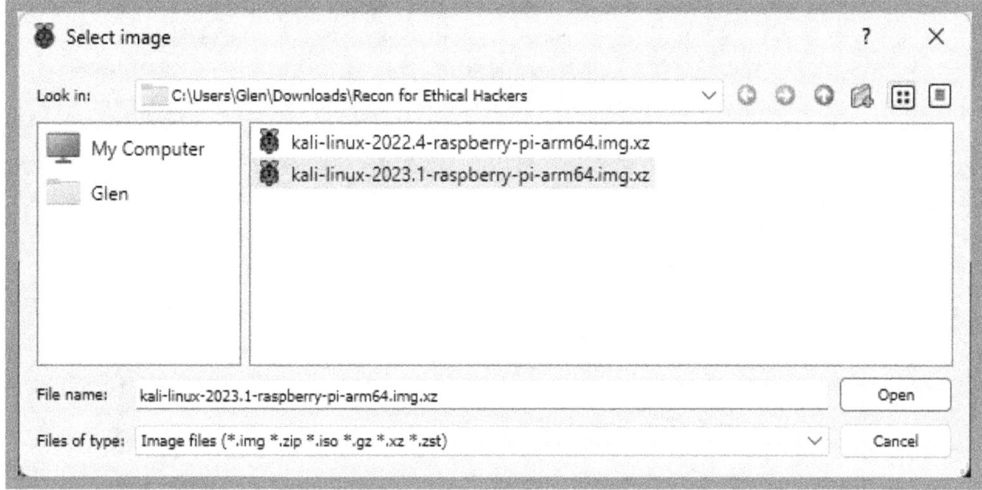

Figure 6.38 – Kali Linux ARM file

7. Next, return to the main menu on **Raspberry Pi Imager**, click on **CHOOSE STORAGE**, and select the microSD card as the **Storage** device. Then, click on the gear icon to open the advanced options:

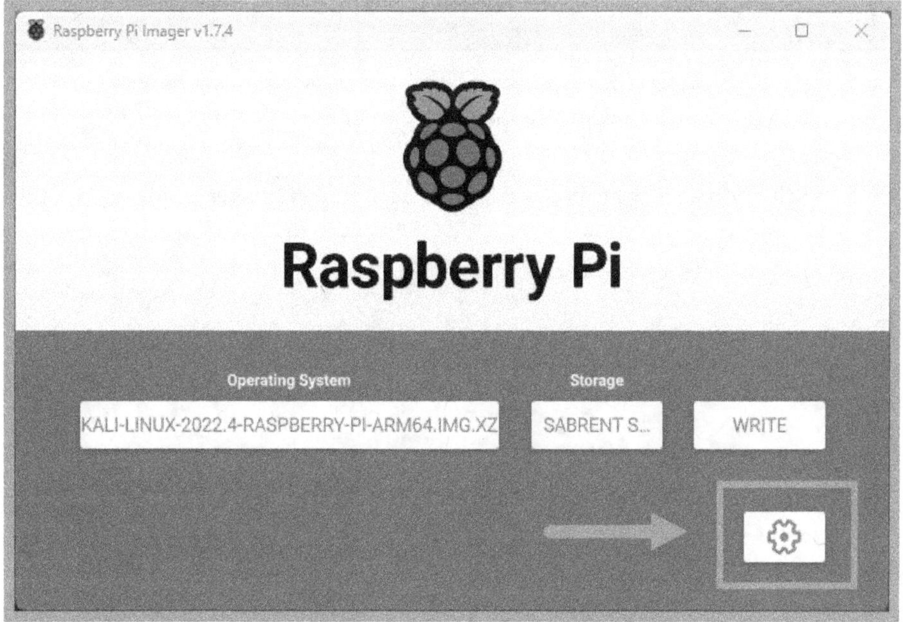

Figure 6.39 – Writing the operating system

8. Next, the **Advanced options** menu will appear. Check the **Enable SSH** box, set a username and password (`kali/kali`), and click on **SAVE**, as shown in the following screenshot:

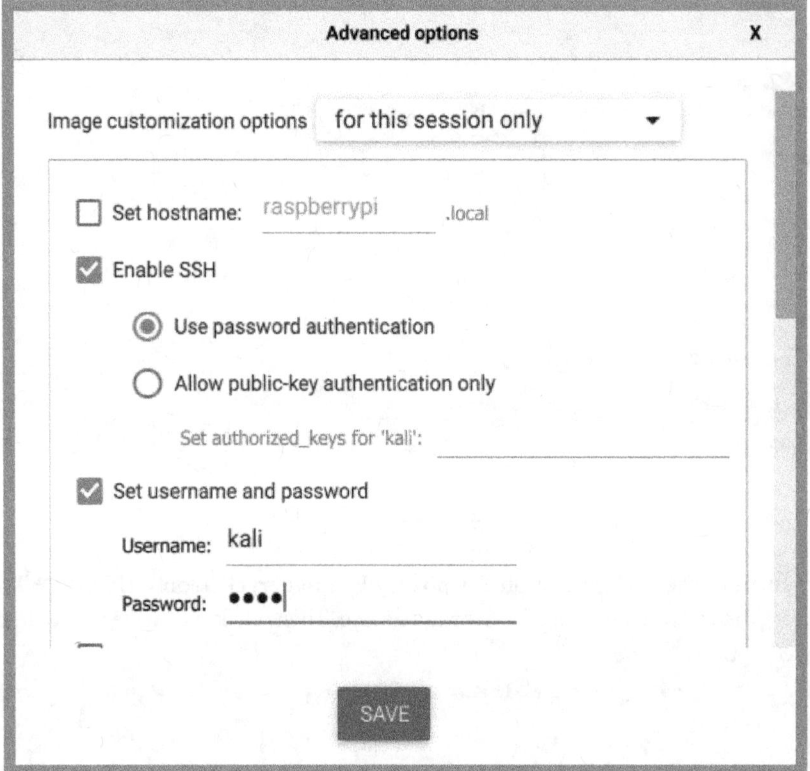

Figure 6.40 – Enabling SSH

9. Next, you will automatically be returned to the **Raspberry Pi Imager** main window. Click on **WRITE** to begin writing the Kali Linux ARM OS image to the microSD card:

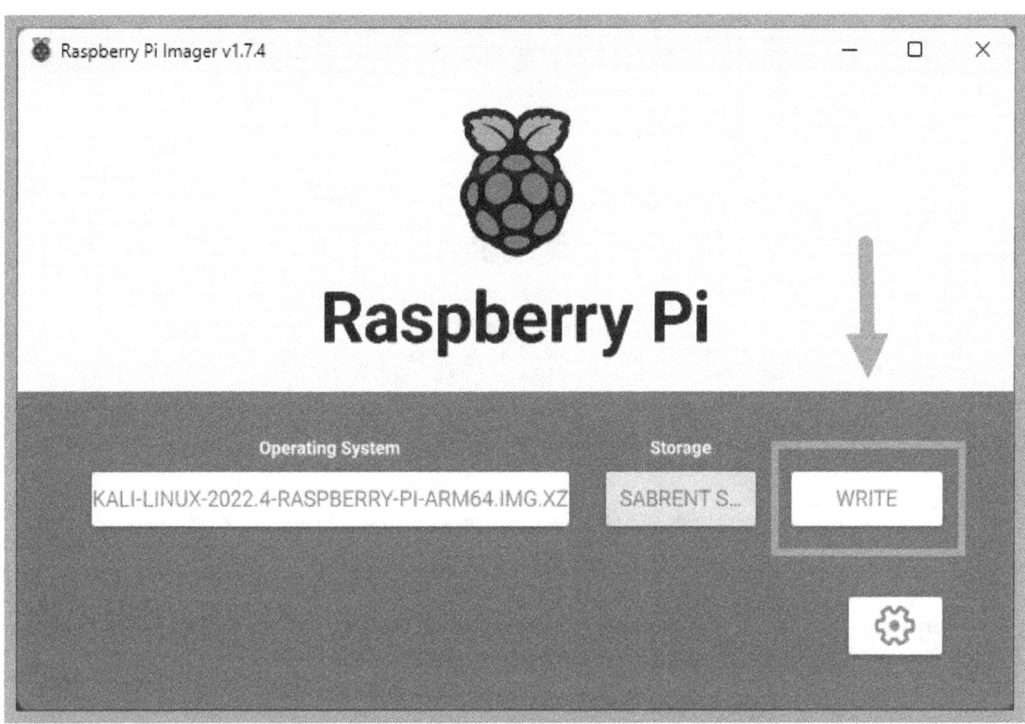

Figure 6.41 – Write button

This process usually takes a few minutes to write and verify.

10. Once the write process is completed, insert the microSD card into Raspberry Pi, connect the GPS dongle and wireless network adapter, and power up the device by connecting it to a power source.
11. Next, connect Raspberry Pi to your wired network using a network cable.
12. Log in to your modem, router, or DHCP server to determine the IP address assigned to Raspberry Pi. The IP address will be required for the next steps.

Part 2 – Setting up Kismet with GPS

1. If you are using a **Windows 10** computer, open **Command Prompt** and execute the `powershell` command before proceeding to the next step.
2. If you are using a **Windows 11** or Linux-based operating system, open the **Terminal** and use the `ssh kali@ip-address` syntax, where `ip-address` is the IP address of Raspberry Pi on your network. Log in using the default user credentials of `kali/kali` as the username and password as shown:

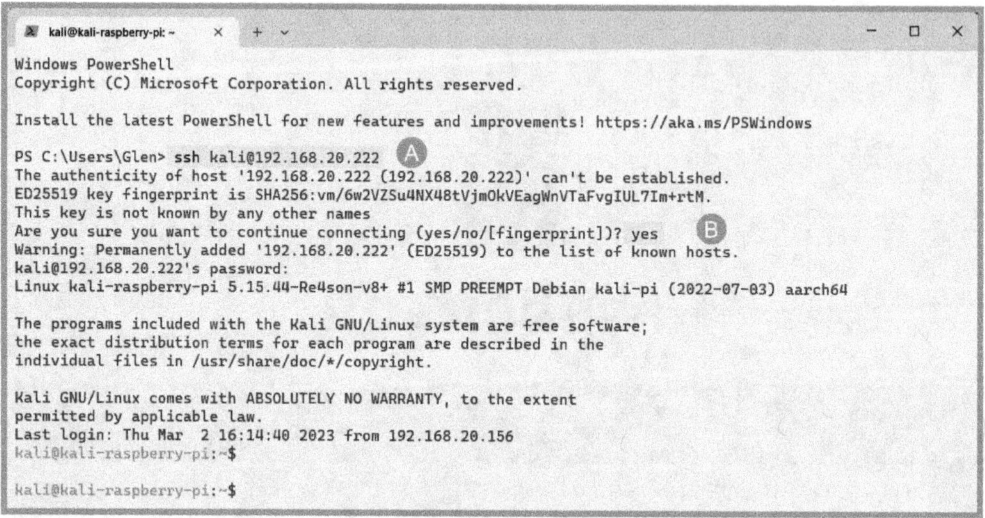

Figure 6.42 – Connecting to Raspberry Pi

As shown in the preceding screenshot, we are using SSH to securely access Raspberry Pi over the network to perform additional configurations and continue the setup process.

> **Tip**
>
> Press *Ctrl* + *P* to change the layout of the Terminal prompt.

3. Next, use the following command to update the software package repository file:

```
kali@kali-raspberry-pi:~$ sudo apt update
```

4. Next, install the latest version of Kismet and the GPS packages on Raspberry Pi using the following command:

```
kali@kali-raspberry-pi:~$ sudo apt install kismet gpsd gpsd-
clients
```

> **Tip**
>
> You can use the `dmesg | grep tty` command to display the kernel messages associated with the **teletypewriter** (**tty**) subsystem. This command helps you to determine whether an attached USB device is detected and recognized by the operating system.

5. Next, you'll need to modify Kismet's configurations to ensure the `gpsd` package is able to connect to the GPS dongle and send data to Kismet. To do this, use the following command to open the `kismet.conf` file in the `nano` command-line text editor:

```
kali@kali-raspberry-pi:~$ sudo nano /etc/kismet/kismet.conf
```

6. Next, use the down arrow on your keyboard to scroll down the text file until you've found the GPS settings. Then, insert the following on a new line:

```
gps=gpsd:host=localhost,port=2947,reconnect=true
```

This should look as follows:

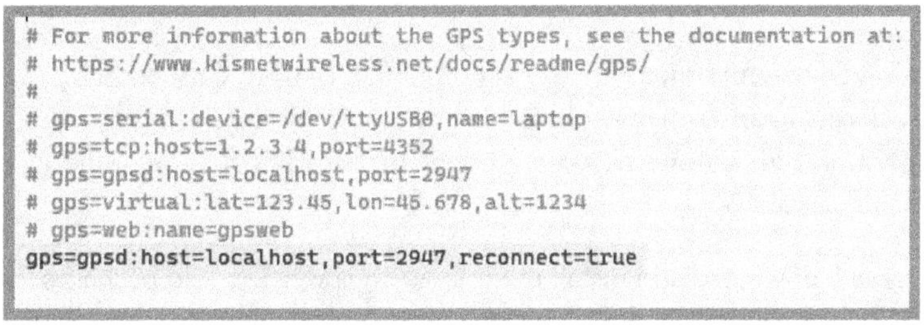

```
# For more information about the GPS types, see the documentation at:
# https://www.kismetwireless.net/docs/readme/gps/
#
# gps=serial:device=/dev/ttyUSB0,name=laptop
# gps=tcp:host=1.2.3.4,port=4352
# gps=gpsd:host=localhost,port=2947
# gps=virtual:lat=123.45,lon=45.678,alt=1234
# gps=web:name=gpsweb
gps=gpsd:host=localhost,port=2947,reconnect=true
```

Figure 6.43 – Modifying the GPS settings in Kismet

7. Next, to save the modifications to the `kismet.conf` file, press *Ctrl* + *X*, then *Y*, and hit *Enter*.
8. Next, configure `gpsd` to work with the GPS receiver on USB0 with the following command:

```
kali@kali-raspberry-pi:~$ sudo gpsd -b /dev/ttyUSB0
```

9. Reboot Raspberry Pi using the `sudo reboot` command and log in again.
10. Next, having ensured the GPS dongle has a clear vertical line of sight to the sky, use either the `cgps -s` or `gpsmon` commands to receive GPS data from orbital satellites, as shown in the following screenshot:

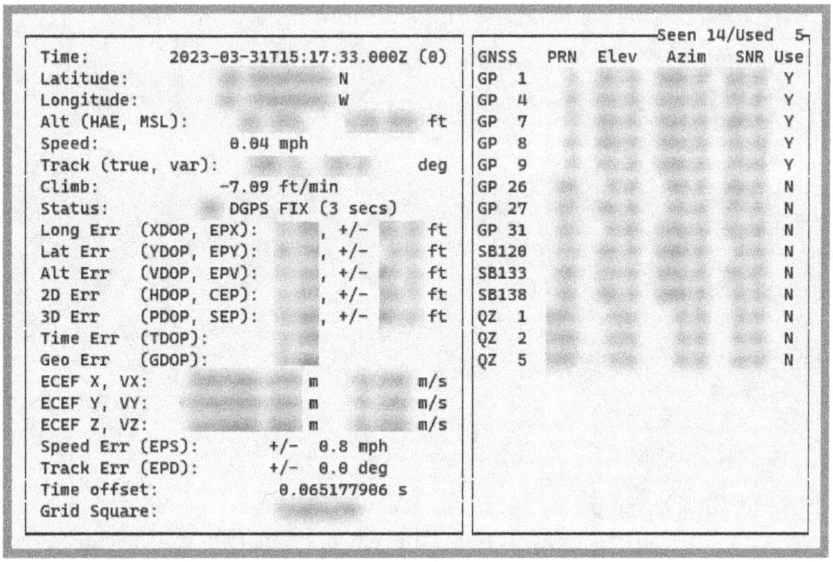

```
Time:            2023-03-31T15:17:33.000Z (0)   GNSS  PRN  Elev  Azim  SNR Use
Latitude:                     N                 GP  1                      Y
Longitude:                    W                 GP  4                      Y
Alt (HAE, MSL):                          ft     GP  7                      Y
Speed:                  0.04 mph                GP  8                      Y
Track (true, var):                       deg    GP  9                      Y
Climb:                 -7.09 ft/min             GP 26                      N
Status:                 DGPS FIX (3 secs)       GP 27                      N
Long Err  (XDOP, EPX):      , +/-        ft     GP 31                      N
Lat Err   (YDOP, EPY):      , +/-        ft     SB120                      N
Alt Err   (VDOP, EPV):      , +/-        ft     SB133                      N
2D Err    (HDOP, CEP):      , +/-        ft     SB138                      N
3D Err    (PDOP, SEP):      , +/-        ft     QZ  1                      N
Time Err  (TDOP):                               QZ  2                      N
Geo Err   (GDOP):                               QZ  5                      N
ECEF X, VX:                  m          m/s
ECEF Y, VY:                  m          m/s
ECEF Z, VZ:                  m          m/s
Speed Err (EPS):        +/-  0.8 mph
Track Err (EPD):        +/-  0.0 deg
Time offset:             0.065177906 s
Grid Square:
                                                Seen 14/Used 5
```

Figure 6.44 – cgps data

As shown in the preceding screenshot, the GPS software and dongle were able to identify and connect to a few satellites (right column). Additionally, Raspberry Pi was able to retrieve its current GPS coordinates (blurred for privacy) and other data for positioning and tracking. Press *Ctrl* + *C* to stop **cgps** from running.

Part 3 – setting up an access point on Raspberry Pi

Setting up Raspberry Pi to function as an access point enables us to connect wirelessly to it using a mobile device such as a laptop or smartphone, hence removing the need for a wired connection. This makes Raspberry Pi fully autonomous and mobile for war-driving, war-walking, or war-flying exercises. Let's do this as follows:

1. On Raspberry Pi, download and install the `dnsmasq` and `hostapd` packages for configuring wireless and IP services:

```
kali@kali-raspberry-pi:~$ sudo apt install dnsmasq hostapd
```

2. Next, use the following commands to stop the `dnsmasq` and `hostapd` services until they're needed later:

```
kali@kali-raspberry-pi:~$ sudo systemctl stop dnsmasq
kali@kali-raspberry-pi:~$ sudo systemctl stop hostapd
```

3. Next, we need to set the parameters for the wireless network that will be broadcasted from the `wlan0` interface. Use the following command to edit the `hostapd.conf` file:

```
kali@kali-raspberry-pi:~$ sudo nano /etc/hostapd/hostapd.conf
```

Then, insert the following configurations into the `hostapd.conf` file:

```
interface=wlan0
driver=nl80211
ssid=MyNetwork
hw_mode=g
channel=7
macaddr_acl=0
auth_algs=1
ignore_broadcast_ssid=0
wpa=2
wpa_passphrase=Password123
wpa_key_mgmt=WPA-PSK
wpa_pairwise=TKIP
rsn_pairwise=CCMP
```

Keep in mind, you can change `MyNetwork` to any wireless network name of your choice, and `Password123` to a stronger password for the network. The following screenshot shows the preceding configurations within the `hostapd.conf` file:

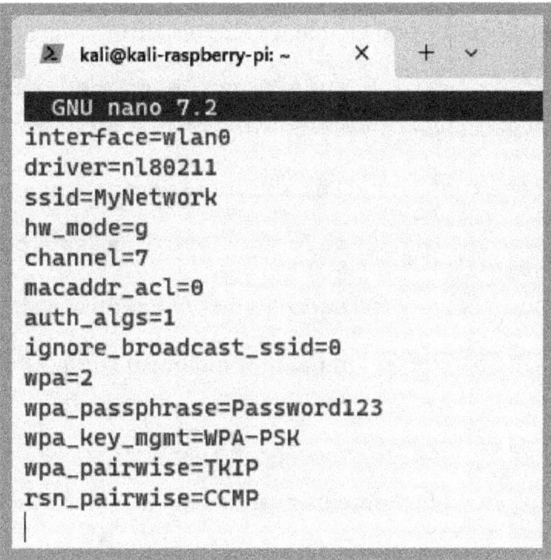

Figure 6.45 – Hostapd configurations

Press *Ctrl* + *X*, then press *Y*, and hit *Enter* to save the file.

4. Next, install the `dhcpcd` package, which we will use to configure a static IP address on the `wlan0` interface of Raspberry Pi:

```
kali@kali-raspberry-pi:~$ sudo apt install dhcpcd
```

5. Next, modify the **dhcpcd** configuration file to assign the `192.168.4.1` address to the `wlan0` wireless adapter using the following command:

```
kali@kali-raspberry-pi:~$ sudo nano /etc/dhcpcd.conf
```

Insert the following parameters at the end of the `dhcpcd.conf` file:

```
interface wlan0
     static ip_address=192.168.4.1/24
     nohook wpa_supplicant
```

The following screenshot shows the preceding parameters inserted at the end of the file:

```
interface wlan0
    static ip_address=192.168.4.1/24
    nohook wpa_supplicant
```

Figure 6.46 – dhcpcd file

Press *Ctrl* + *X*, then press *Y*, and hit *Enter* to save the file.

6. Next, restart the **dhcpcd** service and enable it to automatically start on system boot:

```
kali@kali-raspberry-pi:~$ sudo service dhcpcd restart
kali@kali-raspberry-pi:~$ sudo systemctl enable dhcpcd
```

7. Next, back up the `dnsmasq` configuration file and create a new file using the following commands:

```
kali@kali-raspberry-pi:~$ sudo mv /etc/dnsmasq.conf /etc/
dnsmasq.conf.orig
kali@kali-raspberry-pi:~$ sudo nano /etc/dnsmasq.conf
```

Then, insert the following parameters in the new **dnsmasq** configuration file:

```
interface=wlan0
dhcp-range=192.168.4.2,192.168.4.20,255.255.255.0,24h
```

The following screenshot shows the preceding parameters within the **dnsmasq** configuration file:

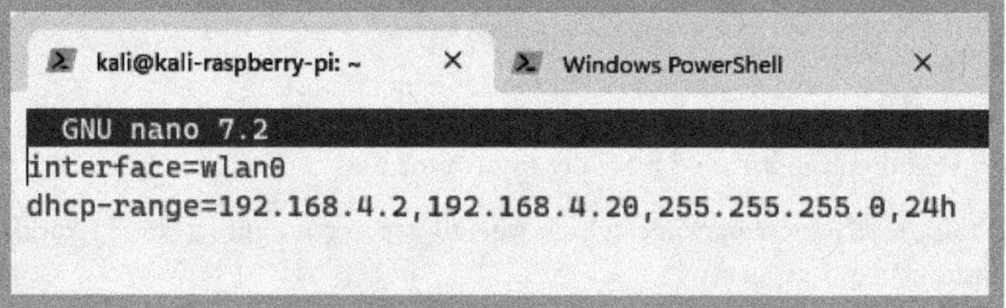

Figure 6.47 – DNSmasq configurations

The configurations shown in the preceding screenshot are used to provide an IP address and subnet mask to any client that connects to the wireless network that's generated from the `wlan0` interface on Raspberry Pi.

8. Next, enable and restart the `dnsmasq` and `hostapd` services with the following commands:

```
kali@kali-raspberry-pi:~$ sudo systemctl enable dnsmasq
kali@kali-raspberry-pi:~$ sudo systemctl restart dnsmasq
kali@kali-raspberry-pi:~$ sudo systemctl unmask hostapd
kali@kali-raspberry-pi:~$ sudo systemctl enable hostapd
kali@kali-raspberry-pi:~$ sudo systemctl restart hostapd
```

9. Lastly, use the `sudo reboot` commands to reboot Raspberry Pi. After the device is rebooted, you will see the `MyNetwork` wireless network is available. Now you can connect to Raspberry Pi using the wireless network, and your device will receive an IP address within the `192.168.4.2` - `192.168.4.20` range. Then, you can access Raspberry Pi over SSH with the `192.168.4.1` address that's statically configured on its `wlan0` interface.

Part 4 – Monitoring and data collection

1. You'll need to change the operating mode of your external wireless network adapter from manage to monitor mode. Use the `iwconfig` command to view the list of wireless adapters on Raspberry Pi as shown:

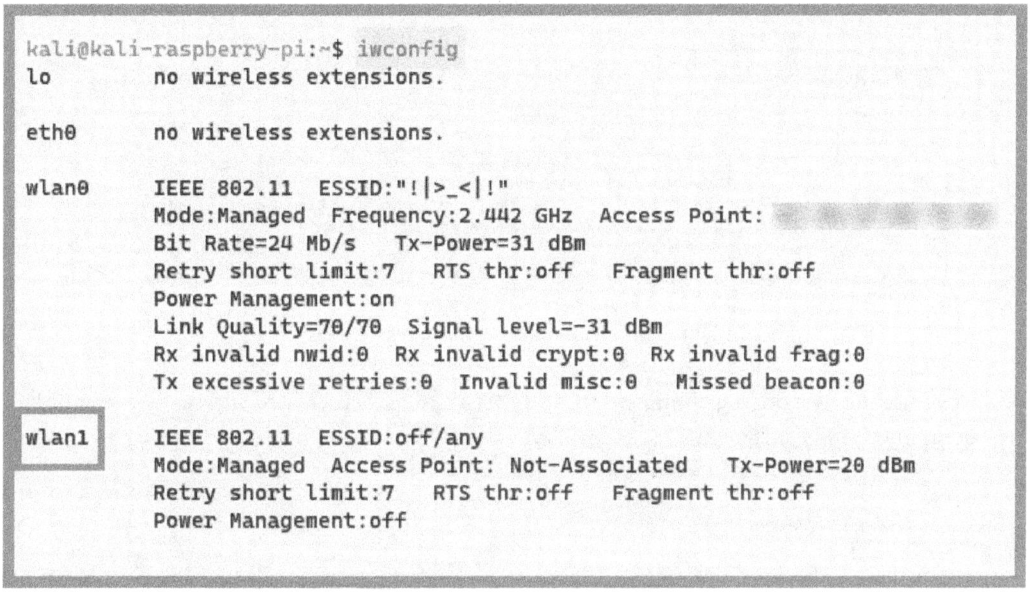

```
kali@kali-raspberry-pi:~$ iwconfig
lo          no wireless extensions.

eth0        no wireless extensions.

wlan0       IEEE 802.11  ESSID:"!|>_<|!"
            Mode:Managed  Frequency:2.442 GHz  Access Point:
            Bit Rate=24 Mb/s   Tx-Power=31 dBm
            Retry short limit:7   RTS thr:off   Fragment thr:off
            Power Management:on
            Link Quality=70/70  Signal level=-31 dBm
            Rx invalid nwid:0  Rx invalid crypt:0  Rx invalid frag:0
            Tx excessive retries:0  Invalid misc:0   Missed beacon:0

wlan1       IEEE 802.11  ESSID:off/any
            Mode:Managed  Access Point: Not-Associated   Tx-Power=20 dBm
            Retry short limit:7   RTS thr:off   Fragment thr:off
            Power Management:off
```

Figure 6.48 – Wireless network adapters

As shown in the preceding screenshot, `wlan0` is the embedded wireless adapter that's used for the access point on Raspberry Pi and `wlan1` is the external adapter.

2. Next, use `airmon-ng` to enable monitoring mode on the `wlan1` adapter:

```
kali@kali-raspberry-pi:~$ sudo airmon-ng start wlan1
```

The following screenshot shows the expected results when creating `wlan1mon`, the adapter with monitor mode:

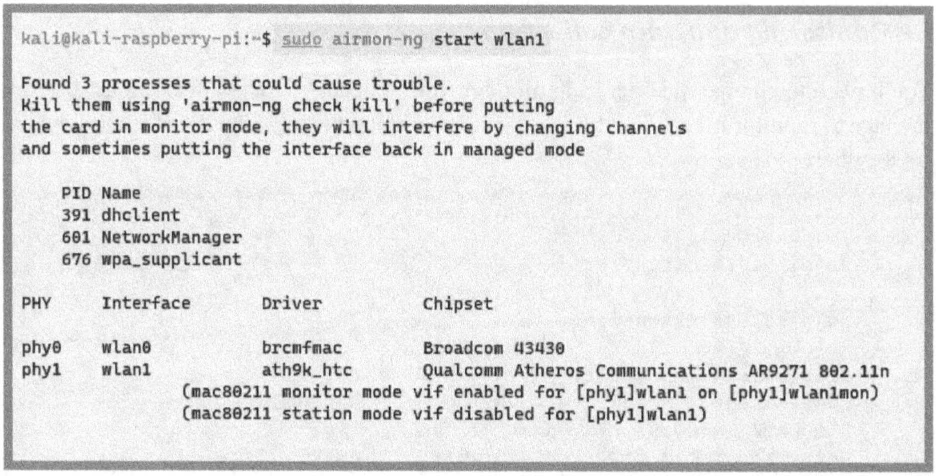

```
kali@kali-raspberry-pi:~$ sudo airmon-ng start wlan1

Found 3 processes that could cause trouble.
Kill them using 'airmon-ng check kill' before putting
the card in monitor mode, they will interfere by changing channels
and sometimes putting the interface back in managed mode

    PID Name
    391 dhclient
    601 NetworkManager
    676 wpa_supplicant

PHY     Interface       Driver          Chipset

phy0    wlan0           brcmfmac        Broadcom 43430
phy1    wlan1           ath9k_htc       Qualcomm Atheros Communications AR9271 802.11n
                (mac80211 monitor mode vif enabled for [phy1]wlan1 on [phy1]wlan1mon)
                (mac80211 station mode vif disabled for [phy1]wlan1)
```

Figure 6.49 – Enabling monitor mode

3. Next, use the `iwconfig` command to verify `wlan1mon` was created and that it's operating in monitor mode as shown:

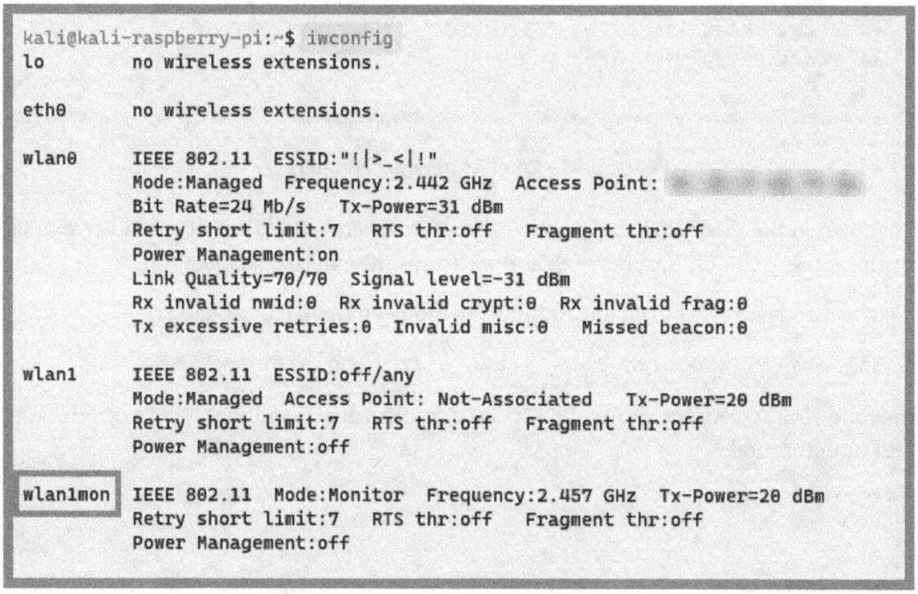

```
kali@kali-raspberry-pi:~$ iwconfig
lo        no wireless extensions.

eth0      no wireless extensions.

wlan0     IEEE 802.11  ESSID:"!|>_<|!"
          Mode:Managed  Frequency:2.442 GHz  Access Point:
          Bit Rate=24 Mb/s   Tx-Power=31 dBm
          Retry short limit:7   RTS thr:off   Fragment thr:off
          Power Management:on
          Link Quality=70/70  Signal level=-31 dBm
          Rx invalid nwid:0  Rx invalid crypt:0  Rx invalid frag:0
          Tx excessive retries:0  Invalid misc:0   Missed beacon:0

wlan1     IEEE 802.11  ESSID:off/any
          Mode:Managed  Access Point: Not-Associated   Tx-Power=20 dBm
          Retry short limit:7   RTS thr:off   Fragment thr:off
          Power Management:off

wlan1mon  IEEE 802.11  Mode:Monitor  Frequency:2.457 GHz  Tx-Power=20 dBm
          Retry short limit:7   RTS thr:off   Fragment thr:off
          Power Management:off
```

Figure 6.50 – Verifying the adapter

4. Next, create a new directory called `wardrive` to store the collected data:

```
kali@kali-raspberry-pi:~$ mkdir wardrive
kali@kali-raspberry-pi:~$ cd wardrive
```

Tip

When Kismet is running, you'll be required to stay connected to the Kismet terminal. Any disruption of the connection will stop Kismet. To prevent this issue, we'll use **Tmux**, which is a multiplexer terminal that lets us send terminal windows to the background without losing the session. Therefore, we can start a Tmux session, disconnect from Raspberry Pi, and then reconnect and regain access to the Tmux session. To learn more about Tmux, please see `https://github.com/tmux/tmux`.

5. Next, start a **Tmux** session using the following command:

   ```
   kali@kali-raspberry-pi:~/wardrive$ tmux
   ```

6. Once the new Tmux session has opened, use the following commands to run **Kismet** and attach the monitor-mode wireless adapter to it:

   ```
   kali@kali-raspberry-pi:~/wardrive$ kismet -c wlan1mon
   ```

   The following screenshot shows Kismet running with the GPS daemon attached, and our signals intelligence monitoring has started:

```
kali@kali-raspberry-pi: ~/war
KISMET - Point your browser to http://localhost:2501 (or the address of this system) for the Kismet UI
INFO: (GPS) Connected to gpsd server [::1]:2947
INFO: (GPS) Connected to a JSON-enabled GPSD (3.22), enabling JSON mode
INFO: Could not open system plugin directory (/usr/lib/aarch64-linux-gnu/ki
      smet/), skipping: No such file or directory
INFO: Did not find a user plugin directory (/home/kali/.kismet//plugins/),
      skipping: No such file or directory
INFO: Found type 'linuxwifi' for 'wlan1mon'
INFO: wlan1mon interface 'wlan1mon' is already in monitor mode
INFO: wlan1mon finished configuring wlan1mon, ready to capture
INFO: Data source 'wlan1mon' launched successfully
INFO: Detected new 802.11 Wi-Fi access point 38:4C:4F:58:EC:AC
INFO: 802.11 Wi-Fi device 38:4C:4F:58:EC:AC advertising SSID
      'Digicel_WiFi_T28R'
INFO: 802.11 Wi-Fi device 38:4C:4F:58:EC:AC advertising SSID
```

Figure 6.51 – Starting Kismet

   As shown in the preceding screenshot, you can access the Kismet web interface by going to `http://ip-address-rasp-pi:2501`.

7. Next, to send the current Tmux session to the background, press *Ctrl* + *B*, then press *D*.

At this point, if your Raspberry Pi is connected to a portable power bank, you can also terminate the SSH session and start war-driving, war-walking, or war-flying to collect signals intelligence within an area.

Part 5 – Using the Kismet web interface

1. While you're still connected to the access point from Raspberry Pi, open your web browser and go to `http://ip-address-rasp-pi:2501` to access the Kismet web interface.

2. Next, set your preferred login credentials to access Kismet and log in. The following screenshot shows the main dashboard of Kismet:

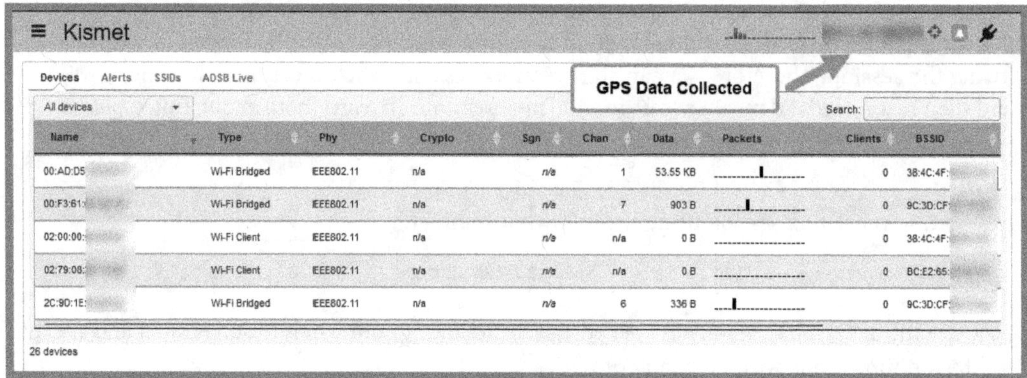

Figure 6.52 – Kismet web interface

As shown in the preceding screenshot, beacons and probes are captured and analyzed by Kismet. Beacons are WLAN frames sent from access points and probes are sent from clients. Additionally, in the top-right corner, you will notice your GPS coordinates will appear.

3. Next, to display all access points within the vicinity, click on **Devices | Wi-Fi Access Points** as shown:

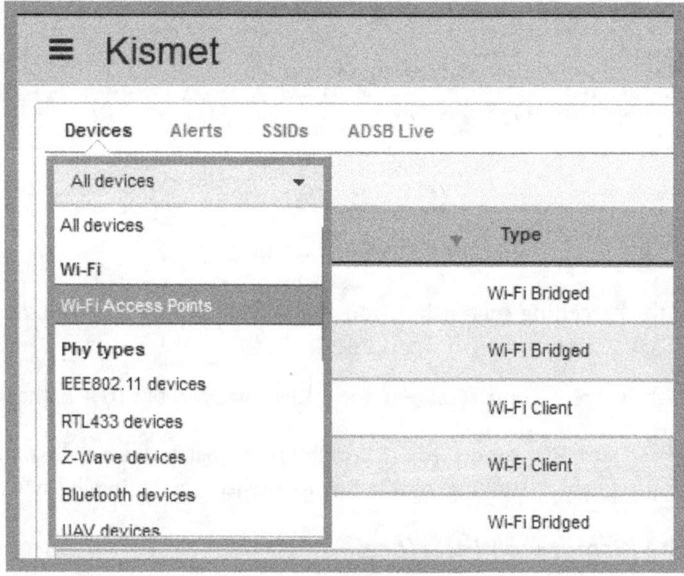

Figure 6.53 – Filtering access points

4. Now, the dashboard will display only access points. Using the interface, you can identify their signal strength, view the amount of data transmitted, see the wireless security features, and more. Click on any wireless access point to view specific data, as shown in the following screenshot:

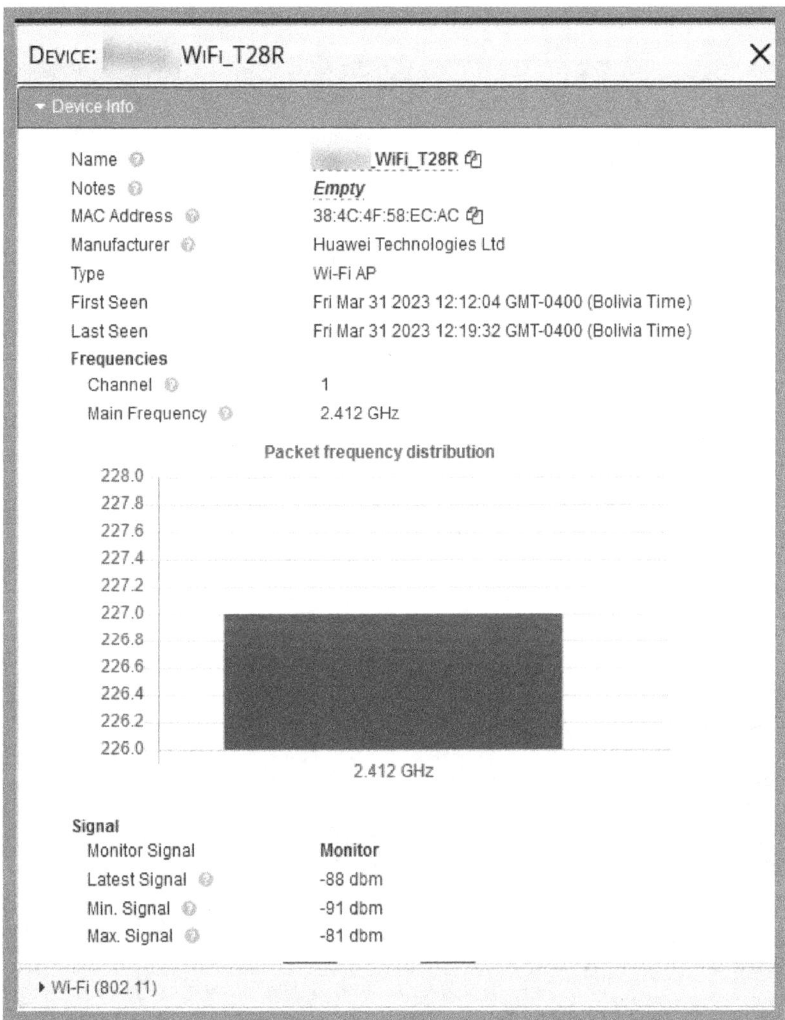

Figure 6.54 – Access point intelligence

As shown in the preceding screenshot, we are able to identify the specific 2.4-GHz channel and the frequency used by the target, along with the signal strength, MAC address of the access point, and manufacturer. Such information helps you to better profile the target device and determine the operating channel and approximate distance between your device and the targeted access point.

5. Next, expand the **Wi-Fi (802.11)** tab to view the wireless network settings as shown:

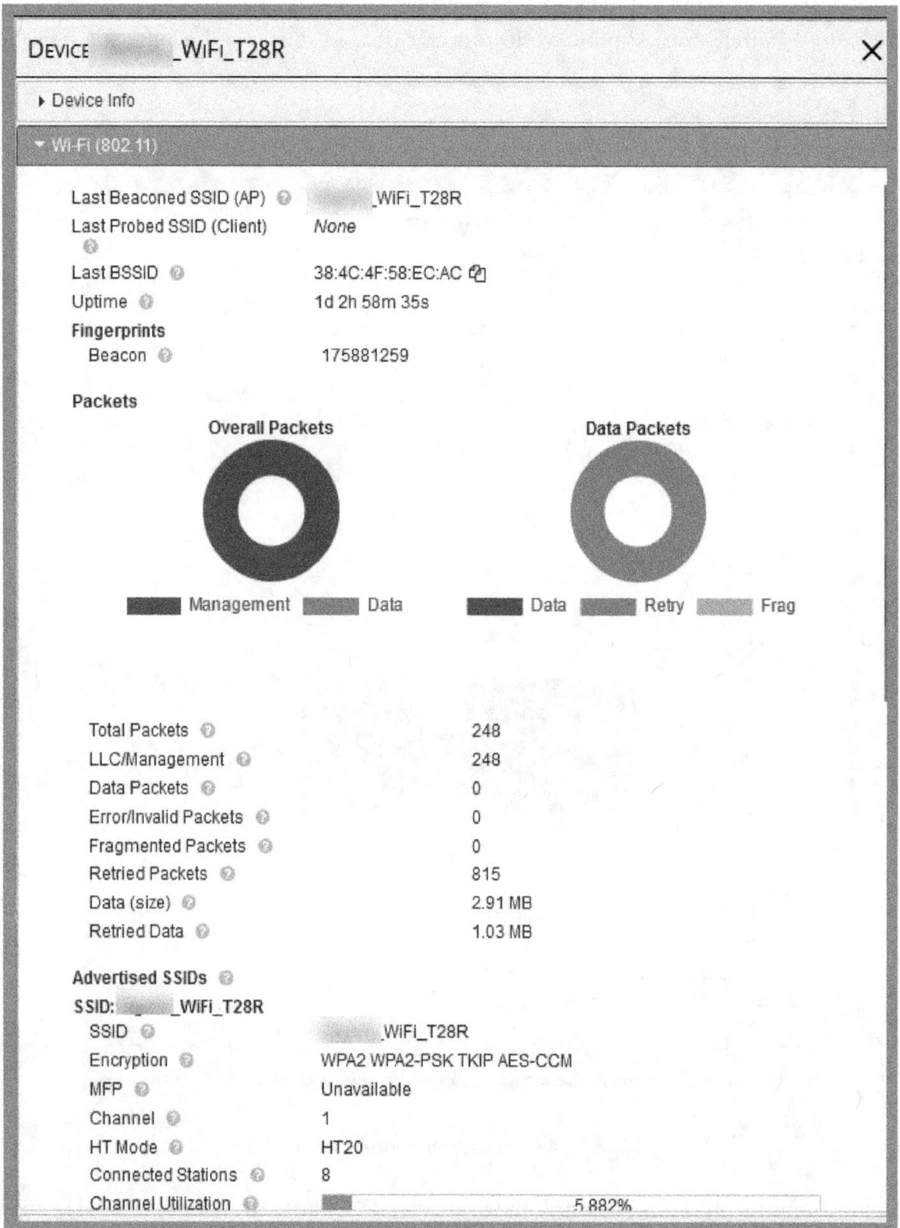

Figure 6.55 – Wireless settings

As shown in the preceding screenshot, Kismet is able to identify the amount of WLAN frames (traffic) traveling through the targeted access point, advertised **Service Set Identifiers** (**SSIDs**), and whether there are multiple SSIDs advertised from the same access point.

6. Next, on the same **Wi-Fi (802.11)** tab, you can see a list of stations (clients) that are associated with the access point:

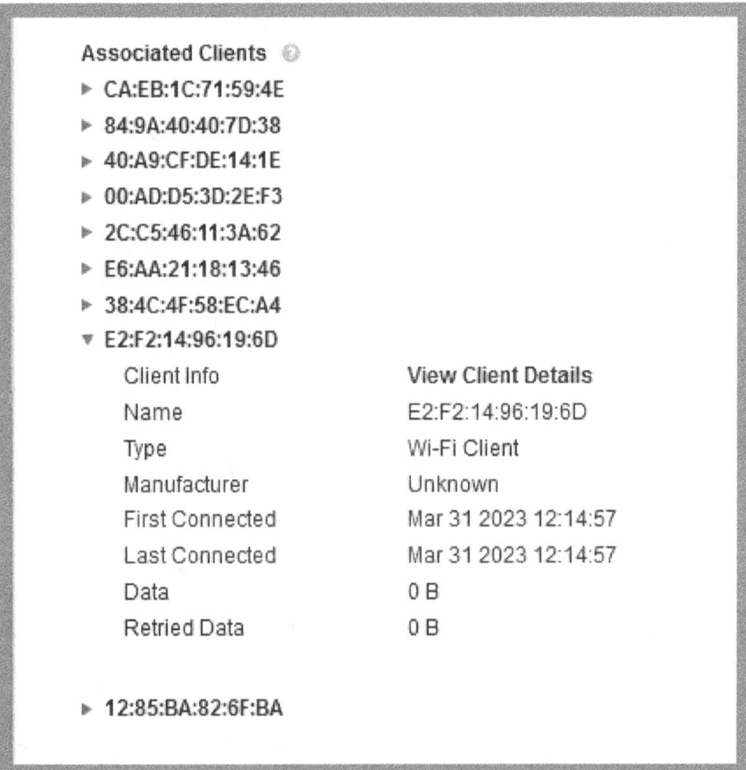

Figure 6.56 – Stations

The stations are represented by their MAC addresses. Clicking on a station provides additional details such as data transmitted, connection status, and times of first and last connected. If an access point is configured with MAC address filtering, obtaining a list of authorized MAC addresses will be useful during a wireless network penetration testing assessment.

7. To view more details about a station, click on **View Client Details** to open a new window as shown in the following screenshot:

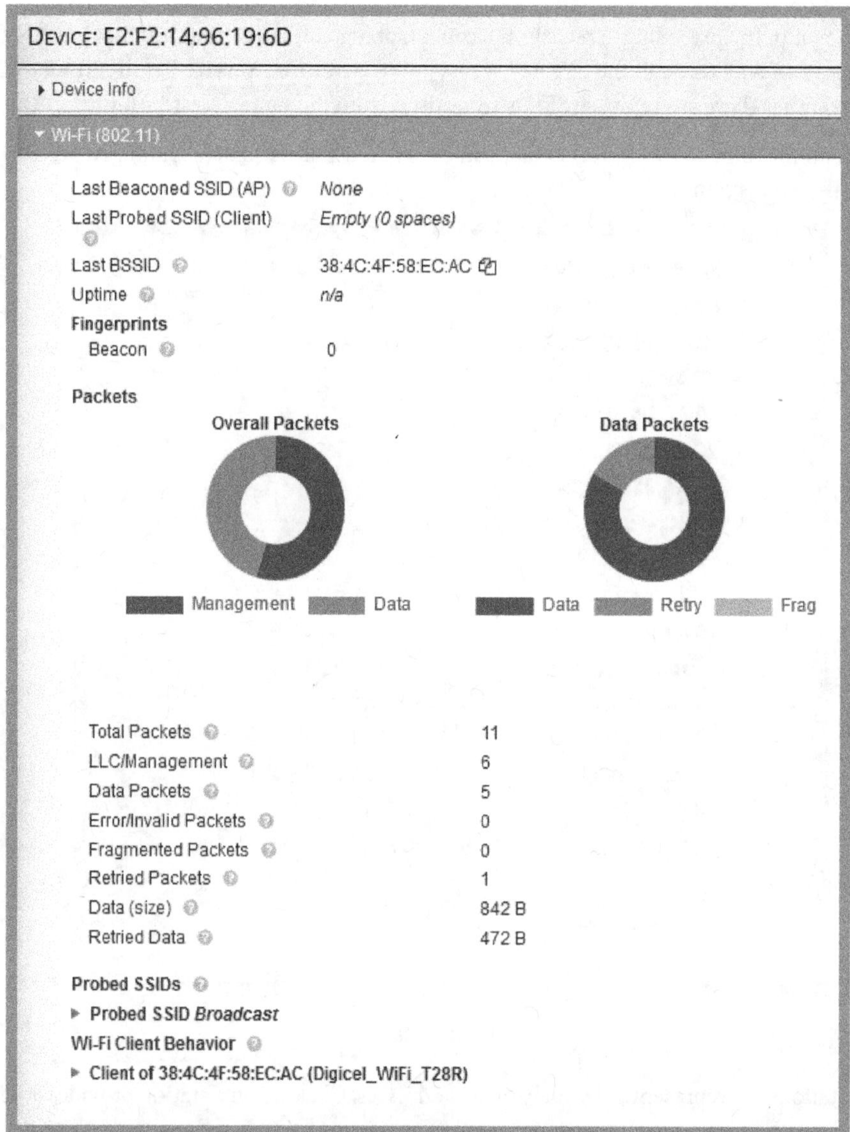

Figure 6.57 – Viewing a station's details

As shown in the preceding screenshot, Kismet helps us determine the associated wireless network of a client, and whether the client is sending probes for any additional wireless networks. As an ethical hacker, you can set up a rogue access point to advertise the SSIDs found within probes. This technique can be used to trick the target into connecting to the fake network and capturing the WPA or WPA2 handshake, which can then be used to perform password cracking.

Part 6 – Mapping the data onto Google Earth

1. After you've collected the data, use the `tmux ls` command to view the Tmux session ID for the background session:

   ```
   kali@kali-raspberry-pi:~/wardrive$ tmux ls
   0: 1 windows (created Fri Mar 31 16:09:42 2023)
   ```

   As shown here, session ID `0` is the background session that's running Kismet.

2. To re-attach the background session to Tmux, use the following commands:

   ```
   kali@kali-raspberry-pi:~/wardrive$ tmux attach -t 0
   ```

   The following screenshots the Kismet session re-attached and running as expected:

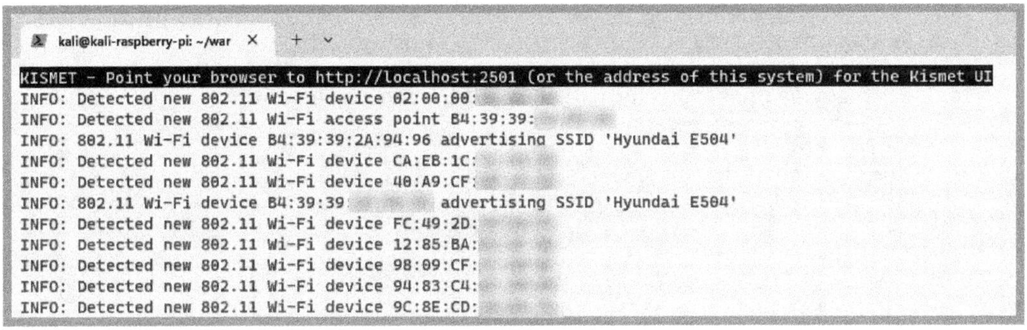

```
kali@kali-raspberry-pi: ~/war
KISMET - Point your browser to http://localhost:2501 (or the address of this system) for the Kismet UI
INFO: Detected new 802.11 Wi-Fi device 02:00:00:
INFO: Detected new 802.11 Wi-Fi access point B4:39:39:
INFO: 802.11 Wi-Fi device B4:39:39:2A:94:96 advertising SSID 'Hyundai E504'
INFO: Detected new 802.11 Wi-Fi device CA:EB:1C:
INFO: Detected new 802.11 Wi-Fi device 40:A9:CF:
INFO: 802.11 Wi-Fi device B4:39:39            advertising SSID 'Hyundai E504'
INFO: Detected new 802.11 Wi-Fi device FC:49:2D:
INFO: Detected new 802.11 Wi-Fi device 12:85:BA:
INFO: Detected new 802.11 Wi-Fi device 98:09:CF:
INFO: Detected new 802.11 Wi-Fi device 94:83:C4:
INFO: Detected new 802.11 Wi-Fi device 9C:8E:CD:
```

Figure 6.58 – Kismet

3. To gracefully stop Kismet and save the collected data, press *Ctrl* + *C* to stop the process:

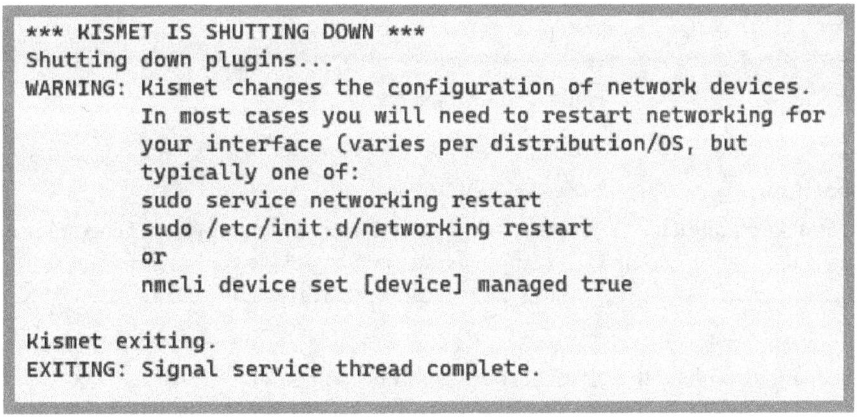

```
*** KISMET IS SHUTTING DOWN ***
Shutting down plugins...
WARNING: Kismet changes the configuration of network devices.
         In most cases you will need to restart networking for
         your interface (varies per distribution/OS, but
         typically one of:
         sudo service networking restart
         sudo /etc/init.d/networking restart
         or
         nmcli device set [device] managed true

Kismet exiting.
EXITING: Signal service thread complete.
```

Figure 6.59 – Stopping Kismet

Once Kismet is stopped, a `.kismet` file is automatically created containing all the collected data:

```
kali@kali-raspberry-pi:~/wardrive$ ls
Kismet-20230302-16-15-13-1.kismet
```

Figure 6.60 – Kismet file

4. Next, type `exit` to leave the Tmux terminal and return to the default shell on the Linux terminal:

```
kali@kali-raspberry-pi:~/wardrive$ exit
```

5. Next, use the following commands to convert the `.kismet` file into a `.kml` file:

```
kali@kali-raspberry-pi:~/wardrive$ sudo kismetdb_to_kml --in
Kismet-20230302-16-15-13-1.kismet --out wardrive1.kml
```

The following screenshot shows the output during the conversion process:

```
kali@kali-raspberry-pi:~/wardrive$ sudo kismetdb_to_kml --in Kismet-20230302-16-15-13-1.kismet --out wardrive1.kml
WARNING:  No packets with GPS info for 'B8:8A:60:        ', skipping
WARNING:  No packets with GPS info for '80:30:49:        ', skipping
WARNING:  No packets with GPS info for '44:1E:98:        ', skipping
WARNING:  No packets with GPS info for 'E0:37:17:        ', skipping
WARNING:  No packets with GPS info for 'DA:CD:74:        ', skipping
WARNING:  No packets with GPS info for '6E:63:9C:        ', skipping
WARNING:  No packets with GPS info for '7C:1C:68:        ', skipping
WARNING:  No packets with GPS info for 'DA:FC:8F:        ', skipping
WARNING:  No packets with GPS info for '40:0D:10:        ', skipping
WARNING:  No packets with GPS info for '40:0D:10:        ', skipping
WARNING:  No packets with GPS info for '4A:78:5E:        ', skipping
WARNING:  No packets with GPS info for '22:96:2B:        ', skipping
WARNING:  No packets with GPS info for '44:1E:98:        ', skipping
WARNING:  No packets with GPS info for '9E:9A:9B:        ', skipping
WARNING:  No packets with GPS info for '46:91:91:        ', skipping
WARNING:  No packets with GPS info for 'A4:77:33:        ', skipping

kali@kali-raspberry-pi:~/wardrive$ ls
Kismet-20230302-16-15-13-1.kismet  wardrive1.kml
```

A

B

Figure 6.61 – Creating a KML file

> **Important note**
>
> **Keyhole Markup Language** (KML) is a data format that is used to represent geographic data and visualization within 2D and 3D mapping systems such as **Google Earth**.

6. Next, reconnect Raspberry Pi to your network and use a tool such as **WinSCP** to securely connect and download the `.kml` file onto your host computer.
7. Next, download and install **Google Earth Pro** from `https://www.google.com/earth/versions/`.
8. Next, launch the **Google Earth Pro** application on your computer and click on **File | Open...** to attach the `.kml` file as shown:

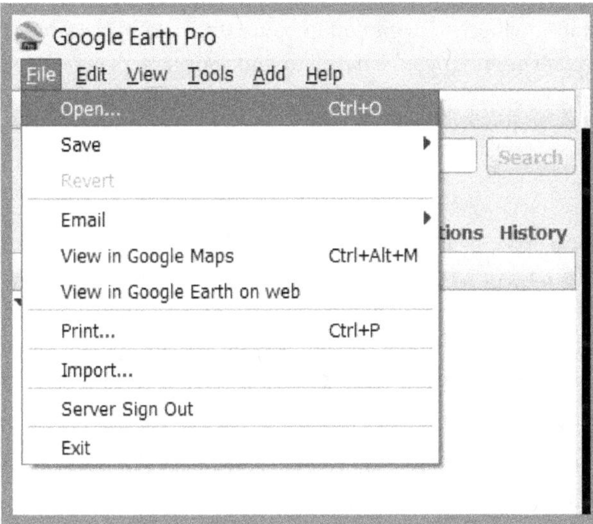

Figure 6.62 – Google Earth Pro File menu

After the `.kml` file is loaded, Google Earth Pro shows yellow pins that represent the locations of wireless devices such as access points and stations discovered using Kismet, as shown in the following screenshot:

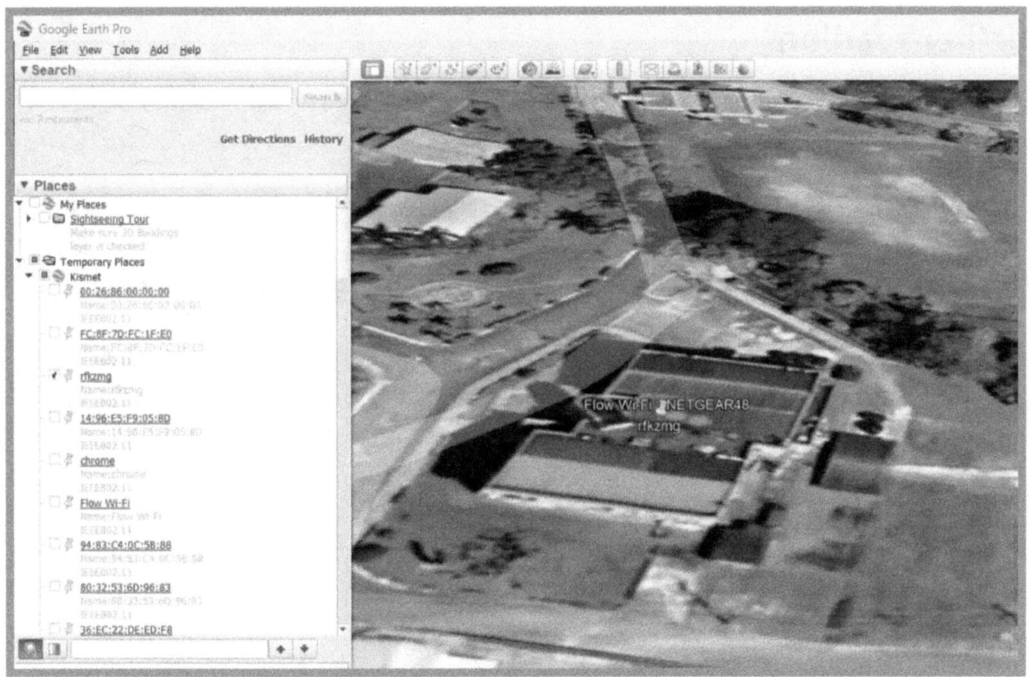

Figure 6.63 – Mapping wireless devices

As an ethical hacker, signals intelligence helps you to better understand the wireless landscape around your target, which is especially useful if you're trying to find your target's wireless networks and clients. Having completed this section, you have learned how to perform wireless signals intelligence.

Summary

During this chapter, you learned how images contain a lot of data that can be leveraged by adversaries to identify the location of a target and their whereabouts. In addition, you gained the skills to extract EXIF data from images and use it with global mapping systems to identify a target's geo-location at the time the picture was taken. Furthermore, you learned how to collect and analyze intelligence on people and search data breaches to identify user credentials for gaining access to a target's account, system, or network.

Lastly, you learned about the importance of performing wireless SIGINT as an ethical hacker and gained the hands-on skills for constructing a mobile wireless SIGINT infrastructure for war-driving, war-walking, and war-flying.

I hope this chapter has been informative and helpful for your journey in the cybersecurity industry. In the next chapter, *Working with Active Reconnaissance*, you will learn how to determine open ports and running services, profile operating systems, and enumerate resources on hosts.

Further reading

- OSINT attack surface diagrams: `https://www.osintdojo.com/diagrams/main`
- OSINT tools and tutorials: `https://osint.tools/`
- SIGINT: `https://www.techtarget.com/whatis/definition/SIGINT-signals-intelligence`
- Understanding location data: `https://www.quadrant.io/resources/location-data`

Part 2: Scanning and Enumeration

In this section, you will learn how to perform advanced scanning techniques to discover hosts, identify security vulnerabilities on systems, enumerate common network services, collect intelligence on websites, and identify reconnaissance activities in a network.

This part has the following chapters:

7
Working with Active Reconnaissance

The more information known about a target, the more ethical hackers are able to identify security vulnerabilities, which can be exploited to get unauthorized access to the systems and networks of organizations. Active reconnaissance helps ethical hackers and penetration testers to collect and analyze information about a target that's not accessible or available during passive reconnaissance. Since passive reconnaissance relies on **Open Source Intelligence** (**OSINT**) and publicly available information from multiple online sources, the information may not be up to date, and inaccuracy affects planning future operations such as weaponization and exploitation of the target. With active reconnaissance, ethical hackers collect data in real time to better identify security vulnerabilities.

In this chapter, you will learn how to efficiently scan hosts on a network to identify their operating systems, open ports, and running services using automated tools and techniques. In addition, you will gain hands-on experience in spoofing your identity and evading threat detection systems, such as firewalls and **Intrusion Detection Systems** (**IDSes**), while performing host discovery on a network. Lastly, you will learn how to enumerate sensitive data from vulnerable systems on a network and perform wireless reconnaissance.

In this chapter, we will cover the following topics:

- Automating reconnaissance
- Spoofing your identity on a network
- Discovering live hosts on a network
- Using evasion techniques
- Enumerating network services
- Wireless reconnaissance

Let's dive in!

Technical requirements

To follow along with the exercises in this chapter, please ensure that you have met the following hardware and software requirements:

- Kali Linux: `https://www.kali.org/get-kali/`
- Metasploitable 3
- An Alfa AWUS036NHA – Wireless B/G/N USB adapter

Active reconnaissance

Unlike passive reconnaissance, which leverages OSINT to help ethical hackers create a profile of a target, active reconnaissance uses a more direct approach by sending probes to the targeted systems and networks to collect sensitive and specific details. For instance, during passive reconnaissance, the threat actor depends on the accuracy of the collected data from multiple online sources. If the data is outdated or inaccurate, it can affect both the weaponization and exploitation phases and prevent ethical hacking from gaining access to the target. With active reconnaissance, the collection of data is performed in real time, enabling ethical hacking to collect and analyze current data to help improve future operations.

> **Important note**
> Scanning is intrusive and illegal in many countries, so ensure that you obtain legal permission from the authorities before scanning any system or network you do not own.

Active reconnaissance plays an important role during ethical hacking and penetration testing, as it helps cybersecurity professionals to collect data about live systems on a network, and such data can be leveraged to identify the attack surface, which consists of the vulnerable points of entry that can be exploited to gain a foothold into a target. The following are the various types of data that can be collected to better profile the target or the targeted systems:

- The number of live hosts on a network
- Open service ports on systems
- Running services (banner grabbing)
- The operating systems of host devices
- Identifying security vulnerabilities
- The network topology

- The placement of networking devices
- Identifying security appliances

To identify live hosts, an ethical hacker sends specially crafted probes on the network. This helps the ethical hacker to determine whether a targeted host is online or not before attempting any future actions, such as enumeration and vulnerability analysis. This is commonly referred to as **network scanning** or **host discovery**.

After the targeted host(s) are discovered on the network, the next step is to identify any open service port on the device. These ports are the entry and exit points within an operating system, and they are associated with an application-layer protocol within the **Transmission Control Protocol/Internet Protocol** (**TCP/IP**) networking model. Before a client sends packets onto a network, it opens an ephemeral/dynamic port for the outgoing traffic. When a server runs an application such as Apache HTTP Server, it opens either port `80` or `443` to listen for incoming packets.

The following diagram shows the source and destination addressing information that will be included within a packet from a client to a web server over a network:

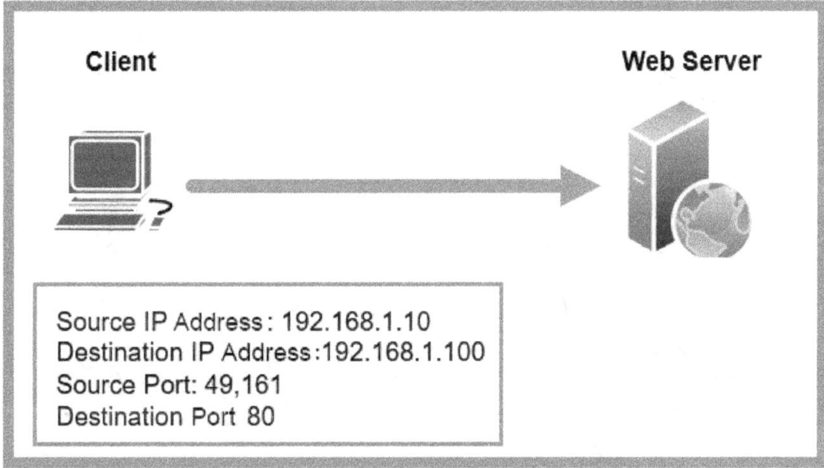

Figure 7.1 – The HTTP request message

The following diagram shows how the source information from the client is used to send a response from the server:

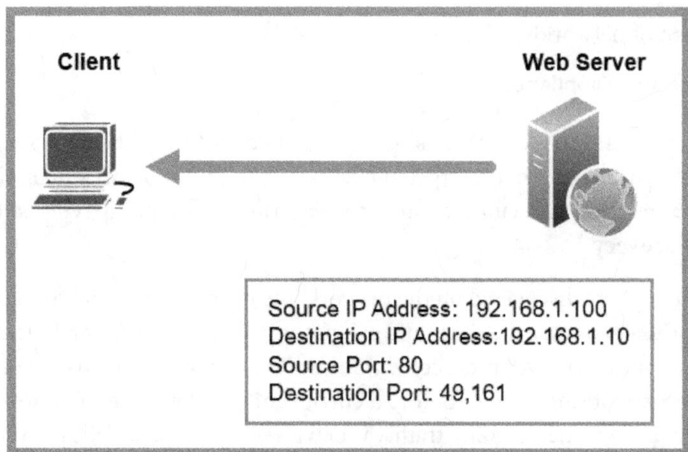

Figure 7.2 – The HTTP response

As an ethical hacker, identifying open ports helps us to determine the running services on a targeted system. In addition, you can attempt to perform banner grabbing to identify the service versions of all running services. The service version can be leveraged to find known security vulnerabilities and their severity for a specific service on the targeted system.

The following table shows the categories of port numbers:

Port Range	Category
0 - 1,023	Well-known ports
1,024 - 49,151	Registered ports
49,152 - 65,535	Private/Dynamic ports

Figure 7.3 – Port number ranges

As shown in the preceding table, ports `1-1023` are commonly referred to as well-known ports, as these ports are associated with common application-layer protocols such as **Hypertext Transfer Protocol Secure** (**HTTPS**), **Domain Name System** (**DNS**), and **Simple Mail Transfer Protocol** (**SMTP**). Registered ports are simply network service ports that are designated to specific application-layer protocols or services, and private/dynamic ports are temporarily open to send packets from a client to other systems on a network.

Next, it's important to identify the operating system of targeted hosts, as it helps ethical hackers to identify possible security vulnerabilities that can be exploited to gain a foothold on the target. In addition, different operating systems have different security controls and mechanisms, and architectures that can make them more or less secure to various types of cyberattacks and threats. Therefore, such information can be leveraged by ethical hackers to customize their exploits and payloads for specific security vulnerabilities on a specific operating system.

For instance, if your target runs outdated versions of an operating system such as Microsoft Windows 7, you can research known security vulnerabilities and how they can be exploited. For example, **Microsoft Security Bulletin MS17-010** describes a known security vulnerability within Microsoft Windows 7–Windows Server 2016 operating systems, which enables an attacker to perform **Remote Code Execution** (**RCE**) by exploiting the Microsoft **Server Message Block** (**SMB**) version 1 service on the host. Such information is useful when performing vulnerability analysis and developing a payload for the exploitation phase.

When scanning a network, ethical hackers can determine the layout of the network topology, such as identifying the placement of network switches and routers, and security appliances such as firewalls. This information helps you to understand what type of devices are between your attacker machine and the targeted host. If there are routers on the network, these routers may be configured with packet-filtering capabilities such as **Access Control Lists** (**ACLs**) to filter specific layer 3 and layer 4 traffic between a source and destination. If firewalls are present on the network, they may be configured to perform **Deep Packet Inspection** (**DPI**) to inspect application-layer traffic for malicious code and advanced filtering between networks. Hence, understanding the network topology helps you to better plan your attacks and future operations.

Having completed this section, you have learned the importance of active reconnaissance and how it helps ethical hackers during their security assessments. Next, you will learn how to spoof your device's identity on a network.

Spoofing your identity on a network

Ethical hackers commonly use the same **Tactics, Techniques, and Procedures** (**TTPs**) as real adversaries and threat actors, with the intention of efficiently discovering hidden security vulnerabilities on a target's systems and network before a real hacker is able to discover and exploit them. A common technique used during ethical hacking, penetration testing, and red teaming is concealing your identity by spoofing your device's IP address and **Media Access Control** (**MAC**) address.

A device can be assigned an IP address via a **Dynamic Host Configuration Protocol** (**DHCP**) server on a network or statically assigned by a user. Any of these methods allow an IP address to be assigned to the device's **Network Interface Card** (**NIC**). There's a unique 48-bit hexadecimal MAC address that's embedded into the firmware of the network adapter by its manufacturer. Whenever a device sends a frame (message) on a network, the source MAC address of the sender and the destination MAC address of the target are inserted into the frame header. The first 24 bits of a MAC address are unique to the manufacturer, which is known as the **Organizational Unique Identifier** (**OUI**), and anyone can perform a vendor lookup on any MAC address using various online databases to determine the manufacturer and type of device. Cybersecurity and networking professionals can capture and analyze packets on a network to identify unauthorized devices based on their IP and MAC addresses.

The following are online databases to perform MAC vendor lookup:

- Wireshark OUI Lookup tool: `https://www.wireshark.org/tools/oui-lookup.html`
- MAC Vendors: `https://macvendors.com/`
- MACaddress.io: `https://macaddress.io/`

As an ethical hacker, consider changing the MAC address of your attacker machine to something that is very common on a typical network within an organization, such as an IP phone, network printer, or a popular brand of devices. The tactic is to ensure that your device does not stand out or attract the attention of the security team during your ethical hacking and penetration testing exercises. Therefore, if your device is assigned the MAC address of a network printer or an IP phone, an inexperienced security professional may not notice your device sending probes to other systems on the network to collect information.

To get started with spoofing the identity of your attacker machine, please use the following instructions:

1. Firstly, power on your **Kali Linux** virtual machine and log in.
2. Next, open the Terminal and use the following commands to verify the MAC address of the `eth1` network adapter that's connected to the `172.30.1.0/24` (*PentestNet*) network within our lab:

   ```
   kali@kali:~$ ip address show eth1
   ```

   The following snippet shows that the `eth1` network adapter is assigned the `08:00:27:7d:98:8e` MAC address:

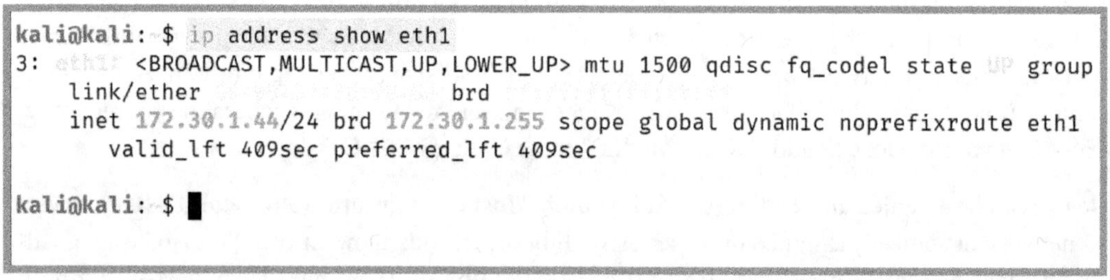

```
kali@kali:~$ ip address show eth1
3: eth1: <BROADCAST,MULTICAST,UP,LOWER_UP> mtu 1500 qdisc fq_codel state UP group
    link/ether                    brd
    inet 172.30.1.44/24 brd 172.30.1.255 scope global dynamic noprefixroute eth1
       valid_lft 409sec preferred_lft 409sec

kali@kali:~$
```

Figure 7.4 – Verifying the MAC address

3. Next, change the `eth1` network adapter status to down:

   ```
   kali@kali:~$ sudo ifconfig eth1 down
   ```

4. Next, we can use the pre-installed `macchanger` tool to view the different options for altering the MAC address on the network adapter:

   ```
   kali@kali:~$ sudo macchanger -h
   ```

   The following snippet shows there are various options to change the MAC address on the network adapter, such as randomly generating one, statically setting the MAC address of your preference, and assigning an address based on a vendor:

```
kali@kali: $ sudo macchanger -h
GNU MAC Changer
Usage: macchanger [options] device

  -h,  --help                  Print this help
  -V,  --version               Print version and exit
  -s,  --show                  Print the MAC address and exit
  -e,  --ending                Don't change the vendor bytes
  -a,  --another               Set random vendor MAC of the same kind
  -A                           Set random vendor MAC of any kind
  -p,  --permanent             Reset to original, permanent hardware MAC
  -r,  --random                Set fully random MAC
  -l,  --list[=keyword]        Print known vendors
  -b,  --bia                   Pretend to be a burned-in-address
  -m,  --mac=XX:XX:XX:XX:XX:XX
       --mac XX:XX:XX:XX:XX:XX  Set the MAC XX:XX:XX:XX:XX:XX
```

Figure 7.5 – The MAC changer menu

5. Next, assign a randomly generated MAC address to the `eth1` network adapter:

   ```
   kali@kali:~$ sudo macchanger -A eth1
   ```

   The following snippet shows the current (original) MAC address and the new randomly generated address on the network adapter:

```
kali@kali: $ sudo macchanger -A eth1
Current MAC:   08:00:27:7d:98:8e (CADMUS COMPUTER SYSTEMS)
Permanent MAC: 08:00:27:7d:98:8e (CADMUS COMPUTER SYSTEMS)
New MAC:       00:e0:cd:97:15:74 (SAAB SENSIS CORPORATION)
```

Figure 7.6 – The spoofed MAC address

6. Next, use the following command to re-enable the `eth1` network adapter:

   ```
   kali@kali:~$ sudo ifconfig eth1 up
   ```

7. Next, use either the `ip address show eth1` or `ifconfig eth1` commands to verify the `eth1` network adapter is up and assigned the spoofed MAC address, as shown here:

```
kali@kali: $ ip address show eth1
3: eth1: <BROADCAST,MULTICAST,UP,LOWER_UP> mtu 1500 qdisc fq_codel state UP group
    link/ether                    brd                    permaddr
    inet 172.30.1.49/24 brd 172.30.1.255 scope global dynamic noprefixroute eth1
       valid_lft 587sec preferred_lft 587sec

kali@kali: $ ifconfig eth1
eth1: flags=4163<UP,BROADCAST,RUNNING,MULTICAST>  mtu 1500
        inet 172.30.1.49  netmask 255.255.255.0  broadcast 172.30.1.255
        ether 00:e0:cd:97:15:74  txqueuelen 1000  (Ethernet)
        RX packets 15  bytes 5140 (5.0 KiB)
        RX errors 0  dropped 0  overruns 0  frame 0
        TX packets 24  bytes 4153 (4.0 KiB)
        TX errors 0  dropped 0 overruns 0  carrier 0  collisions 0
```

Figure 7.7 – Verifying the interface status

Any frames that originate from the `eth1` network adapter of your Kali Linux machine will be assigned the spoofed MAC address as the new source MAC address.

Having completed this section, you have learned how to spoof the MAC address of a network adapter to spoof the identity of your device. In the next section, you will learn how to discover and profile hosts on a network.

Discovering live hosts on a network

Host discovery is an important part of ethical hacking and penetration testing, as it enables an ethical hacker to identify which systems on a network are discoverable and live within an organization. If your targeted host is offline, you won't be able to identify security vulnerabilities and send an exploit to it.

There are various techniques and tools that help ethical hackers efficiently discover live hosts on a network and identify open ports and running services. This section will help you develop the skills and knowledge to perform host discovery as an ethical hacker.

Performing passive scanning with Netdiscover

Netdiscover is a pre-installed tool within Kali Linux that enables ethical hackers to either actively scan a network range or passively listen and analyze network packets to discover live hosts. Active scanning sends probes to each device within the network to determine which hosts are online. When using passive-mode, network packets that are captured by the attacker's machine will be analyzed to identify the IP and MAC addresses of other devices. However, passive-mode within Netdiscover reduces your threat detection level on the network unless the security team actively scans their network infrastructure to identify unauthorized devices.

To get started with using Netdiscover in passive-mode, please use the following instructions:

1. Firstly, power on the **Kali Linux** virtual machine and log in.
2. Next, power on the **Metasploitable 3** virtual machine. This machine will be used to generate traffic on the `PentestNet` network.
3. Next, open the Terminal on Kali Linux and use the following commands to verify the IP address and subnet mask of the `eth1` network adapter:

```
kali@kali:~$ ip addr show eth1
```

As shown in the following snippet, `eth1` is connected to the `172.30.1.0/24` network (*PentestNet*):

```
kali@kali: $ ip addr show eth1
3:       <BROADCAST,MULTICAST,UP,LOWER_UP> mtu 1500 qdisc fq_codel state    group
    link/ether                   brd                      permaddr
    inet 172.30.1.49/24 brd 172.30.1.255 scope global dynamic noprefixroute eth1
       valid_lft 120sec preferred_lft 120sec
```

Figure 7.8 – Checking the interface status

It's important to verify which network adapter on your attacker machine is connected to the targeted network. If you start scanning the wrong network or range of IP addresses, you can land yourself in some legal trouble, as scanning is considered illegal, and you'll be breaching the legal penetration testing agreement.

4. Next, enable **Netdiscover** to operate in passive-mode and listen on the `eth1` interface for network traffic:

```
kali@kali:~$ sudo netdiscover -p -i eth1
```

The following snippet shows Netdiscover was able to identify the MAC and IP addresses of devices that transmitted messages on the `172.30.1.0/24` network:

```
 Currently scanning: (passive)   |   Screen View: Unique Hosts

 5 Captured ARP Req/Rep packets, from 2 hosts.   Total size: 300
 _____________________________________________________________________________
   IP            At MAC Address     Count     Len  MAC Vendor / Hostname
 -----------------------------------------------------------------------------
 172.30.1.45     08:00:27:bf:3e:9c      3     180  PCS Systemtechnik GmbH
 172.30.1.1      08:00:27:2f:d8:4a      2     120  PCS Systemtechnik GmbH
```

Figure 7.9 – Passive scanning

> Passive-mode in Netdiscover can only analyze network traffic if it's detected on the `eth1` network adapter on a Kali Linux virtual machine. If you're not seeing any results appearing, simply log in to the Metasploitable 3 virtual machine with the `Administrator/vagrant` user credentials and send **ping** messages to the Kali Linux virtual machine.

While passive scanning may reduce your threat detection by the target, it may not provide the expected results needed to move forward with your ethical hacking and penetration testing methodology. Next, you will learn how to perform a ping sweep across an entire network.

Performing a ping sweep

A **ping sweep** is a common technique used by IT and security professionals to easily identify live hosts on a network. It involves sending **ICMP ECHO Request** messages to each host on the network; when a live host receives the ICMP ECHO Request message, it will respond by sending an **ICMP ECHO Reply** message to the sender, which indicates that the host is online. Within client and server operating systems, there's a **Ping** utility, which leverages **Internet Control Message Protocol** (**ICMP**), a network protocol that helps networking professionals to identify common network issues.

The following snippet shows a Wireshark packet capture of ICMP messages between a client and Google's DNS server:

No.	Time	Source	Destination	Protocol	Length	Info
163	9.878252	172.16.17.65	8.8.8.8	ICMP	74	Echo (ping) request
164	9.927742	8.8.8.8	172.16.17.65	ICMP	74	Echo (ping) reply
166	10.895600	172.16.17.65	8.8.8.8	ICMP	74	Echo (ping) request
167	10.944807	8.8.8.8	172.16.17.65	ICMP	74	Echo (ping) reply
169	11.904669	172.16.17.65	8.8.8.8	ICMP	74	Echo (ping) request
170	11.953887	8.8.8.8	172.16.17.65	ICMP	74	Echo (ping) reply
172	12.922175	172.16.17.65	8.8.8.8	ICMP	74	Echo (ping) request
173	12.971113	8.8.8.8	172.16.17.65	ICMP	74	Echo (ping) reply

Figure 7.10 – Ping messages

If you want to perform a ping sweep across the *PentestNet* network within our lab environment, you will need to ping each host on the `172.30.1.0/24` network that ranges from `172.30.1.1-172.30.1.254`. This is usually a manual, mundane, and time-consuming task for many. However, as an aspiring ethical hacker, you can automate the ping sweep process using various programming languages and even the native **Bourne Again Shell** (**BASH**) in Linux-based operating systems.

To automate the ping sweep process on Kali Linux using BASH, please use the following instructions:

1. Firstly, power on both the **Kali Linux** and **Metasploitable 3** Windows virtual machines.
2. Next, log in to **Kali Linux** and use the following commands within the Terminal to create a new BASH script file, using the `nano` command-line text editor:

```
kali@kali:~$ sudo nano ping-sweep.sh
```

3. Next, when Nano opens on the Terminal, type the following code to create the script:

```
#!/bin/bash

echo "Enter the network ID you want to scan (e.g. 192.168.1.0):"
read network

echo "Enter the subnet mask in CIDR notation (e.g. 24):"
read mask

if [[ $mask =~ ^[0-9]+$ ]] && [ $mask -le 32 ]; then
  wildcard=$(( 2**(32-$mask) - 1 ))
  network_address=$(echo $network | awk -F'.' '{print
$1"."$2"."$3"."0}')
  for host in $(seq 1 254); do
    ip=$(echo $network_address | awk -F'.' '{print
$1"."$2"."$3"."'$host'}')
    ping -c1 -W1 $ip > /dev/null
    if [ $? -eq 0 ]; then
      echo "$ip is up"
    fi
  done
else
  echo "Invalid subnet mask"
fi
```

When the preceding code executes, it will prompt the user to enter the network ID and subnet mask of the targeted network. It will calculate the range of host IP addresses and send one ICMP ECHO Request message to each host IP address.

4. To save the file, press the *Ctrl* + *X* keys, then press *Y*, and hit *Enter* to confirm and exit.

5. Next, add execution privileges to the `ping-sweep.sh` file:

```
kali@kali:~$ sudo chmod +x ping-sweep.sh
```

6. Next, use the following commands to execute the script:

```
kali@kali:~$ ./ping-sweep.sh
```

The following snippet shows that the script has prompted the user input of the network ID and subnet mask, and then it proceeds to check whether the hosts are online:

```
kali@kali:~$ ./ping-sweep.sh
Enter the network ID you want to scan (e.g. 192.168.1.0):
172.30.1.0
Enter the subnet mask in CIDR notation (e.g. 24):
24
172.30.1.1 is up
172.30.1.45 is up
172.30.1.49 is up
```

Figure 7.11 – A ping sweep

As shown in the following screenshot, the Ping utility and the preceding script used the ICMP network protocol to perform the ping sweep across the `172.30.1.0/24` network:

*eth1

File Edit View Go Capture Analyze Statistics Telephony Wireless Tools Help

icmp

No.	Time	Source	Destination	Protocol	Length	Info
3	0.000226479	172.30.1.49	172.30.1.1	ICMP	98	Echo (ping) request
4	0.000524646	172.30.1.1	172.30.1.49	ICMP	98	Echo (ping) reply
134	43.290568291	172.30.1.49	172.30.1.45	ICMP	98	Echo (ping) request
137	43.291630131	172.30.1.45	172.30.1.49	ICMP	98	Echo (ping) reply

Figure 7.12 – Wireshark packet capture

While the preceding method seems efficient for discovering live hosts on a network, system administrators and security professionals often disable ICMP responses from critical assets, such as servers and firewalls, to reduce the likelihood of a novice hacker identifying these systems. In addition, sending too many ICMP messages on a network can generate a lot of *noise*, which will increase the detection level of the ethical hacker on the network. Next, you will learn how to automate and improve your scanning techniques while reducing the detection level, using a powerful network scanner.

Host discovery with Nmap

Network Mapper (**Nmap**) is a powerful network scanning tool with lots of advanced capabilities to help ethical hackers and penetration testers to efficiently profile their targets. Nmap enables ethical hackers to discover live hosts and identify open **Transmission Control Protocol** (**TCP**) and **User Datagram Protocol** (**UDP**) ports, running services and their versions, and the operating system of a target.

To get started with using Nmap for host discovery, follow these steps:

1. Firstly, power on both the **Kali Linux** and **Metasploitable 3** virtual machines. As always, the attacker machine will be Kali Linux and the target will be Metasploitable 3.

2. On **Kali Linux**, open the Terminal and use either the `ip address` or `ifconfig` command to identify the IP address that's assigned to the `eth1` interface, which is connected to the `172.30.1.0/24` network (*PentestNet*).

   If important to identify the IP address of your attacker machine (Kali Linux) so that you do not scan your own device during security assessments; otherwise, you'll need to remove its details from the results.

3. Next, perform a ping sweep on the entire `172.30.1.0/24` network while excluding the IP address of the Kali Linux machine, using the following commands:

   ```
   kali@kali:~$ nmap -sn 172.30.1.0/24 --exclude 172.30.1.49
   ```

   As shown in the following snippet, the `-sn` syntax specifies performing a ping scan on the target, while `--exclude` allows us to exclude a host, a range of addresses, or a network during the scan:

```
kali@kali:~$ nmap -sn 172.30.1.0/24 --exclude 172.30.1.49
Starting Nmap 7.93 ( https://nmap.org ) at 2023-04-06 11:36 AST
Nmap scan report for 172.30.1.45
Host is up (0.0035s latency).
Nmap done: 255 IP addresses (1 host up) scanned in 18.96 seconds
```

Figure 7.13 – A ping scan using Nmap

   As shown in the preceding snippet, the ping scan was able to identify that the `172.30.1.45` host is online.

   Nmap does not send typical ICMP messages during a ping scan to a target. If the targeted device is configured not to respond to ICMP messages, then a typical ping sweep will fail to identify the status of the host on the network. Instead, Nmap sends specially crafted TCP **Synchronization** (**SYN**) packets to the targeted host, with the intention of triggering a TCP **Acknowledgement** (**ACK**) as a response from a live/online host.

   The following snippet shows a packet capture within Wireshark of the Nmap ping scan:

```
*eth1
File Edit View Go Capture Analyze Statistics Telephony Wireless Tools Help
ip.addr==172.30.1.49 && ip.addr==172.30.1.45
No.  Time          Source       Destination  Protocol Length Info
177  18.896530093  172.30.1.49  172.30.1.45  TCP      74     36038 → 443 [SYN] Seq=0
181  18.897178127  172.30.1.45  172.30.1.49  TCP      60     443 → 36038 [RST, ACK]
603  20.867170284  172.30.1.49  172.30.1.45  TCP      74     36044 → 443 [SYN] Seq=0
604  20.867712274  172.30.1.45  172.30.1.49  TCP      60     443 → 36044 [RST, ACK]
```

Figure 7.14 – Wireshark packet capture

As shown in the preceding snippet, Kali Linux (`172.30.1.49`) sent a **TCP SYN** message to the target (`172.30.1.45`) on service port `443` (HTTPS), and the target responded with a **TCP ACK/RST** packet to acknowledge and reset the connection. This response indicates that port `443` is open and the target is live. While devices may be configured to not respond to ICMP messages, TCP/IP is designed to respond to **TCP SYN** packets on an open service port, and Nmap uses this technique to better determine whether a target is online or not.

> **Tip**
>
> To learn more about how TCP establishes a connection with a three-way handshake, please see `https://hub.packtpub.com/understanding-network-port-numbers-tcp-udp-and-icmp-on-an-operating-system/`.

4. Next, perform a port scan to identify the top 1,000 service ports that are open on Metasploitable 3:

```
kali@kali:~$ nmap 172.30.1.45
```

As shown in the following snippet, Nmap was able to identify the top 1,000 service ports that opened on the target and determined their service:

```
kali@kali:~$ nmap 172.30.1.45
Starting Nmap 7.93 ( https://nmap.org ) at 2023-04-06 11:51 AST
Nmap scan report for 172.30.1.45
Host is up (0.0017s latency).
Not shown: 981 closed tcp ports (conn-refused)
PORT      STATE SERVICE
21/tcp    open  ftp
22/tcp    open  ssh
80/tcp    open  http
135/tcp   open  msrpc
139/tcp   open  netbios-ssn
445/tcp   open  microsoft-ds
3306/tcp  open  mysql
3389/tcp  open  ms-wbt-server
4848/tcp  open  appserv-http
7676/tcp  open  imqbrokerd
```


Figure 7.15 – Identifying the top 1,000 service ports

As shown in the preceding snippet, Nmap helps us to better identify which ports are open and the associated services for each open port. Such information can be leveraged to identify the role, function, and the attack surface of a targeted device on a network. For instance, the preceding snippet shows that port `22` is open and running **Secure Shell** (**SSH**), a remote access service that can be compromised by an ethical hacker to gain a foothold into a system. It's important to research the services for each open port to identify security vulnerabilities and how a real attacker can compromise the target.

> **Important note**
>
> By default, Nmap scans TCP ports only.

5. Next, use the `-p-` syntax to specify all 65,535 ports and the `-T4` syntax to increase the scanning speed, as shown here:

```
kali@kali:~$ nmap -T4 -p- 172.30.1.45
```

The following snippet shows that Nmap has identified additional open service ports on the target system:

```
kali@kali: $ nmap -T4 -p- 172.30.1.45
Starting Nmap 7.93 ( https://nmap.org ) at 2023-04-06 11:53 AST
Nmap scan report for 172.30.1.45
Host is up (0.0015s latency).
Not shown: 65497 closed tcp ports (conn-refused)
PORT      STATE SERVICE
21/tcp    open  ftp
22/tcp    open  ssh
80/tcp    open  http
135/tcp   open  msrpc
139/tcp   open  netbios-ssn
445/tcp   open  microsoft-ds
1617/tcp  open  nimrod-agent
3306/tcp  open  mysql
3389/tcp  open  ms-wbt-server
3700/tcp  open  lrs-paging
4848/tcp  open  appserv-http
```

Scanned all 65,535 TCP ports

Figure 7.16 – Scanning all ports

As an ethical hacker, it's important to scan all service ports on a targeted device to identify whether there are additional running services and determine whether they are vulnerable to cyber-attacks and threats. In addition, if your goal is to gain a foothold into the targeted system, you may be able to exploit a security vulnerability found on a running service.

> **Tip**
>
> Nmap can scan a range of host addresses on a network by using the `nmap 172.30.1.10-100` syntax. Additionally, you can scan the selected host by using the `nmap 172.30.1.10,172.30.1.20,172.30.1.40-50` command. `-T<0-5>` allows you to set the timing template on an Nmap scan; higher is faster. Keep in mind that faster scans increase the amount of noise generated on a network and can raise the suspicions of the security team.

6. By default, Nmap performs a ping scan on a targeted host to determine whether it's online, and then it proceeds to scan. However, if the targeted host is configured to block ping probes from Nmap, we can perform a scan without pinging the target with the following commands:

```
kali@kali:~$ nmap -PN 172.30.1.45
```

The following snippet shows the execution of the preceding commands:

```
kali@kali: $ nmap -PN 172.30.1.45
Starting Nmap 7.93 ( https://nmap.org ) at 2023-04-06 12:12 AST
Nmap scan report for 172.30.1.45
Host is up (0.0041s latency).
Not shown: 980 closed tcp ports (conn-refused)
PORT      STATE SERVICE
21/tcp    open  ftp
22/tcp    open  ssh
80/tcp    open  http
135/tcp   open  msrpc
139/tcp   open  netbios-ssn
445/tcp   open  microsoft-ds
3306/tcp  open  mysql
3389/tcp  open  ms-wbt-server
```

Don't Ping Scan

Figure 7.17 – Don't ping scan

1. To scan the top 1,000 UDP service ports on a targeted system, use the following command:

```
kali@kali:~$ sudo nmap -sU 172.30.1.45
```

The following snippet shows Nmap was able to identify open UDP ports on the target:

```
kali@kali: $ sudo nmap -sU 172.30.1.45
Starting Nmap 7.93 ( https://nmap.org ) at 2023-04-06 12:16 AST
Nmap scan report for 172.30.1.45
Host is up (0.00069s latency).
Not shown: 993 closed udp ports (port-unreach)
PORT     STATE         SERVICE
137/udp  open          netbios-ns
138/udp  open|filtered netbios-dgm
161/udp  open          snmp
500/udp  open|filtered isakmp
4500/udp open|filtered nat-t-ike
5353/udp open|filtered zeroconf
5355/udp open|filtered llmnr
MAC Address: 08:00:27:BF:3E:9C (Oracle VirtualBox virtual NIC)

Nmap done: 1 IP address (1 host up) scanned in 1105.24 seconds
```

UDP scan

Figure 7.18 – UDP port scanning

Since Nmap scans the top 1,000 ports by default, you can use the `nmap -sU -p- 172.30.1.45` command to scan all 65,535 UDP ports on the target. Additionally, you can use the `-p` syntax to specify a specific port or a range of ports – for instance, `-p22` and `-p1-65535`. The `-U` syntax allows you to specify a UDP port scan with the `-p U:53` syntax.

> **Important note**
>
> Filtered ports indicate whether there's a firewall or another security mechanism that's blocking the port, and Nmap is unable to determine whether the port is open or closed on the target.

2. To identify the service versions of running services on a targeted device, use the following commands:

```
kali@kali:~$ sudo nmap -sV 172.30.1.45
```

The following snippet shows Nmap was able to identify the service versions of each running service:

```
kali@kali: $ sudo nmap -sV 172.30.1.45
Starting Nmap 7.93 ( https://nmap.org ) at 2023-04-06 12:38 AST
Nmap scan report for 172.30.1.45
Host is up (0.00092s latency).
Not shown: 980 closed tcp ports (reset)
PORT      STATE SERVICE            VERSION
21/tcp    open  ftp                Microsoft ftpd
22/tcp    open  ssh                OpenSSH 7.1 (protocol 2.0)
80/tcp    open  http               Microsoft IIS httpd 7.5
135/tcp   open  msrpc              Microsoft Windows RPC
139/tcp   open  netbios-ssn        Microsoft Windows netbios-ssn
445/tcp   open  microsoft-ds       Microsoft Windows Server 2008 R2 - 2012 microsoft-ds
3306/tcp  open  mysql              MySQL 5.5.20-log
3389/tcp  open  ms-wbt-server?
```

Figure 7.19 – Identifying the service versions

Identifying the service versions helps us to determine whether the running application or service is outdated, and whether there are any known security vulnerabilities that can be exploited to gain authorized access. If you recall, one of the reasons to perform reconnaissance is to determine the attack surface of a target, such as discovering security vulnerabilities and points of entry.

3. To perform operating system detection on a targeted host, use the `-O` syntax, as shown here:

```
kali@kali:~$ sudo nmap -O 172.30.1.45
```

The following snippet shows Nmap was able to identify the target host as running Microsoft Windows 7, 8, or Server 2008:

```
Device type: general purpose
Running: Microsoft Windows 7|2008|8.1
OS CPE: cpe:/o:microsoft:windows_7::- cpe:/o:microsoft:windows_7::sp1 cpe:/o:microsoft:windows_server_2008::sp1 cpe:/o:microso
ft:windows_server_2008:r2 cpe:/o:microsoft:windows_8 cpe:/o:microsoft:windows_8.1
OS details: Microsoft Windows 7 SP0 - SP1, Windows Server 2008 SP1, Windows Server 2008 R2, Windows 8, or Windows 8.1 Update 1
Network Distance: 1 hop

OS detection performed. Please report any incorrect results at https://nmap.org/submit/ .
Nmap done: 1 IP address (1 host up) scanned in 2.65 seconds
```

Figure 7.20 – Identifying the operating system

As shown in the preceding snippet, identifying the operating system architecture and build number helps us to better plan future attacks during ethical hacking and penetration testing. For instance, you can research known security vulnerabilities and develop and test exploits for this specific operating system version, before preceding to the exploitation phase.

4. Lastly, Nmap enables us to perform operating system detection, service version detection, script scanning, and traceroute by using the `-A` syntax, as shown here:

```
kali@kali:~$ sudo nmap -A 172.30.1.45
```

As shown in the following snippet, the `-A` syntax enables Nmap to perform multiple types of scanning techniques at the same time:

```
kali@kali: $ sudo nmap -A 172.30.1.45
Starting Nmap 7.93 ( https://nmap.org ) at 2023-04-06 12:44 AST
Nmap scan report for 172.30.1.45
Host is up (0.00066s latency).
Not shown: 980 closed tcp ports (reset)
PORT      STATE SERVICE              VERSION
21/tcp    open  ftp                  Microsoft ftpd
| ftp-syst:
|_  SYST: Windows_NT
22/tcp    open  ssh                  OpenSSH 7.1 (protocol 2.0)
| ssh-hostkey:
|   2048 a5702aacb0ab784fbc1efc053623de38 (RSA)
|_  521 4aa1db4e864943fe0eb86a337fa97882 (ECDSA)
80/tcp    open  http                 Microsoft IIS httpd 7.5
| http-methods:
|_  Potentially risky methods: TRACE
|_http-title: Site doesn't have a title (text/html).
|_http-server-header: Microsoft-HTTPAPI/2.0
135/tcp   open  msrpc                Microsoft Windows RPC
139/tcp   open  netbios-ssn          Microsoft Windows netbios-ssn
445/tcp   open  microsoft-ds         Windows Server 2008 R2 Standard 7601 Service Pack 1
3306/tcp  open  mysql                MySQL 5.5.20-log
```

Figure 7.21 – Multiple scans

As shown in the preceding snippet, the `-A` syntax further enables Nmap to perform banner grabbing on each running service. Banner grabbing enables us to collect additional information about a service, such as the header details of a Microsoft IIS service.

In addition, the following is the remaining portion of the scan, which shows the SMB version, operating system version, computer name/hostname, and workgroup or domain:

```
Host script results:
| smb2-time:
|   date: 2023-04-06T16:46:39
|_  start_date: 2023-04-06T15:46:29
| smb-security-mode:
|   account_used: <blank>
|   authentication_level: user
|   challenge_response: supported
|_  message_signing: disabled (dangerous, but default)
|_clock-skew: mean: 1h45m00s, deviation: 3h30m00s, median: 0s
|_nbstat: NetBIOS name: VAGRANT-2008R2, NetBIOS user: <unknown>, NetBIOS MAC: 080027bf3e9c
| smb2-security-mode:
|   210:
|_    Message signing enabled but not required
| smb-os-discovery:
|   OS: Windows Server 2008 R2 Standard 7601 Service Pack 1 (Windows Server 2008 R2 Standard 6.1)
|   OS CPE: cpe:/o:microsoft:windows_server_2008::sp1
|   Computer name: vagrant-2008R2
|   NetBIOS computer name: VAGRANT-2008R2\x00
|   Workgroup: WORKGROUP\x00
|_  System time: 2023-04-06T09:46:37-07:00
```

Figure 7.22 – OS and SMB detection

As you can see, Nmap is a powerful network scanning tool, with many features and capabilities to help ethical hackers and penetration testers collect sensitive information during active reconnaissance of their targets.

> **Tip**
> To learn more about Nmap and its scanning capabilities, please see `https://nmap.org/book/man.html`. In addition, you can use the `man nmap` command on the Terminal to view the Nmap manual page.

Having completed this section, you have learned how to perform host discovery in active reconnaissance. In the next section, you will learn how to perform evasion techniques during network scanning.

Using evasion techniques

When a device such as a computer sends a packet to a network, the sender's IP and MAC addresses are inserted into the packet header. As such, when ethical hackers perform host discovery on a network, their IP and MAC addresses are also included within each packet from their machine to the targeted hosts. Ethical hackers and penetration testers always try to remain undetected during security assessments, as part of simulating real-world cyberattacks on an organization's network infrastructure.

There are various evasion techniques that can be used during host discovery to reduce the risk of being detected and traced by threat detection systems and the security team of a target. To get started, please use the following instructions:

1. Power on both the **Kali Linux** and **Metasploitable 3** virtual machines. Ensure you're logged in to Kali Linux.
2. Ensure you identify the IP address of Metasploitable 3, as this virtual machine will be the target of this exercise.
3. Nmap enables us to perform a scan using decoys to trick the target into thinking the source of the scan originates from different sources, rather than a single device. The `-D` syntax allows you to specify any number of decoy addresses, as shown here:

   ```
   kali@kali:~$ nmap 172.30.1.45 -D 172.30.1.20,
   172.30.1.21,172.30.1.22
   ```

   Before Nmap performs the scan on the target, it checks each decoy address on the network to ensure that these systems are online. If a decoy address is not reachable by Nmap, it will be excluded from the scan. Therefore, I recommend performing a status check on the decoy systems before performing a scan.
4. Nmap enables you to automatically spoof your MAC address when scanning a target by using the `--spoof-mac` syntax, as shown here:

   ```
   kali@kali:~$ sudo nmap -sT -PN --spoof-mac 0 172.30.1.45
   ```

   The `-ST` syntax specifies performing a **TCP Connect** scan by establishing the TCP three-way handshake, `-PN` specifies not pinging the host, and `--spoof-mac 0` indicates spoofing a random MAC address. The following snippet shows the random spoof MAC address during the scan:

   ```
   kali@kali:~$ sudo nmap -sT -PN --spoof-mac 0 172.30.1.45
   Starting Nmap 7.93 ( https://nmap.org ) at 2023-04-06 13:09 AST
   Spoofing MAC address AB:A4:7A:56:A5:E3 (No registered vendor)
   You have specified some options that require raw socket access.
   These options will not be honored for TCP Connect scan.
   Nmap scan report for 172.30.1.45
   Host is up (0.0012s latency).
   Not shown: 981 closed tcp ports (conn-refused)
   PORT      STATE SERVICE
   21/tcp    open  ftp
   22/tcp    open  ssh
   80/tcp    open  http
   135/tcp   open  msrpc
   139/tcp   open  netbios-ssn
   445/tcp   open  microsoft-ds
   3306/tcp  open  mysql
   ```

   Figure 7.23 – Spoofing the random MAC address

5. If you want to spoof a specific vendor's MAC address, Nmap allows you to insert the vendor's name, as shown in the following command:

```
kali@kali:~$ sudo nmap -sT -PN --spoof-mac dell 172.30.1.45
```

The following snippet shows that Nmap uses a spoof MAC address from Dell:

```
kali@kali:~$ sudo nmap -sT -PN --spoof-mac dell 172.30.1.45
Starting Nmap 7.93 ( https://nmap.org ) at 2023-04-06 13:10 AST
Spoofing MAC address 00:00:97:47:11:CD (Dell EMC)
You have specified some options that require raw socket access.
These options will not be honored for TCP Connect scan.
Nmap scan report for 172.30.1.45
Host is up (0.0012s latency).
Not shown: 981 closed tcp ports (conn-refused)
PORT      STATE SERVICE
21/tcp    open  ftp
22/tcp    open  ssh
80/tcp    open  http
135/tcp   open  msrpc
```

Figure 7.24 – Spoofing the vendor address

6. If there's a specific MAC address you want to spoof, it can be included, as shown in the following command:

```
kali@kali:~$ sudo nmap -sT -PN --spoof-mac 12:34:56:78:9A:BC
172.30.1.45
```

The following snippet shows that the custom MAC address was used as the spoof address during the Nmap scan:

```
kali@kali:~$ sudo nmap -sT -PN --spoof-mac 12:34:56:78:9A:BC 172.30.1.45
Starting Nmap 7.93 ( https://nmap.org ) at 2023-04-06 13:11 AST
Spoofing MAC address 12:34:56:78:9A:BC (No registered vendor)
You have specified some options that require raw socket access.
These options will not be honored for TCP Connect scan.
Nmap scan report for 172.30.1.45
Host is up (0.0018s latency).
Not shown: 981 closed tcp ports (conn-refused)
PORT      STATE SERVICE
21/tcp    open  ftp
22/tcp    open  ssh
80/tcp    open  http
135/tcp   open  msrpc
139/tcp   open  netbios-ssn
```

Figure 7.25 – Using a specific spoof address

7. Lastly, Nmap provides the capability to send probes as smaller fragments on the network to reduce detection by network security systems. The following command is an example:

```
kali@kali:~$ sudo nmap -f 172.30.1.45
```

The following snippet shows a Wireshark capture of Kali Linux (`172.30.1.44`) sending multiple fragmented IP packets of 42 bytes to Metasploitable 3 (`172.30.1.45`):

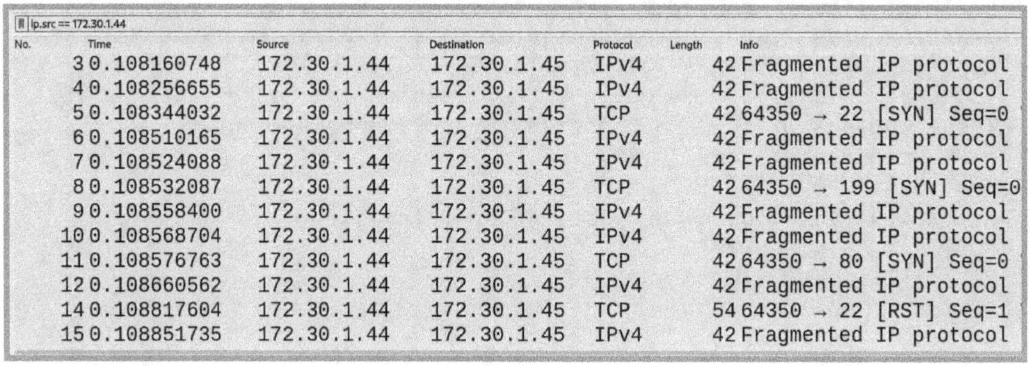

ip.src == 172.30.1.44

No.	Time	Source	Destination	Protocol	Length	Info
3	0.108160748	172.30.1.44	172.30.1.45	IPv4	42	Fragmented IP protocol
4	0.108256655	172.30.1.44	172.30.1.45	IPv4	42	Fragmented IP protocol
5	0.108344032	172.30.1.44	172.30.1.45	TCP	42	64350 → 22 [SYN] Seq=0
6	0.108510165	172.30.1.44	172.30.1.45	IPv4	42	Fragmented IP protocol
7	0.108524088	172.30.1.44	172.30.1.45	IPv4	42	Fragmented IP protocol
8	0.108532087	172.30.1.44	172.30.1.45	TCP	42	64350 → 199 [SYN] Seq=0
9	0.108558400	172.30.1.44	172.30.1.45	IPv4	42	Fragmented IP protocol
10	0.108568704	172.30.1.44	172.30.1.45	IPv4	42	Fragmented IP protocol
11	0.108576763	172.30.1.44	172.30.1.45	TCP	42	64350 → 80 [SYN] Seq=0
12	0.108660562	172.30.1.44	172.30.1.45	IPv4	42	Fragmented IP protocol
14	0.108817604	172.30.1.44	172.30.1.45	TCP	54	64350 → 22 [RST] Seq=1
15	0.108851735	172.30.1.44	172.30.1.45	IPv4	42	Fragmented IP protocol

Figure 7.26 – Fragmenting packets

Important note

Using the `-sS` syntax enables Nmap to perform a **stealth scan** (sometimes referred to as a *half-open scan*) on a target. The stealth scan does not establish a full TCP three-way handshake with the target. However, the Nmap stealth scan can be easily detected by seasoned cybersecurity professionals. To learn more about the Nmap stealth scan, please see `https://nmap.org/book/synscan.html`.

As you learned in this section, there are various techniques that are used by ethical hackers to evade detection during host discovery. In the next section, you will explore various enumeration techniques.

Enumerating network services

During the host discovery phase of ethical hacking, many open ports and running services can be found. The information collected during the host discovery phase is used to perform additional research to find known security vulnerabilities within those running services and how they can be exploited to gain unauthorized access to a target. Enumerating network services on targeted systems helps ethical hackers and penetration testers to better understand the target's network infrastructure and attack surfaces and which attack vectors can be used to deliver a payload to the target.

NetBIOS and SMB enumeration

Network Basic Input/Output System (**NetBIOS**) is an old legacy network protocol that's commonly used within Windows-based environments for file and print sharing, and name resolution services on a **Local Area Network** (**LAN**). SMB is a newer network protocol that's also used within Windows-based environments for file and print-sharing services. As an ethical hacker, performing NetBIOS and SMB enumeration helps you identify which systems on a network have network shares. A misconfigured system enables an ethical hacker to access sensitive directories and files over a network.

To get started with NetBIOS and SMB enumeration, please use the following instruction:

1. Firstly, power on the **Kali Linux** and **Metasploitable 3** virtual machines. Kali Linux will be the attacker machine and Metasploitable 3 will operate as the target.
2. Next, use `nbtscan` to retrieve the NetBIOS name and MAC address of Metasploitable 3:

   ```
   kali@kali:~$ nbtscan -r 172.30.1.45
   ```

   The `-r` syntax is used to specify a remote host on the network, as shown in the following snippet:

```
kali@kali:~$ nbtscan -r 172.30.1.45
Doing NBT name scan for addresses from 172.30.1.45

IP address       NetBIOS Name     Server    User             MAC address
------------------------------------------------------------------------------
172.30.1.45      VAGRANT-2008R2   <server>  <unknown>        08:00:27:bf:3e:9c
```

Figure 7.27 – A NetBIOS scan

   If there are any logged-in users on the target, their usernames will appear within the `User` field.

3. Next, use **Nmap** to identify whether SMB runs the Metasploitable 3 by scanning port `445`, using the following commands:

   ```
   kali@kali:~$ nmap -A -T4 -Pn -n -p 445 172.30.1.45
   ```

   As shown in the following snippet, port `445` is open, and Nmap was able to profile the host operating system:

```
kali@kali:~$ nmap -A -T4 -Pn -n -p 445 172.30.1.45
Starting Nmap 7.93 ( https://nmap.org ) at 2023-04-14 12:28 AST
Nmap scan report for 172.30.1.45
Host is up (0.00037s latency).

PORT    STATE SERVICE      VERSION
445/tcp open  microsoft-ds Windows Server 2008 R2 Standard 7601 Service Pack 1 microsoft-ds
Service Info: OS: Windows Server 2008 R2 - 2012; CPE: cpe:/o:microsoft:windows
```

Figure 7.28 – Identifying SMB on a target

4. Next, use Nmap to perform a NetBIOS scan of Metasploitable 3 with the following command:

```
kali@kali:~$ sudo nmap -sV --script nbstat.nse 172.30.1.45
```

As shown in the following snippet, Nmap was able to verify the operating system of the host:

```
MAC Address: 08:00:27:BF:3E:9C (Oracle VirtualBox virtual NIC)
Service Info: OSs: Windows, Windows Server 2008 R2 - 2012; CPE: cpe:/o:microsoft:windows

Host script results:
| nbstat: NetBIOS name: VAGRANT-2008R2, NetBIOS user: <unknown>, NetBIOS MAC: 080027bf3e9c
| Names:
|   VAGRANT-2008R2<00>    Flags: <unique><active>
|   WORKGROUP<00>         Flags: <group><active>
|_  VAGRANT-2008R2<20>    Flags: <unique><active>

Service detection performed. Please report any incorrect results at https://nmap.org/submit/ .
Nmap done: 1 IP address (1 host up) scanned in 108.31 seconds
```

Figure 7.29 – Identifying the target operating system

5. Since SMB is found running on Metasploitable 3, use the `smbclient` tool to enumerate network scans on the target; `-U` specifies the username:

```
kali@kali:~$ smbclient -L //172.30.1.45 -U vagrant
```

When prompted for a password, use `vagrant` to allow `smbclient` to enumerate the shares, as shown here:

```
kali@kali:~$ smbclient -L //172.30.1.45 -U vagrant
Password for [WORKGROUP\vagrant]:

        Sharename       Type      Comment
        ---------       ----      -------
        ADMIN$          Disk      Remote Admin
        C$              Disk      Default share
        IPC$            IPC       Remote IPC
Reconnecting with SMB1 for workgroup listing.
do_connect: Connection to 172.30.1.45 failed (Error NT_STATUS_RESOURCE_NAME_NOT_FOUND)
Unable to connect with SMB1 -- no workgroup available
```

Figure 7.30 – Network shares

As shown in the preceding snippet, there are multiple network shares on the target.

> **Tip**
>
> To access a network share, use the `smbclient -L //172.30.1.45/sharename -U vagrant` command. Ensure you adjust the IP address, share name, and username accordingly for your target.

6. Next, use the smbmap tool to identify the permissions on each network share:

```
kali@kali:~$ smbmap -H 172.30.1.45 -u vagrant -p vagrant
```

The -u and -p syntax allows us to specify the username and password to access the target, as shown here:

```
kali@kali:~$ smbmap -H 172.30.1.45 -u vagrant -p vagrant
[+] IP: 172.30.1.45:445 Name: 172.30.1.45
        Disk                                                    Permissions     Comment
        ----                                                    -----------     -------
        ADMIN$                                                  READ, WRITE     Remote Admin
        C$                                                      READ, WRITE     Default share
        IPC$                                                    NO ACCESS       Remote IPC
        Users                                                   READ, WRITE
```

Figure 7.31 – Share permissions

As shown in the preceding snippet, some shares enable network users to both read and write to the directory. Access to sensitive directories is useful during ethical hacking and penetration testing.

7. Next, use the following commands to perform SMB enumeration of shares with Nmap:

```
kali@kali:~$ sudo nmap -p139,445 --script=smb-enum-shares.nse
172.30.1.45
```

The following snippet shows that Nmap attempted to find and enumerate shares on the target:

```
kali@kali:~$ sudo nmap -p139,445 --script=smb-enum-shares.nse 172.30.1.45
Starting Nmap 7.93 ( https://nmap.org ) at 2023-04-15 11:01 AST
Nmap scan report for 172.30.1.45
Host is up (0.00015s latency).

PORT    STATE SERVICE
139/tcp open  netbios-ssn
445/tcp open  microsoft-ds
MAC Address: 08:00:27:BF:3E:9C (Oracle VirtualBox virtual NIC)

Host script results:
| smb-enum-shares:
|   note: ERROR: Enumerating shares failed, guessing at common ones (NT_STATUS_ACCESS_DENIED)
|   account_used: <blank>
|   \\172.30.1.45\ADMIN$:
|     warning: Couldn't get details for share: NT_STATUS_ACCESS_DENIED
|     Anonymous access: <none>
|   \\172.30.1.45\C$:
|     warning: Couldn't get details for share: NT_STATUS_ACCESS_DENIED
|     Anonymous access: <none>
|   \\172.30.1.45\IPC$:
|     warning: Couldn't get details for share: NT_STATUS_ACCESS_DENIED
|     Anonymous access: READ
|   \\172.30.1.45\USERS:
|     warning: Couldn't get details for share: NT_STATUS_ACCESS_DENIED
|_    Anonymous access: <none>
```

Figure 7.32 – Nmap SMB enumeration script

8. Lastly, we can use Nmap to identify SMB vulnerabilities on a target using the following commands:

```
kali@kali:~$ sudo nmap -p139,445 --script=smb-vuln-* 172.30.1.45
```

As shown in the following snippet, Nmap was able to identify a security vulnerability on the target system and provide us with reference information:

```
Host script results:
|_smb-vuln-ms10-054: false
|_smb-vuln-ms10-061: NT_STATUS_ACCESS_DENIED
| smb-vuln-ms17-010:
|   VULNERABLE:
|   Remote Code Execution vulnerability in Microsoft SMBv1 servers (ms17-010)
|     State: VULNERABLE
|     IDs:  CVE:CVE-2017-0143
|     Risk factor: HIGH
|       A critical remote code execution vulnerability exists in Microsoft SMBv1
|        servers (ms17-010).
|
|     Disclosure date: 2017-03-14
|     References:
|       https://technet.microsoft.com/en-us/library/security/ms17-010.aspx
|       https://cve.mitre.org/cgi-bin/cvename.cgi?name=CVE-2017-0143
|_      https://blogs.technet.microsoft.com/msrc/2017/05/12/customer-guidance-for-wannacrypt-attacks/

Nmap done: 1 IP address (1 host up) scanned in 5.45 seconds
```

Figure 7.33 – SMB vulnerability discovery

Having completed this section, you have learned how to use various tools and techniques to perform service enumeration. In the next section, you will learn how to perform wireless reconnaissance.

Wireless reconnaissance

Wireless reconnaissance enables ethical hackers and penetration testers to identify their target's wireless network and determine associated clients, network settings, operating frequencies and channels, and the approximate distance between you and the access point.

When an access point is powered on, it sends **beacons** to advertise its presence and network information to nearby wireless clients. Within these beacons, the access point inserts the network name or **Service Set Identifier** (**SSID**), which helps clients to identify one wireless network from the other. Once a client is connected (associated) with a wireless network, it automatically saves the network information and password within its **Preferred Network List** (**PNL**). From then on, when wireless capabilities are enabled on the client, it will send **probes** to seek any of the saved networks from the PNL; once a network is found and within range, the client will attempt to automatically establish an association. As an ethical hacker, capturing and analyzing beacons and probes from wireless networks and clients helps us to identify whether the targeted wireless network is vulnerable to wireless-based attacks.

> **Tip**
> You can use **Wigle.net** to find wireless networks around the world. The website also allows users to upload Kismet data and helps ethical hackers to locate their target networks.

To get started with wireless reconnaissance, follow the following instructions.

Part 1 – attaching a wireless network adapter

To attach a wireless network adapter, follow these steps.

1. Connect the Alfa AWUS036NHA – Wireless B/G/N USB adapter to your host computer.
2. Next, open **VirtualBox Manager**, select the **Kali Linux** virtual machine, and click on **Settings**, as shown here:

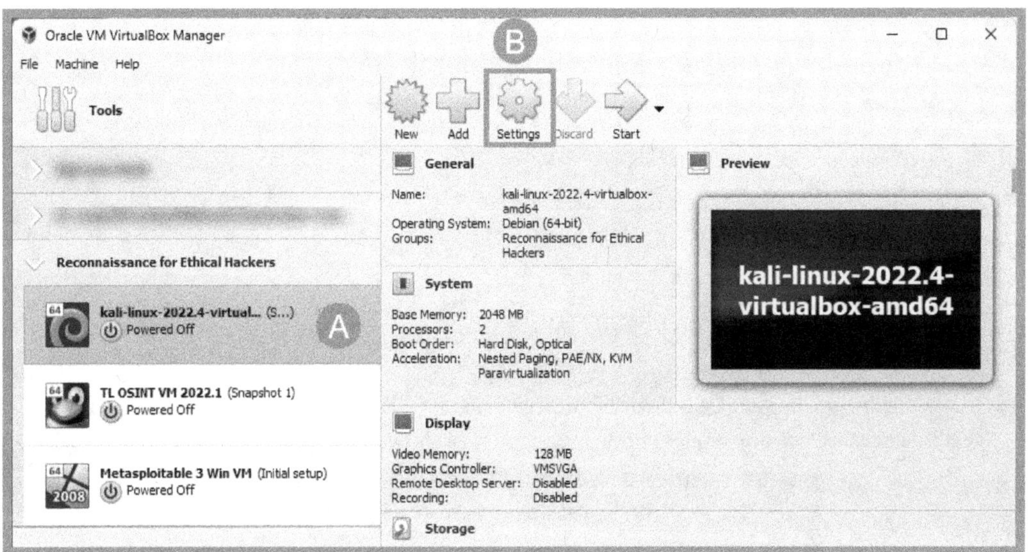

Figure 7.34 – VirtualBox Manager

3. Next, select the **USB** category from the left menu, and then select either the USB 2.0 or 3.0 controller, depending on your USB interface types on your host computer, and click on the **Add USB device** icon, as shown here:

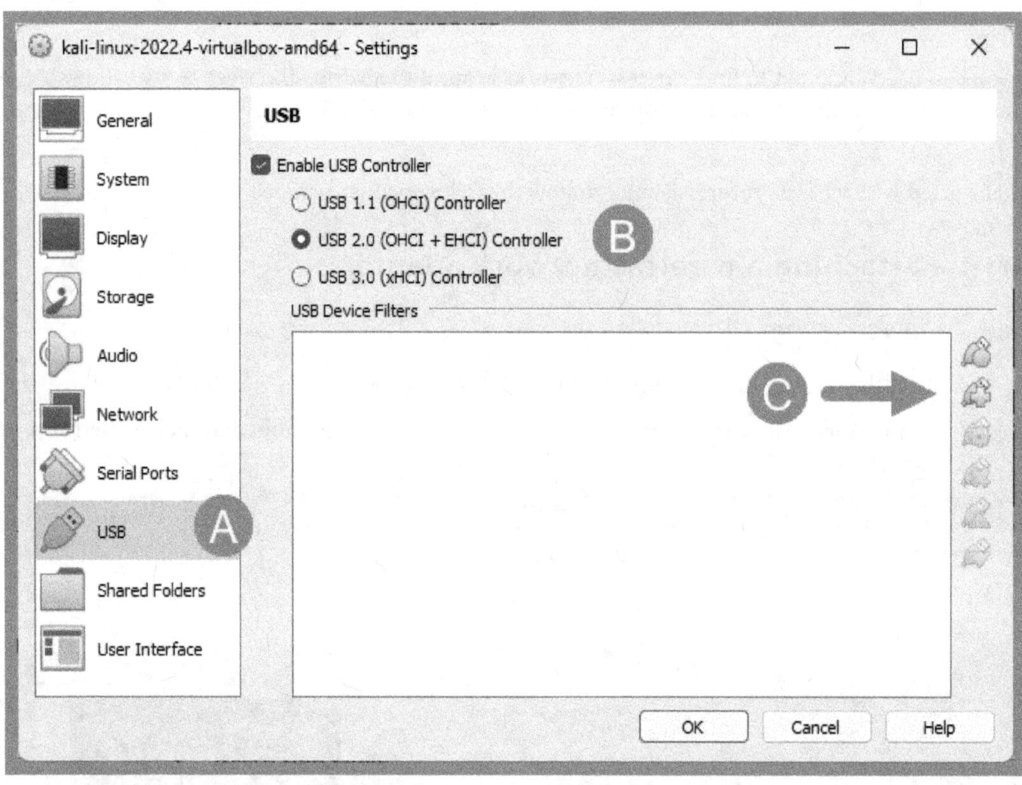

Figure 7.35 – The Settings menu

4. After clicking on the **Add USB device** icon, a side menu will appear, displaying all your connected USB devices. Select the wireless network adapter, as shown here:

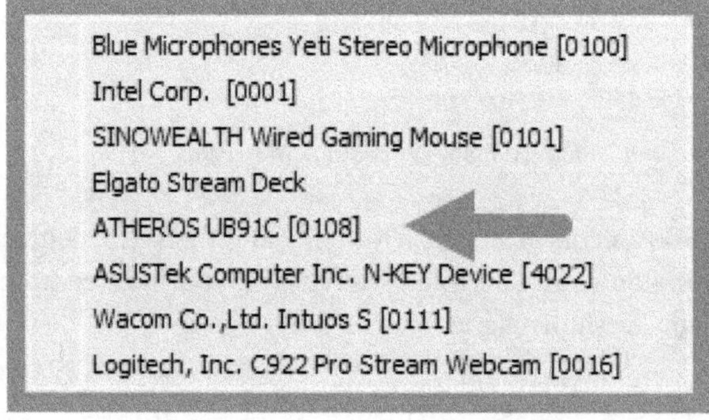

Figure 7.36 – Selecting the wireless network adapter

5. Next, you will automatically return to the main **Settings** menu. Click on **OK** to save:

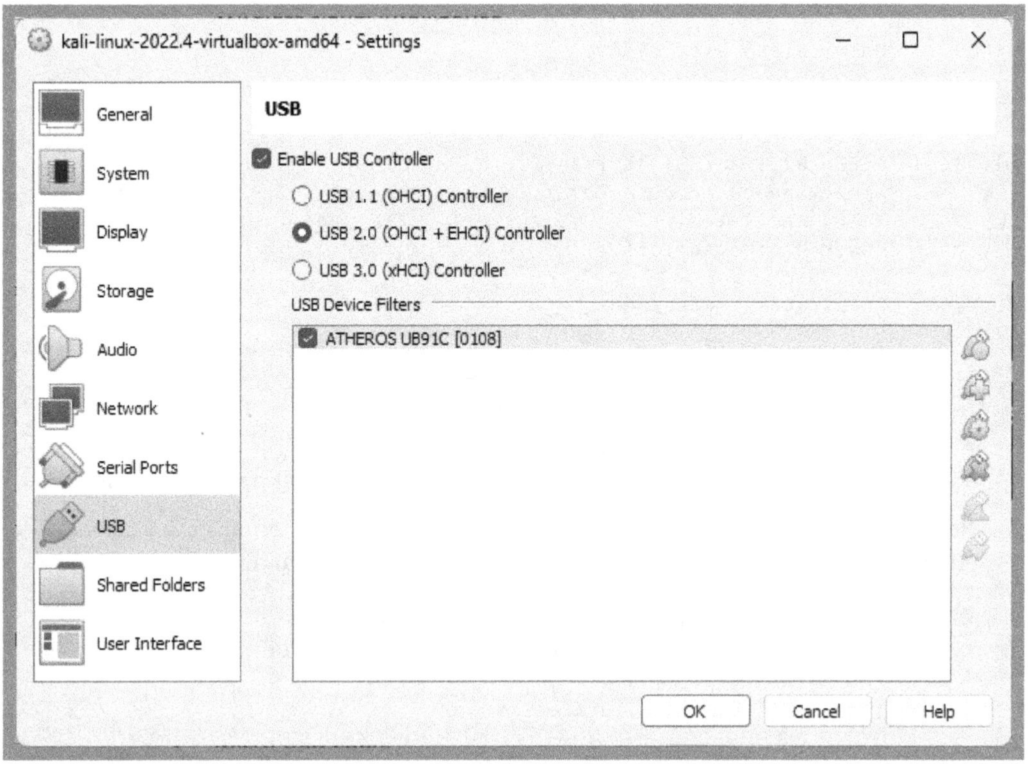

Figure 7.37 – The wireless network adapter

As shown in the preceding snippet, the wireless network adapter is now connected to the Kali Linux virtual machine as a USB device.

Part 2 – enabling monitor mode

To enable monitor mode, follow these steps:

1. Power on the **Kali Linux** virtual machine and log in.
2. Next, open the Terminal and use the `iwconfig` command to identify the wireless network adapter, as shown here:

```
kali@kali:~$ iwconfig
lo        no wireless extensions.

eth0      no wireless extensions.

eth1      no wireless extensions.

wlan0     IEEE 802.11  ESSID:off/any
          Mode:Managed  Access Point: Not-Associated   Tx-Power=20 dBm
          Retry short limit:7   RTS thr:off   Fragment thr:off
          Power Management:off

docker0   no wireless extensions.
```

Figure 7.38 – Verifying the network adapter

As shown in the preceding snippet, the wireless network adapter is displayed as the `wlan0` interface.

3. Next, use the `sudo airmon-ng check kill` command to terminate any conflicting processes that may prevent changing the `wlan0` interface to monitor mode:

```
kali@kali:~$ sudo airmon-ng check kill
[sudo] password for kali:

Killing these processes:

    PID Name
    771 wpa_supplicant
```

Figure 7.39 – Killing processes

4. Next, use the `sudo airmon-ng start wlan0` command to enable monitor mode on the `wlan0` interface:

```
kali@kali:~$ sudo airmon-ng start wlan0

PHY     Interface       Driver          Chipset

phy0    wlan0           ath9k_htc       Qualcomm Atheros Communications AR9271 802.11n
                (mac80211 monitor mode vif enabled for [phy0]wlan0 on [phy0]wlan0mon)
                (mac80211 station mode vif disabled for [phy0]wlan0)
```

Figure 7.40 – Enabling monitor mode

5. Next, use the `iwconfig` command to verify the new interface. `wlan0mon` is created and available on Kali Linux, as shown here:

```
kali@kali:~$ iwconfig
lo        no wireless extensions.

eth0      no wireless extensions.

eth1      no wireless extensions.

docker0   no wireless extensions.

wlan0mon  IEEE 802.11  Mode:Monitor  Frequency:2.457 GHz  Tx-Power=20 dBm
          Retry short limit:7   RTS thr:off   Fragment thr:off
          Power Management:off
```

Figure 7.41 – Verifying the interface status

As shown in the preceding snippet, `wlan0mon` operates in monitor mode, which enables us to capture beacons and probes on wireless networks.

Part 3 – performing wireless reconnaissance

To perform wireless reconnaissance, follow these steps:

1. Use the **Airodump-ng** tool to monitor (capture and analyze) nearly all the beacons and probes by using the following commands:

```
kali@kali:~$ sudo airodump-ng wlan0mon
```

As shown in the following snippet, airodump-ng was able to identify wireless networks, access points, and clients (stations):

```
CH 14 ][ Elapsed: 1 min ][ 2021-09-12 13:10

BSSID              PWR  Beacons    #Data, #/s  CH   MB   ENC CIPHER  AUTH ESSID

9C:3D:CF:          -25      149        2    0   4  540   WPA2 CCMP   PSK  ! ▷_◁ !
68:7F:74:01:28:E1  -36       76        1    0   6  130   WPA2 CCMP   PSK  Corp_Wi-Fi
38:4C:4F:          -72       52       46    0   1  195   WPA2 CCMP   PSK  Digicel_WiFi_T28R
B4:39:39:          -83       26       73    0  11   65   WPA2 CCMP   PSK  Hyundai E504
2C:9D:1E:          -88        9        3    0   7  195   WPA2 CCMP   PSK  Digicel_WiFi_fh4w
80:02:9C:          -92        1        0    0  11  130   WPA2 CCMP   PSK  WLAN11_113CAD
04:C3:E6:           -1        0        2    0   9   -1   WPA              <length:  0>
38:4C:4F:          -88        2        1    0   1  195   WPA2 CCMP   PSK  Doh Study It
A8:2B:CD:          -88        5        0    0  11  130   WPA2 CCMP   PSK  Digicel_WiFi_94J3

BSSID              STATION            PWR   Rate    Lost    Frames  Notes  Probes

(not associated)   98:09:CF:          -38    0 - 1      0        5
68:7F:74:01:28:E1  D8:50:E6:2F:F9:2B  -27    0 - 6      0        5
68:7F:74:01:28:E1  18:31:BF:1A:92:D1  -40    0 - 1      0       25
38:4C:4F:          2C:C5:46:          -84  24e- 1e   1772      103
38:4C:4F:          B0:C0:90:          -86  24e- 1       0        9
38:4C:4F:          B8:C3:85:          -89  24e- 1       0       36
38:4C:4F:          88:29:9C:          -89    0 - 1      0        2
38:4C:4F:          E4:C8:01:          -90  12e- 1       0        6
```

Figure 7.42 – Wireless networks and clients

The following is a description of each field shown in the preceding snippet:

- `BSSID`: This is the **Basic Service Set Identifier** (**BSSID**) or the MAC address of the access point.
- `PWR`: The power level helps ethical hackers to determine the distance between themselves and an access point or client (station). Lowering the power level means you are further away from the target.
- `Beacons`: This field contains the number of beacons that have been sent from access points.
- `#Data`: This field indicates the amount of captured packets on the network.
- `#/s`: This field helps us to determine the number of packets over a 10-second interval.
- `CH`: This field indicates the operating channel of the wireless network on the access point.
- `MB`: This field helps us to determine the maximum speed of the wireless network.
- `ENC`: This field specifics the wireless security standard for the wireless network.
- `AUTH`: This field indicates the authentication type or protocol of the wireless network.
- `ESSID`: This is the **Extended Service Set Identifier** (**ESSID**), which is equivalent to the SSID or network name.
- `STATION`: This field contains the MAC addresses of nearly wireless clients, either connected or not connected to a wireless network.
- `Probes`: This field displays the saved networks from each client from their PNL.

The following snippet shows the station-to-access point associations by observing the `BSSID` and `STATION` fields:

```
BSSID              STATION            PWR   Rate    Lost    Frames  Notes  Probes

9C:3D:CF:          F8:54:B8:          -45   24e- 1e    0        11
9C:3D:CF:          78:BD:BC:          -34    0 - 1e    0         2         Preferred Network List
68:7F:74:01:28:E1  18:31:BF:1A:92:D1  -31   24e- 1     0        77
38:4C:4F:          B0:C0:90:          -82   24e- 1     0        20
38:4C:4F:          E4:C8:01:          -83    5e- 1     0        47         cwc-4361983,cwc - 4361983,
38:4C:4F:          88:9F:6F:          -84   24e- 1     0        52         Digicel_5G_WiFi_37CS
38:4C:4F:          B8:C3:85:          -89   24e- 1     0       146
38:4C:4F:          2C:C5:46:          -93   24e- 1e    0       359
```

Figure 7.43 – Identifying the wireless clients

2. If the targeted wireless network operates on a specific channel, use the `-c` syntax to listen on a specific channel, as shown here:

```
kali@kali:~$ sudo airodump-ng -c 6 wlan0mon
```

As shown in the following snippet, Airodump-ng listens on channel `6` only:

```
CH  6 ][ Elapsed: 42 s ][ 2021-09-12 13:17

BSSID              PWR RXQ  Beacons    #Data, #/s  CH   MB   ENC CIPHER  AUTH ESSID

9C:3D:CF:          -33  16       69        0    0   4  540   WPA2 CCMP   PSK  !ᐅ_ᐊ!
68:7F:74:01:28:E1  -47  96      430        0    0   6  130   WPA2 CCMP   PSK  Corp_Wi-Fi

BSSID              STATION            PWR   Rate    Lost    Frames  Notes  Probes

68:7F:74:01:28:E1  D8:50:E6:2F:F9:2B  -24    1e- 6     0         5
68:7F:74:01:28:E1  18:31:BF:1A:92:D1  -34    1e- 1     0         3
```

Figure 7.44 – Listening on a specific channel

This technique helps ethical hackers to filter out any unwanted output and focus on their target.

3. To set the focus on a specific target, the `--essid` syntax can be used to specify a targeted ESSID network name by using the following command:

```
kali@kali:~$ sudo airodump-ng -c 6 --essid Corp_Wi-Fi wlan0mon
```

As shown in the following snippet, only the **Corp_Wi-Fi** network and its associated clients are displayed:

```
CH  6 ][ Elapsed: 42 s ][ 2021-09-12 13:22

BSSID              PWR RXQ  Beacons    #Data, #/s  CH   MB   ENC CIPHER  AUTH ESSID

68:7F:74:01:28:E1  -44 100       443       37    0   6  130   WPA2 CCMP   PSK  Corp_Wi-Fi

BSSID              STATION            PWR   Rate    Lost    Frames  Notes  Probes

68:7F:74:01:28:E1  D8:50:E6:2F:F9:2B  -25    0 - 6      0        2
68:7F:74:01:28:E1  18:31:BF:1A:92:D1  -29   24e- 1    134       46
```

Figure 7.45 – Filtering the networks

4. Associated stations that transmit data on a wireless network will be displayed under the `STATION` field. However, there may be idle stations that are associated with the target network but not shown on Airodump-ng.

   As an ethical hacker, you can perform a dis-association attack on the target network to force all associated clients to disconnect and reconnect. This technique enables us to identify all associated clients for a specific network. To perform this action, open a new Terminal and use **Aireplay-ng** to send 100 de-authentication wireless frames to the target:

   ```
   kali@kali:~$ sudo aireplay-ng -0 100 -e Corp_Wi-Fi wlan0mon
   ```

   The following snippet shows aireplay-ng sending the de-authentication attack to the target:

```
kali@kali:~$ sudo aireplay-ng -0 100 -e Corp_Wi-Fi wlan0mon
13:28:15  Waiting for beacon frame (ESSID: Corp_Wi-Fi) on channel 6
Found BSSID "68:7F:74:01:28:E1" to given ESSID "Corp_Wi-Fi".
NB: this attack is more effective when targeting
a connected wireless client (-c <client's mac>).
13:28:15  Sending DeAuth (code 7) to broadcast -- BSSID: [68:7F:74:01:28:E1]
13:28:16  Sending DeAuth (code 7) to broadcast -- BSSID: [68:7F:74:01:28:E1]
13:28:16  Sending DeAuth (code 7) to broadcast -- BSSID: [68:7F:74:01:28:E1]
13:28:17  Sending DeAuth (code 7) to broadcast -- BSSID: [68:7F:74:01:28:E1]
13:28:18  Sending DeAuth (code 7) to broadcast -- BSSID: [68:7F:74:01:28:E1]
13:28:18  Sending DeAuth (code 7) to broadcast -- BSSID: [68:7F:74:01:28:E1]
13:28:19  Sending DeAuth (code 7) to broadcast -- BSSID: [68:7F:74:01:28:E1]
13:28:19  Sending DeAuth (code 7) to broadcast -- BSSID: [68:7F:74:01:28:E1]
13:28:20  Sending DeAuth (code 7) to broadcast -- BSSID: [68:7F:74:01:28:E1]
13:28:20  Sending DeAuth (code 7) to broadcast -- BSSID: [68:7F:74:01:28:E1]
```

Figure 7.46 – The de-authentication attack

5. Lastly, by checking the **Airodump-ng** window, you will see all other associated stations and their PNL, as shown here:

```
CH  6 ][ Elapsed: 2 mins ][ 2021-09-12 13:30 ][ PMKID found: 68:7F:74:01:28:E1

BSSID              PWR RXQ  Beacons    #Data, #/s  CH   MB   ENC CIPHER  AUTH ESSID

68:7F:74:01:28:E1  -31 100     1675       139    0   6  130   WPA2 CCMP   PSK  Corp_Wi-Fi

BSSID              STATION            PWR   Rate    Lost    Frames  Notes  Probes

68:7F:74:01:28:E1  D8:50:E6:2F:F9:2B  -28    1e- 1      0       78  PMKID  Corp_Wi-Fi
68:7F:74:01:28:E1  18:31:BF:1A:92:D1  -30    1e- 1      0      123  PMKID
```

Figure 7.47 – Observing stations

Having completed this section, you have learned how to perform active reconnaissance on wireless networks.

Summary

During this chapter, you learned about the importance of active reconnaissance and how it helps ethical hackers to discover live hosts and identify open ports and running services. In addition, you gained the skills to perform reconnaissance on wireless network infrastructure to identify targeted access points and clients, as well as network settings.

I hope this chapter has been informative for you and helpful in your journey in the cybersecurity industry. In the next chapter, *Performing Vulnerability Assessments*, you will learn how to set up and work with vulnerability assessment tools.

Further reading

- How port numbers work: `https://www.techtarget.com/searchnetworking/definition/port-number`
- TCP three-way handshake: `https://www.techopedia.com/definition/10339/three-way-handshake`

8
Performing Vulnerability Assessments

The race between cybersecurity professionals and threat actors is a never-ending marathon as hackers are always developing new techniques to discover and exploit security weaknesses in their targets' systems, and security teams are working continuously to find hidden vulnerabilities and implement countermeasures to protect their assets before a real cyber-attack occurs. While software and product vendors frequently push security updates and patches to their applications, there's no one system that's fully secure from threats and attacks. Hence, it's important that organizations perform regular ethical hacking and penetration testing assessments on their systems and networks to find any hidden security flaws and determine whether their security controls are effective in preventing real attacks and threats.

In this chapter, you will learn how vulnerability management helps organizations to improve their security posture and techniques that are commonly used by cybersecurity professionals such as ethical hackers to efficiently discover various security flaws on their target systems. Furthermore, you will learn how to install and set up common vulnerability management tools to help you identify and analyze security flaws on a system as an ethical hacker.

In this chapter, we will cover the following topics:

- The importance of vulnerability management
- Working with Nessus
- Using Greenbone Vulnerability Manager
- Vulnerability discovery with Nmap

Let's dive in!

Technical requirements

To follow along with the exercises in this chapter, please ensure that you have met the following hardware and software requirements:

- Kali Linux: `https://www.kali.org/get-kali/`
- Metasploitable 3
- Nessus Essentials: `https://www.tenable.com/downloads/nessus`

The importance of vulnerability management

A **vulnerability** is a security weakness that exists on a system that can be exploited by anyone, such as a threat actor, to compromise the confidentiality, integrity, and/or availability of the targeted system. Ethical hackers and penetration testers are usually hired by organizations to simulate real-world cyber-attacks on their systems and networks to identify hidden security flaws that are not easily found by the company's internal IT team. Ethical hackers commonly use the same **Tactics, Techniques, and Procedures** (**TTPs**) as real adversaries to efficiently find hidden security weaknesses. By identifying security vulnerabilities within an organization, the security team can better understand how a real attacker can gain unauthorized access to their networks. In addition, the organization can improve their threat detection, implement countermeasures to prevent threats, and improve their incident response plan.

As an ethical hacker, you'll be using various tools and techniques to discover security vulnerabilities, determine the severity and impact if the targeted host were to be compromised, and provide recommendations on how to resolve the security flaws to improve the overall security posture of the organization.

The following are common data sources used by cybersecurity professionals and researchers to collect information about new and emerging threats on the internet:

- **Vendor website**: The vendor website of a product usually contains information about known security issues that directly affect their products, devices, and applications
- **Vulnerability feeds**: These are usually news feeds that provide information about new security vulnerabilities found by the global community
- **Conferences**: These are security-focused conferences where cybersecurity professionals and researchers discuss new vulnerabilities, malware, and hacking techniques
- **Academic journals**: Security researchers within the academic community provide their research findings on various cyber-attacks and threats
- **Threat feeds**: These are data sources that provide real-time information to the public
- **MITRE ATT&CK**: The MITRE ATT&CK framework provides information on the TTPs used by adversaries to compromise their targets

In addition, ethical hackers and penetration testers can leverage the information found on the following open vulnerability databases to improve their vulnerability detection and classification:

- **Common Vulnerability Scoring System** (**CVSS**): `https://www.first.org/cvss/`
- **Common Vulnerabilities and Exposures** (**CVE**): `https://cve.mitre.org/`
- **National Vulnerability Database** (**NVD**): `https://nvd.nist.gov/`
- **Common Weakness Enumeration** (**CWE**): `https://cwe.mitre.org/`

The following types and classifications of security vulnerabilities can be found on a system:

- **Default settings**: These are the configurations on systems and devices that were implemented by the vendor and shipped to the customer. Default configurations are basic settings on the device to help the customer easily access the device to perform additional device configurations.
- **Weak encryption**: Some systems contain weak encryption protocols and algorithms that can be exploited by a threat actor to gain access to confidential information and unauthorized entry into a targeted device.
- **Unsecure protocols**: There are many unsecure network protocols that transmit data in plaintext. These unsecure protocols do not encrypt any data before placing the packet on a network.
- **Open permission**: There are many files, folders, network shares, and devices that provide full access to anyone. This type of vulnerability can lead to a malicious user leveraging full control of a system or network share.
- **Unsecure root accounts**: If the root account on a Linux-based operating system is not secured using complex password policies and **Multi-Factor Authentication** (**MFA**), a threat actor can gain full administrative control of the system. Similarly, the *administrator* account on Windows-based operating systems should be secured.
- **Open ports and services**: Sometimes a system may be running a lot of unnecessary services. When some applications are running, it operates as a service on the operating system. Some applications on a targeted host may not be fully patched and may contain security vulnerabilities that can be exploited over a network. In addition, some services may open a network port to allow inbound connections from a remote system. A threat actor can launch a remote exploit across the network to take advantage of the vulnerable service on a targeted system.
- **Zero days**: These are vulnerabilities that are discovered and exploited by hackers before the vendor is made aware of the security issue and has time to release a security update to mitigate the threat. Zero days are considered to be a nightmare for security teams as there are no known security fixes until the vendor provides information disclosure about the issue.
- **Legacy systems**: Legacy systems often run **End-of-Life** (**EoL**) operating systems, firmware, applications, and hardware components that are no longer supported by the vendor. As a result

of running legacy systems on a network, they are most vulnerable to modern cyber-attacks and threats.

Vulnerability management life cycle

Vulnerability management is a continuous life cycle of identifying new security flaws in systems to implement mitigation and countermeasures to reduce the attack surface and improve the security posture of the organization.

The following diagram shows the general life cycle of vulnerability management:

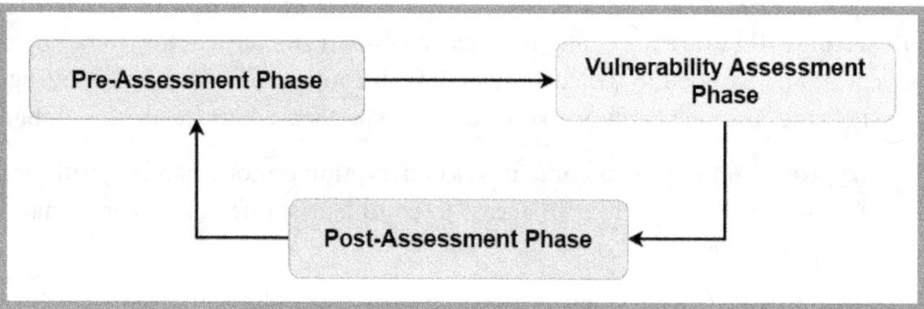

Figure 8.1 – Vulnerability management life cycle

Let's explore these phases in detail.

Pre-assessment phase

The *pre-assessment phase* focuses on identifying all assets owned by the organization. If an organization installs new devices on their network without keeping track of this system, there's the likelihood the IT team may not be managing or monitoring the device for security vulnerabilities, and this creates a risk of a cyber-attack. Additionally, a baseline of normal operations should be created on all critical assets, such as switches, routers, servers, and firewalls. A baseline is used to measure what is considered to be normal and abnormal behavior of a system or network.

During this phase, it's important to gain a solid understanding of the organization and its business processes. Sometimes an organization's business processes may be creating a security risk and increase the chance a hacker is able to compromise the company's systems and network. Identifying security vulnerabilities within business processes can help the leadership team to better understand the risk of doing specific actions.

In addition, the pre-assessment phase helps the ethical hacker to determine what and whether there are any security controls in place to prevent potential attacks and threats from hackers. Such information is useful in better understanding the attack surface, security posture, and cyber defenses of the company. During this phase, it's important to identify whether the organization has implemented any regulatory standards and frameworks in their operating industry and whether these frameworks are enforced.

Lastly, the pre-assessment phase is used to better understand the scope of the vulnerability assessment and targets and obtain legal permission from the authorities as vulnerability scanning can be intrusive at times.

Vulnerability assessment phase

The vulnerability assessment phase focuses on performing different types of automated and manual testing on targeted systems within the scope to identify security weaknesses and determine their severity and impact if they are compromised. In addition, it's important for ethical hackers to check each vulnerability to determine whether it's a false positive or false negative during an automated scan.

False positives are simply alerts on a vulnerability scanner that indicate a vulnerability is present on a system but does not actually exist after manual testing is done. A false negative is when the vulnerability scanner does not detect a known or hidden vulnerability on a target, but a security vulnerability was found after manual testing. Hence, it's important to validate the authenticity of each security vulnerability on a system with manual testing. Lastly, during each vulnerability assessment, the ethical hacker will collect sufficient data that will be used for analysis and reporting.

Post-assessment phase

During the post-assessment phase, the ethical hacker uses the information collected during the vulnerability assessment phase to perform a risk assessment on the assets and organization and categorize each risk based on their severity level and impact on the organization. In addition, the analyzed data is used to determine the threat level of the organization and the risk of being compromised by a real threat actor.

The ethical hacker also provides recommendations on how to mitigate each risk by implementing countermeasures and security controls in the organization. It's important to consider giving priority to high-severity risks and allocating resources to resolve them quickly, as these vulnerabilities usually have a higher impact than others.

When a security vulnerability is found, it's good practice to perform a **root cause analysis** to determine the root causes of the security risk and vulnerability on the target. This process usually helps ethical hackers to better identify the appropriate solutions to prevent this problem from re-occurring in the future.

Based on the recommendations, the internal security or IT team of the organization should implement any fixes, updates and patches, and security controls to prevent a real attack from occurring and improve the security posture. Furthermore, it's essential to keep a record of all the lessons learned during each vulnerability assessment as it can be useful in the future.

Lastly, continuous monitoring and verification are required to determine whether the recommendations are enforced and effective in preventing future cyber-attacks and threats from adversaries. This means regular vulnerability scanning and monitoring of logs using security solutions such as **Intrusion**

Detection Systems (**IDSs**), **Intrusion Prevention Systems** (**IPSs**), and **Security Information and Event Management** (**SIEM**).

Let's not forget, the ethical hacker will analyze all the collected data during the vulnerability assessment to generate both an executive and technical report for the organization. The executive report focuses on the high-level overview of the assessment and it's presented to the executive or leadership team of the company, who are usually non-technical employees. Therefore, the executive report should not contain any jargon or too many technical terms that are confusing to non-technical staff members. The technical report contains the specifics about the vulnerability assessment, such as the tools and techniques that were used to identify and test the validity of each vulnerability.

Working with Nessus

Nessus is one of the most popular vulnerability scanners within the cybersecurity industry. It enables system administrators, cybersecurity professionals, and compliance auditors to identify whether a targeted system contains any known security weaknesses and meets various regulatory standards in the industry. Additionally, Nessus can be deployed on a centralized server and configured to perform regular vulnerability scans on targeted systems using its automation capabilities, hence enabling ethical hackers to easily perform a gap analysis between vulnerability assessments to determine whether the security team of an organization is implementing countermeasures to mitigate and prevent potential cyber-attacks.

As an aspiring ethical hacker and penetration tester, it's essential to gain hands-on experience in setting up Nessus and performing vulnerability scans and assessments on a targeted system. To get started with setting up Nessus, please use the following instructions.

> **Important note**
> There were some challenges when setting up Nessus on Kali Linux using Parallels on the M1 Mac (ARM64) chip. However, the setup process works fine on a Windows-based machine.

Part 1 – setting up Nessus

In this exercise, you will learn how to install Nessus Essentials on Kali Linux to perform vulnerability discovery on Metasploitable 3, our targeted system:

1. Firstly, power on the Kali Linux and Metasploitable 3 virtual machines.
2. Next, open a web browser, go to `https://www.tenable.com/products/nessus/nessus-essentials`, and register for a free Nessus Essentials license key. Upon registration, Tenable will send a unique Nessus Essentials license key to the email address you've provided.

3. On Kali Linux, open Terminal and update the software packages repository list using the following command:

```
kali@kali:~$ sudo apt update
```

4. Next, use the following command to download the Nessus software onto Kali Linux:

```
kali@kali:~$ curl --request GET \
  --url 'https://www.tenable.com/downloads/api/v2/pages/nessus/
files/Nessus-10.5.1-debian10_amd64.deb' \
  --output 'Nessus-10.5.1-debian10_amd64.deb'
```

The following figure shows the execution of the preceding commands:

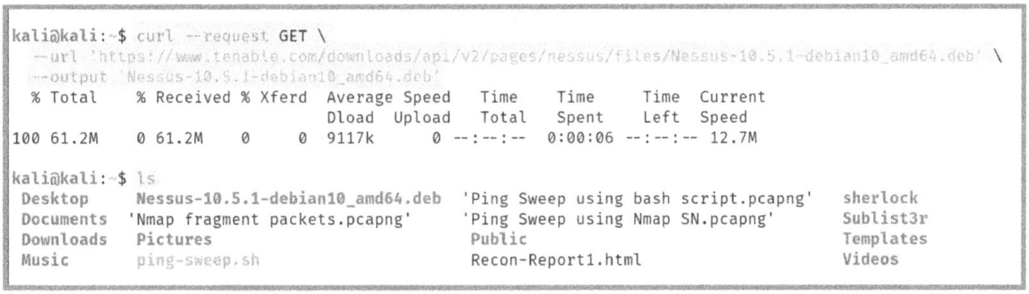

```
kali@kali:~$ curl --request GET \
  --url 'https://www.tenable.com/downloads/api/v2/pages/nessus/files/Nessus-10.5.1-debian10_amd64.deb' \
  --output 'Nessus-10.5.1-debian10_amd64.deb'
  % Total    % Received % Xferd  Average Speed   Time    Time     Time  Current
                                 Dload  Upload   Total   Spent    Left  Speed
100 61.2M    0 61.2M    0     0  9117k      0 --:--:--  0:00:06 --:--:-- 12.7M

kali@kali:~$ ls
 Desktop     Nessus-10.5.1-debian10_amd64.deb  'Ping Sweep using bash script.pcapng'   sherlock
 Documents  'Nmap fragment packets.pcapng'     'Ping Sweep using Nmap SN.pcapng'       Sublist3r
 Downloads   Pictures                           Public                                 Templates
 Music       ping-sweep.sh                      Recon-Report1.html                     Videos
```

Figure 8.2 – Downloading Nessus

Alternatively, you can download the Nessus software from `https://www.tenable.com/downloads/nessus` for **Linux - Debian - AMD64** (Kali Linux).

5. Next, install the Nessus package using the following command:

```
kali@kali:~$ sudo dpkg -i Nessus-10.5.1-debian10_amd64.deb
```

The following figure shows the beginning of the installation process:

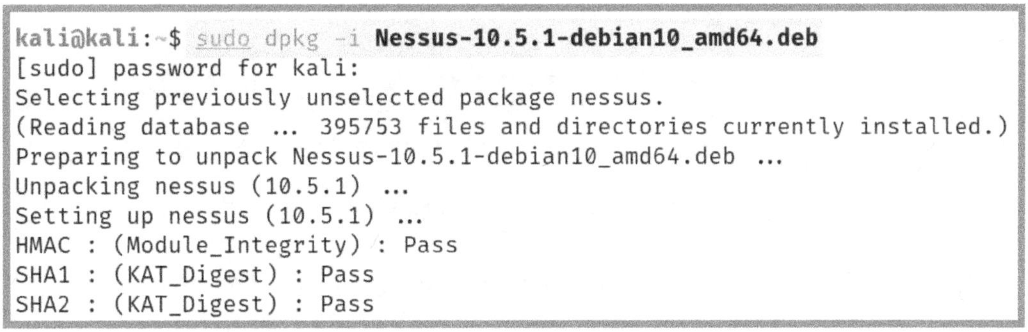

```
kali@kali:~$ sudo dpkg -i Nessus-10.5.1-debian10_amd64.deb
[sudo] password for kali:
Selecting previously unselected package nessus.
(Reading database ... 395753 files and directories currently installed.)
Preparing to unpack Nessus-10.5.1-debian10_amd64.deb ...
Unpacking nessus (10.5.1) ...
Setting up nessus (10.5.1) ...
HMAC : (Module_Integrity) : Pass
SHA1 : (KAT_Digest) : Pass
SHA2 : (KAT_Digest) : Pass
```

Figure 8.3 – Installation process

6. Next, start and restart the Nessus service on Kali Linux:

```
kali@kali:~$ sudo /bin/systemctl start nessusd.service
kali@kali:~$ sudo /bin/systemctl restart nessusd.service
```

7. Once the Nessus service has started, open the web browser within Kali Linux and go to `https://kali:8834/` to access the Nessus web interface and ensure you have accepted the security warnings, as shown:

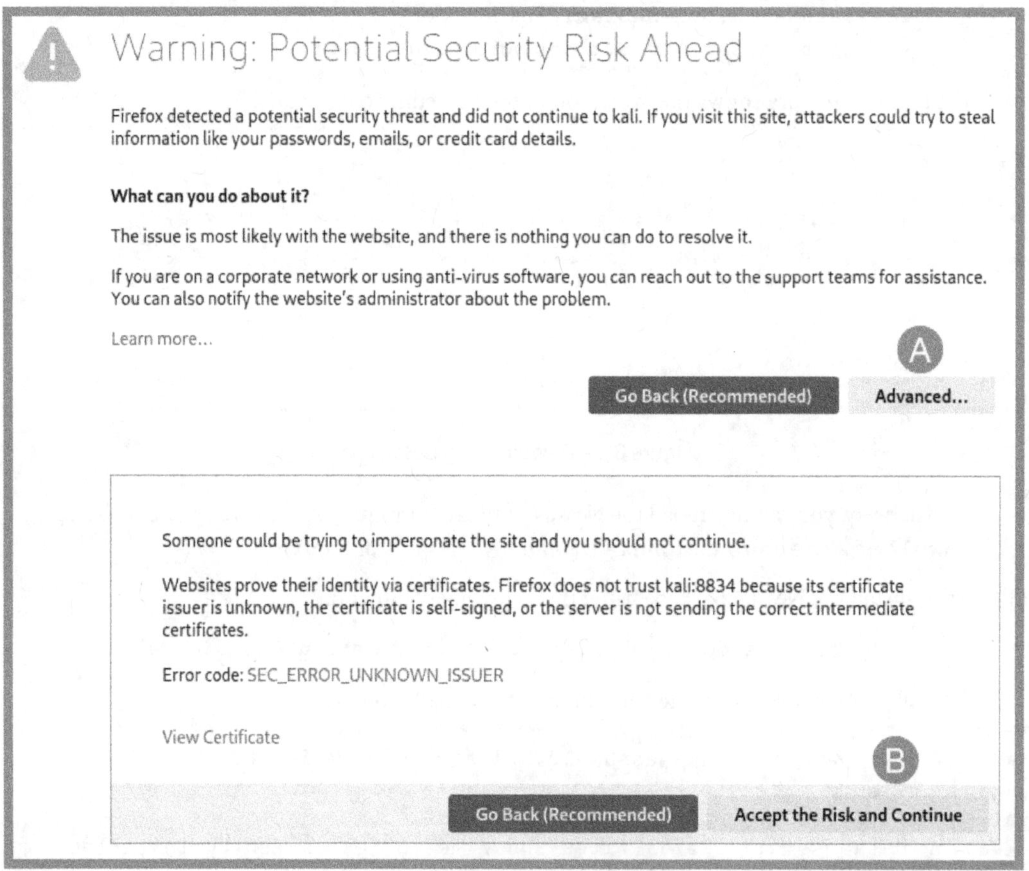

Figure 8.4 – Security warnings

8. Next, you'll be presented with the Nessus initialization page. Click on **Continue**:

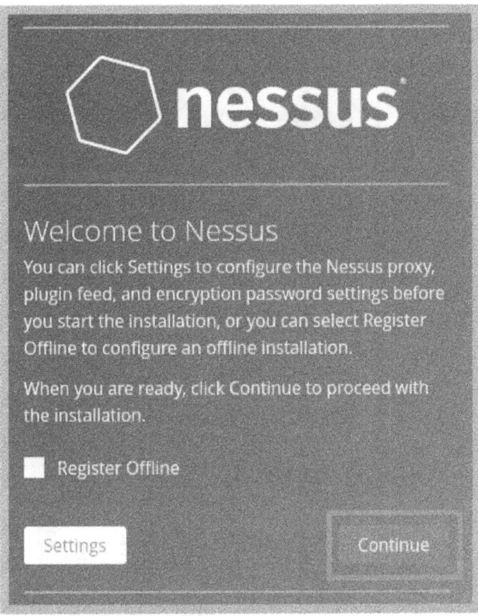

Figure 8.5 – Nessus initialization page

9. Next, select **Register for Nessus Essentials** and click on **Continue**:

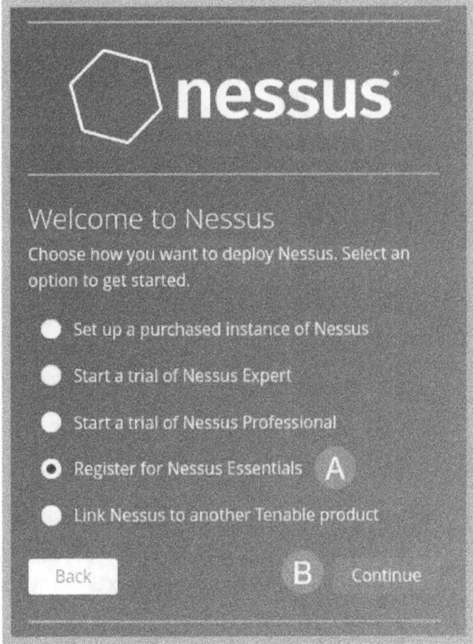

Figure 8.6 – Nessus deployment

10. Since you already registered for a Nessus Essentials license key in *step 2*, click on **Skip** in the registration window shown here:

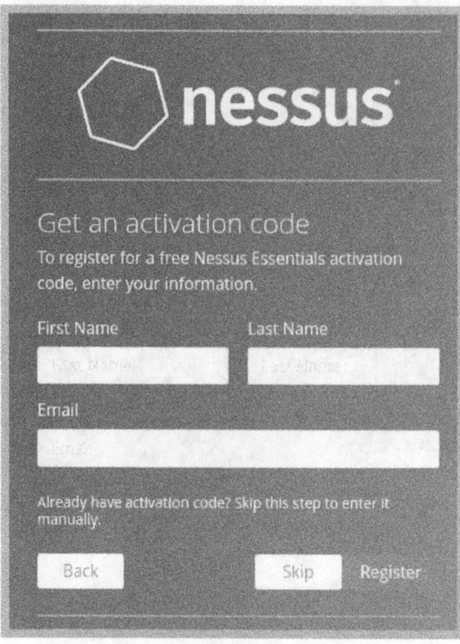

Figure 8.7 – Nessus registration window

11. Next, copy the license key from your email and paste it into the **Activation Code** field, then click on **Continue**, as shown:

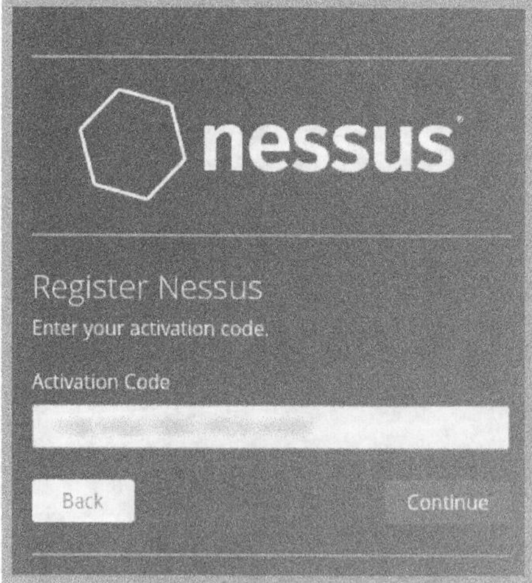

Figure 8.8 – Activation Code

12. Next, Nessus will show you the activation code/license information. Click on **Continue**:

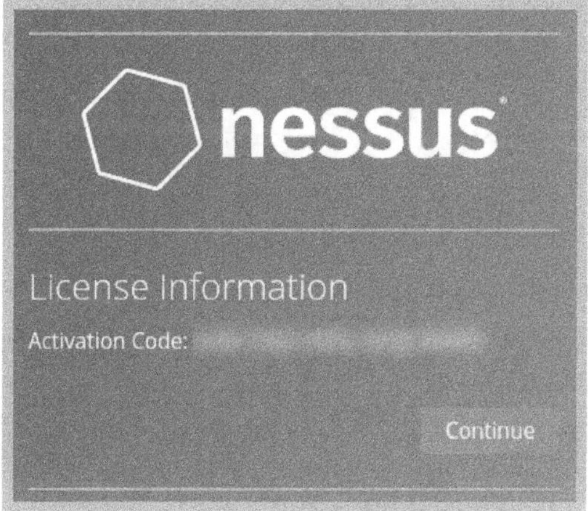

Figure 8.9 – License Information

13. Next, create your user account and click on **Submit**:

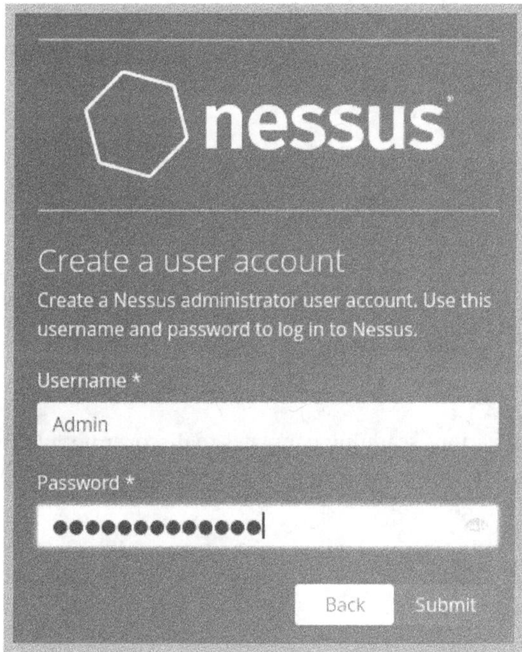

Figure 8.10 – Creating a user account

14. Next, Nessus will begin its initialization process and download additional update plugins. This process usually takes a few minutes depending on your internet service. After the initialization process is completed, Nessus will automatically log in to the user dashboard.
15. Once you log in for the first time, Nessus takes a while to compile all the plugins before you are able to perform any scans.

Part 2 – scanning using Nessus

1. To begin your first vulnerability scan using Nessus Essentials, click on **New Scan**:

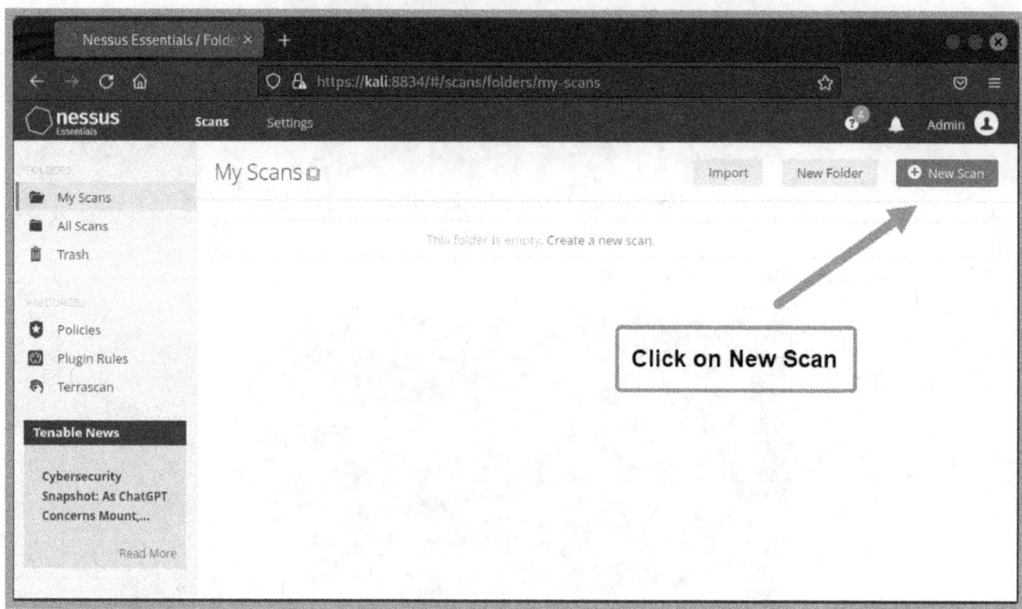

Figure 8.11 – Starting a new scan

2. Next, you'll be shown various scanning templates to discover popular security vulnerabilities, such as PrintNightmare, WannaCry, and Log4Shell, on a targeted system. Click on **Basic Network Scan**:

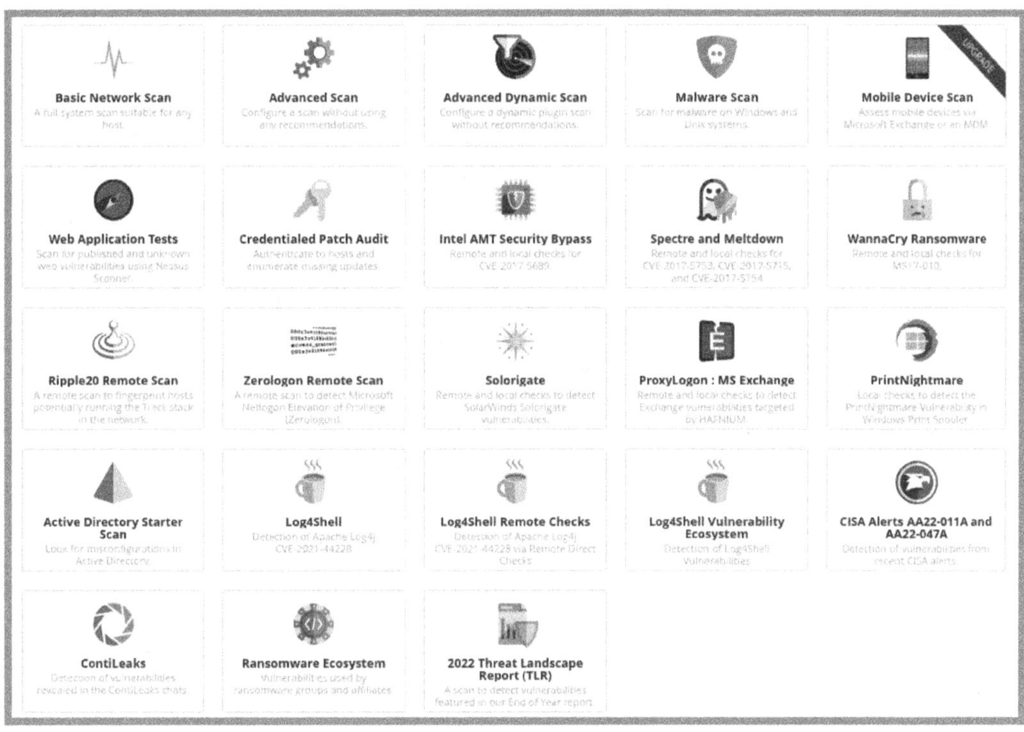

Figure 8.12 – Scanning templates

3. On the **Settings** page, fill in the required fields, such as setting a name for the scan and the IP address of the Metasploitable 3 virtual machine as the targeted system, and click on **Launch**:

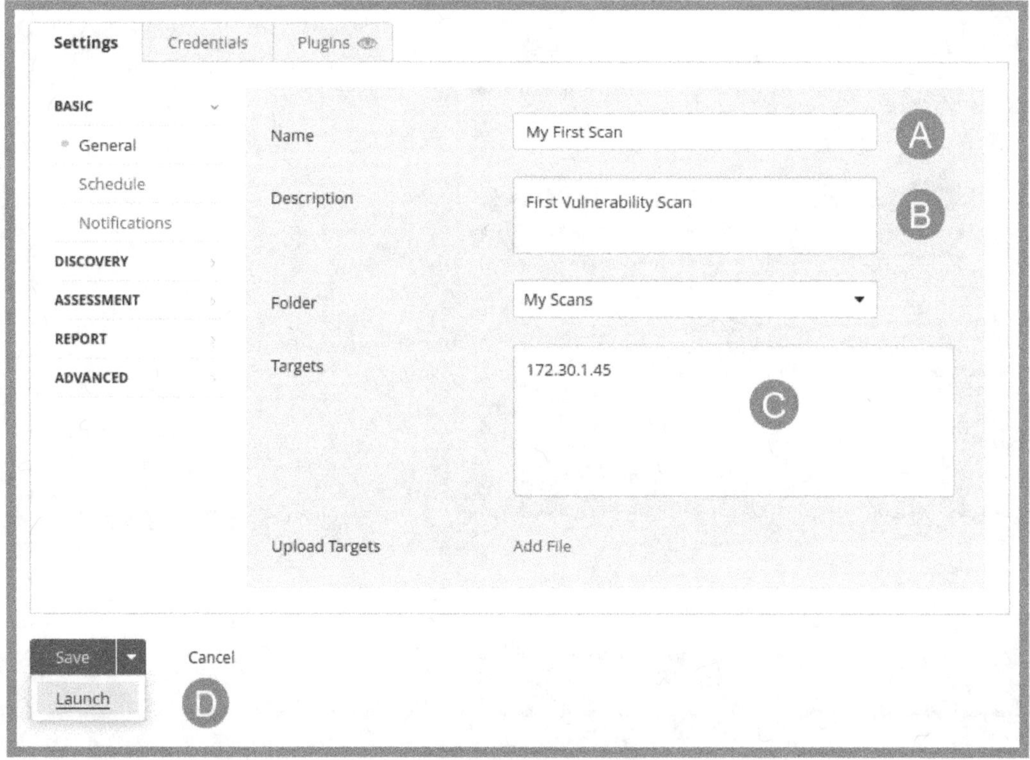

Figure 8.13 – First scan

4. Next, the scan progress will be available on the **My Scans** summary window, as shown:

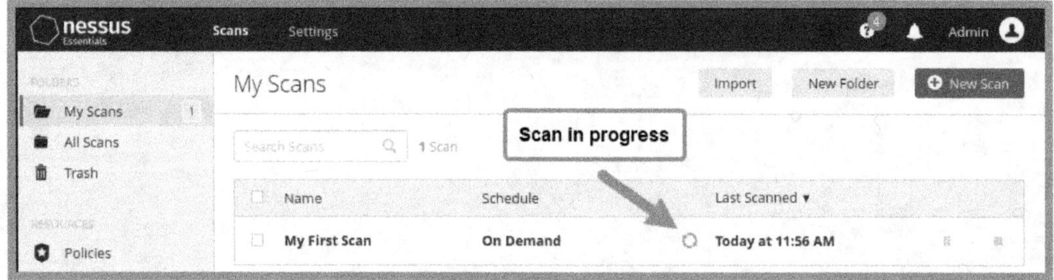

Figure 8.14 – Scan progress

Once the scan is completed, the scan status will automatically update, as shown in the following figure:

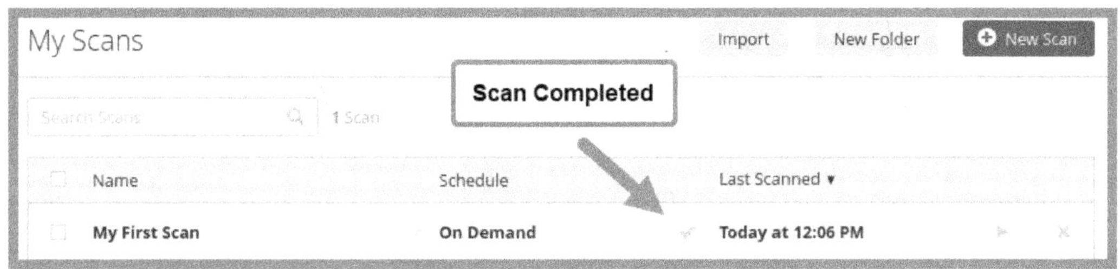

Figure 8.15 – Scan completion

As an ethical hacker, you are often tasked with performing regular vulnerability assessments over multiple assets. Nessus enables you to automate different scan types, allowing you to configure asset and vulnerability discovery techniques, assessment types, and reporting.

Part 3 – vulnerability analysis

1. To view the scan results, click on **My Scan** | **My First Scan** to show the summary:

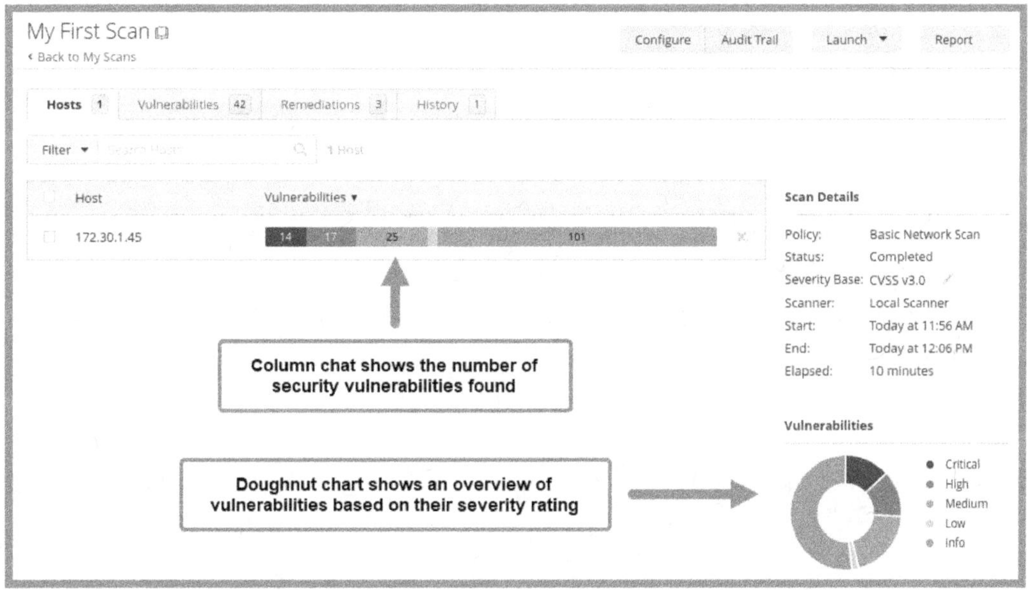

Figure 8.16 – Scan results

As shown in the preceding figure, the column chart shows the number of security vulnerabilities found based on their severity levels. The doughnut chart shows an overview of vulnerabilities.

2. Next, click on the column chart to view a list of all security vulnerabilities on the targeted host:

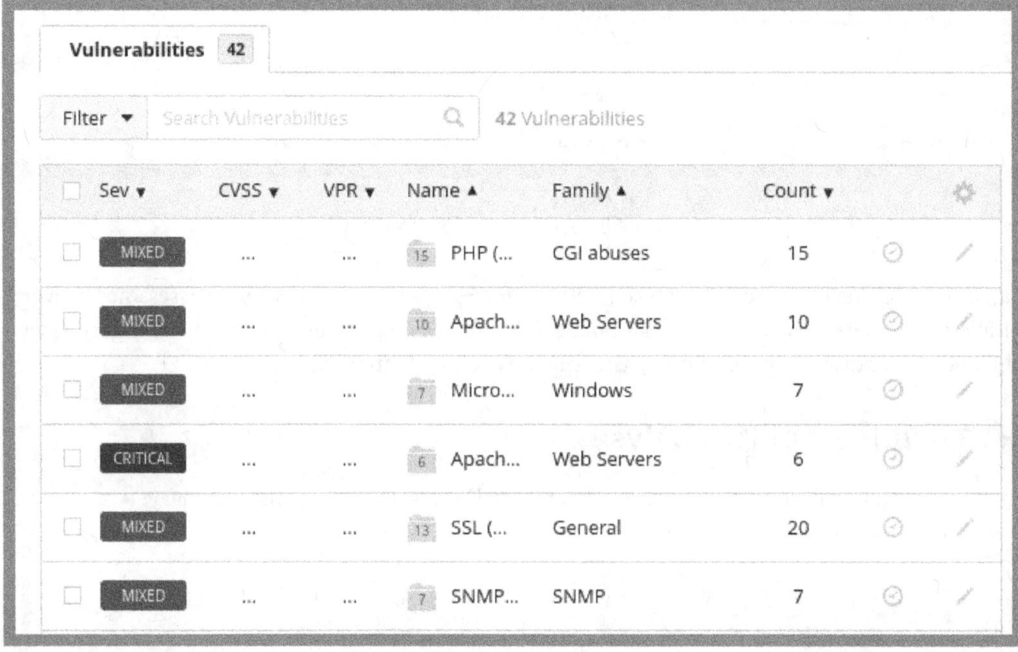

Figure 8.17 – List of vulnerabilities

As shown in the preceding figure, there are multiple security vulnerabilities that are grouped within each severity level.

3. Next, click on the **CRITICAL** severity level to view a list of all critical security vulnerabilities on the host:

Vulnerabilities 42

Search Vulnerabilities 6 Vulnerabilities

Sev ▾	CVSS ▾	VPR ▾	Name ▴	Family ▴	Count ▾
CRITICAL	9.8	9.0	Apache 2.4...	Web Servers	1
CRITICAL	9.8	7.4	Apache < 2...	Web Servers	1
CRITICAL	9.8	7.4	Apache 2.4...	Web Servers	1
CRITICAL	9.8	7.4	Apache 2.4...	Web Servers	1
CRITICAL	9.0	8.1	Apache < 2...	Web Servers	1
CRITICAL	9.0	7.3	Apache 2.4...	Web Servers	1

Figure 8.18 – Critical security vulnerabilities

As shown in the preceding figure, Nessus was able to evaluate each vulnerability severity rating using CVSS, an open scoring calculator that's used by cybersecurity professionals and researchers to determine the severity level and risk of a vulnerability. Similar to CVSS, Tenable uses their **Vulnerability Priority Rating (VPR)** scoring system to determine security risks.

Important note

The severity ratings help cybersecurity professionals during their decision-making processes to prioritize resources toward more severe risks and vulnerabilities that will create a higher impact. Hence, it is important to always determine the severity rating of each security vulnerability.

4. Next, click on any of the critical security vulnerabilities to view more detail:

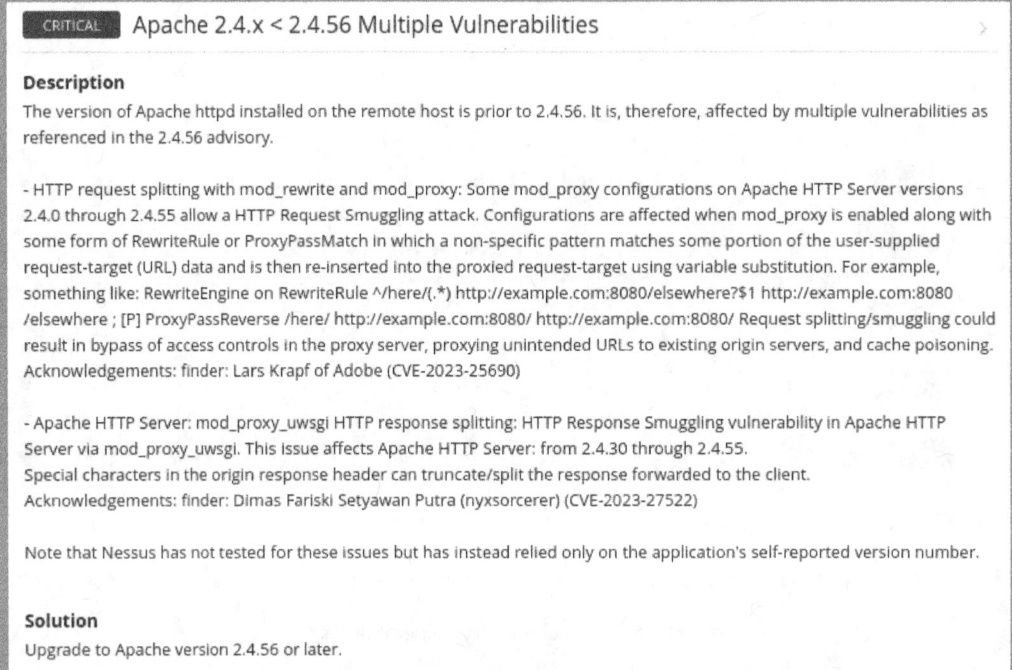

CRITICAL
Apache 2.4.x < 2.4.56 Multiple Vulnerabilities
Description
The version of Apache httpd installed on the remote host is prior to 2.4.56. It is, therefore, affected by multiple vulnerabilities as referenced in the 2.4.56 advisory.
- HTTP request splitting with mod_rewrite and mod_proxy: Some mod_proxy configurations on Apache HTTP Server versions 2.4.0 through 2.4.55 allow a HTTP Request Smuggling attack. Configurations are affected when mod_proxy is enabled along with some form of RewriteRule or ProxyPassMatch in which a non-specific pattern matches some portion of the user-supplied request-target (URL) data and is then re-inserted into the proxied request-target using variable substitution. For example, something like: RewriteEngine on RewriteRule ^/here/(.*) http://example.com:8080/elsewhere?$1 http://example.com:8080 /elsewhere ; [P] ProxyPassReverse /here/ http://example.com:8080/ http://example.com:8080/ Request splitting/smuggling could result in bypass of access controls in the proxy server, proxying unintended URLs to existing origin servers, and cache poisoning. Acknowledgements: finder: Lars Krapf of Adobe (CVE-2023-25690)
- Apache HTTP Server: mod_proxy_uwsgi HTTP response splitting: HTTP Response Smuggling vulnerability in Apache HTTP Server via mod_proxy_uwsgi. This issue affects Apache HTTP Server: from 2.4.30 through 2.4.55.
Special characters in the origin response header can truncate/split the response forwarded to the client.
Acknowledgements: finder: Dimas Fariski Setyawan Putra (nyxsorcerer) (CVE-2023-27522)
Note that Nessus has not tested for these issues but has instead relied only on the application's self-reported version number.
Solution
Upgrade to Apache version 2.4.56 or later.

Figure 8.19 – Security vulnerability detail

As shown in the preceding figure, Nessus provides a full description of the security vulnerability and recommendations on how to mitigate or resolve the issue.

5. Furthermore, the vulnerability details page also shows the metrics that were used to determine the CVSS score, as shown:

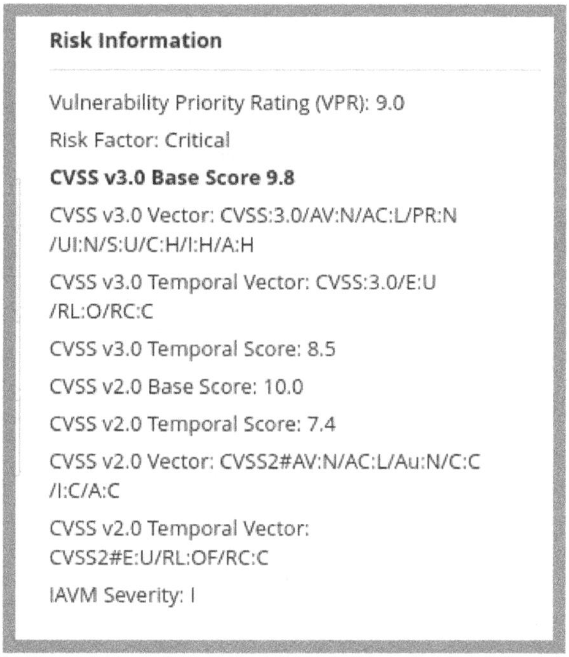

Risk Information

Vulnerability Priority Rating (VPR): 9.0

Risk Factor: Critical

CVSS v3.0 Base Score 9.8

CVSS v3.0 Vector: CVSS:3.0/AV:N/AC:L/PR:N
/UI:N/S:U/C:H/I:H/A:H

CVSS v3.0 Temporal Vector: CVSS:3.0/E:U
/RL:O/RC:C

CVSS v3.0 Temporal Score: 8.5

CVSS v2.0 Base Score: 10.0

CVSS v2.0 Temporal Score: 7.4

CVSS v2.0 Vector: CVSS2#AV:N/AC:L/Au:N/C:C
/I:C/A:C

CVSS v2.0 Temporal Vector:
CVSS2#E:U/RL:OF/RC:C

IAVM Severity: I

Figure 8.20 – CVSS metrics

As shown in the preceding figure, Nessus performed calculations using several different versions of CVSS.

6. Let's use the following CVSS 3.0 vectors to determine how a hacker would be able to compromise the target. Copy the following code:

```
CVSS:3.0/AV:N/AC:L/PR:N/UI:N/S:U/C:H/I:H/A:H
```

7. Next, append the CVSS vectors at the end of the following URL:

   `https://www.first.org/cvss/calculator/3.0#`

   This creates the following final URL with the CVSS vectors:

   `https://www.first.org/cvss/calculator/3.0#CVSS:3.0/AV:N/AC:L/PR:N/UI:N/S:U/C:H/I:H/A:H`

8. Next, go to the final URL using your web browser to view the actual metrics that were used to determine the vulnerability score, as shown:

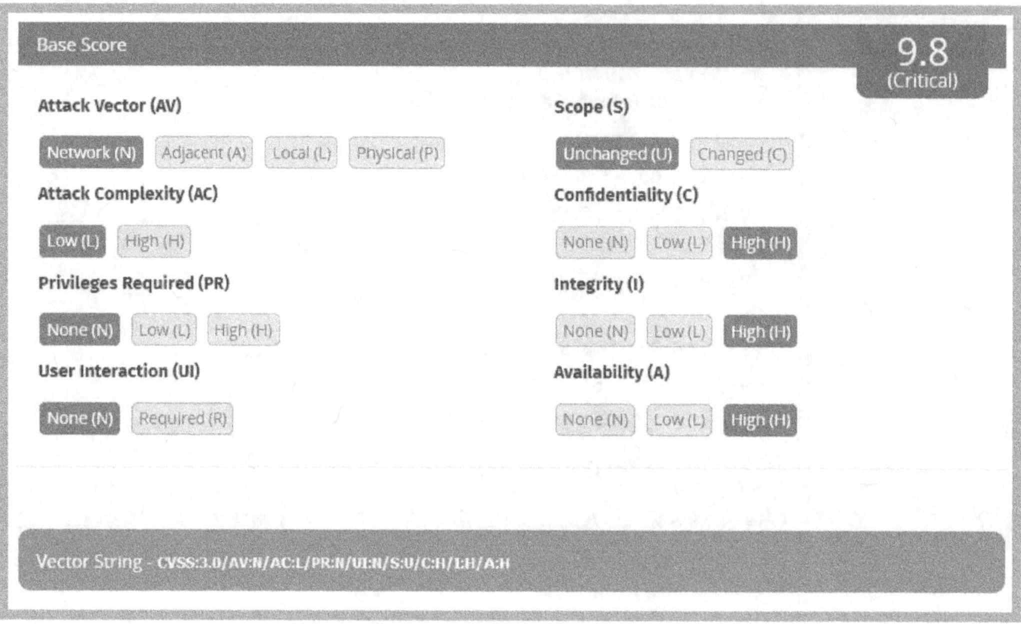

Figure 8.21 – CVSS base score

According to the preceding figure, a hacker will need to design an exploit that's deliverable over a network with **Low (L)** complexity to the target. The exploit requires no higher privileges and no user interactions to be executed on the host. The scope will remain **Unchanged (U)** while the exploit is running, and it will have a **High (H)** impact on the confidentiality, integrity, and availability of the host.

> **Tip**
>
> To learn more about CVSS, please visit `https://www.first.org/cvss/`.

9. To generate a vulnerability report from Nessus, in the top-right corner, click on **Report**.
10. Next, a pop-up window will appear. Select the report format and template and click on **Generate Report**, as shown:

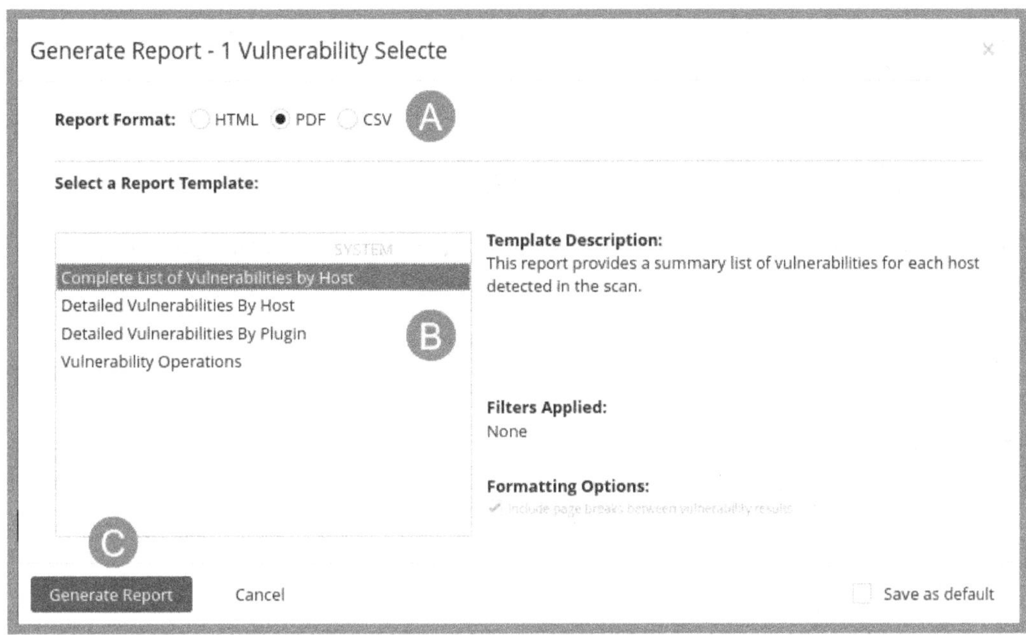

Figure 8.22 – Generate Report

Once the report is generated, save it to your desktop and open it using a PDF reader, as shown:

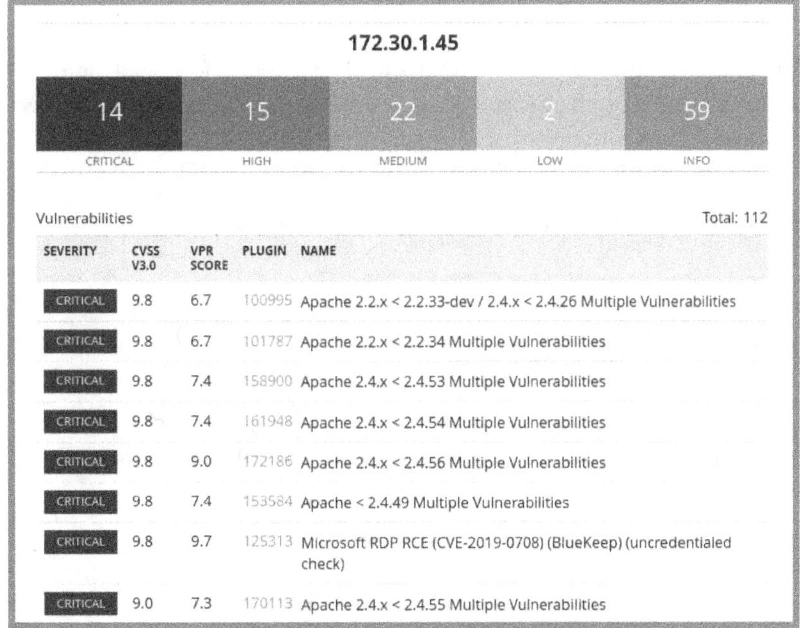

172.30.1.45

14	15	22	2	59
CRITICAL	HIGH	MEDIUM	LOW	INFO

Vulnerabilities Total: 112

SEVERITY	CVSS V3.0	VPR SCORE	PLUGIN	NAME
CRITICAL	9.8	6.7	100995	Apache 2.2.x < 2.2.33-dev / 2.4.x < 2.4.26 Multiple Vulnerabilities
CRITICAL	9.8	6.7	101787	Apache 2.2.x < 2.2.34 Multiple Vulnerabilities
CRITICAL	9.8	7.4	158900	Apache 2.4.x < 2.4.53 Multiple Vulnerabilities
CRITICAL	9.8	7.4	161948	Apache 2.4.x < 2.4.54 Multiple Vulnerabilities
CRITICAL	9.8	9.0	172186	Apache 2.4.x < 2.4.56 Multiple Vulnerabilities
CRITICAL	9.8	7.4	153584	Apache < 2.4.49 Multiple Vulnerabilities
CRITICAL	9.8	9.7	125313	Microsoft RDP RCE (CVE-2019-0708) (BlueKeep) (uncredentialed check)
CRITICAL	9.0	7.3	170113	Apache 2.4.x < 2.4.55 Multiple Vulnerabilities

Figure 8.23 – Nessus PDF report

Having completed this section, you have learned how to set up and use Nessus for vulnerability discovery. In the next section, you will learn how to use an open source vulnerability management tool to discover security flaws in a targeted system.

Using Greenbone Vulnerability Manager

Greenbone Vulnerability Manager (**GVM**) is an open source vulnerability management tool that enables ethical hackers to leverage **Cyber Threat Intelligence** (**CTI**) to efficiently discover security weaknesses in targeted systems.

To get started with setting up GVM, please use the following instructions.

Part 1 – setting up GVM

1. Firstly, power on the Kali Linux and Metasploitable 3 virtual machines.
2. Next, open Terminal on Kali Linux and use the following commands to update the software packages repository list and install GVM:

```
kali@kali:~$ sudo apt update
kali@kali:~$ sudo apt install gvm
```

3. After the installation is completed, use the following commands to initiate the GVM setup process:

```
kali@kali:~$ sudo gvm-setup
```

   This setup process usually takes a few minutes to complete as it downloads additional plugins and updates. Once the setup process is completed, the *admin* account will be created with a random password, as shown:

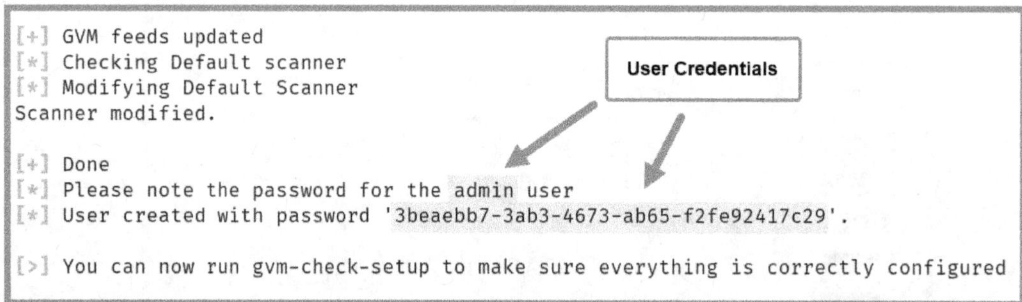

Figure 8.24 – GVM setup

4. Next, execute the `sudo gvm-check-setup` command to ensure GVM is set up correctly.
5. Once everything is configured correctly, open the web browser on Kali Linux, go to `https://127.0.0.1:9392`, and accept the security risks to access the GVM web interface.

6. Ensure you use the default username (*admin*) and the password that was generated at the end of the setup process and click on **Sign In**, as shown:

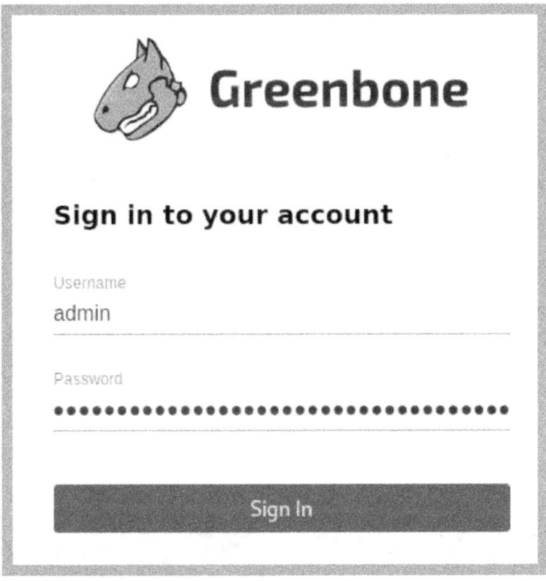

Figure 8.25 – GVM sign-in page

7. Once you're logged in to GVM, click on **Administration** | **Feed status**, as shown:

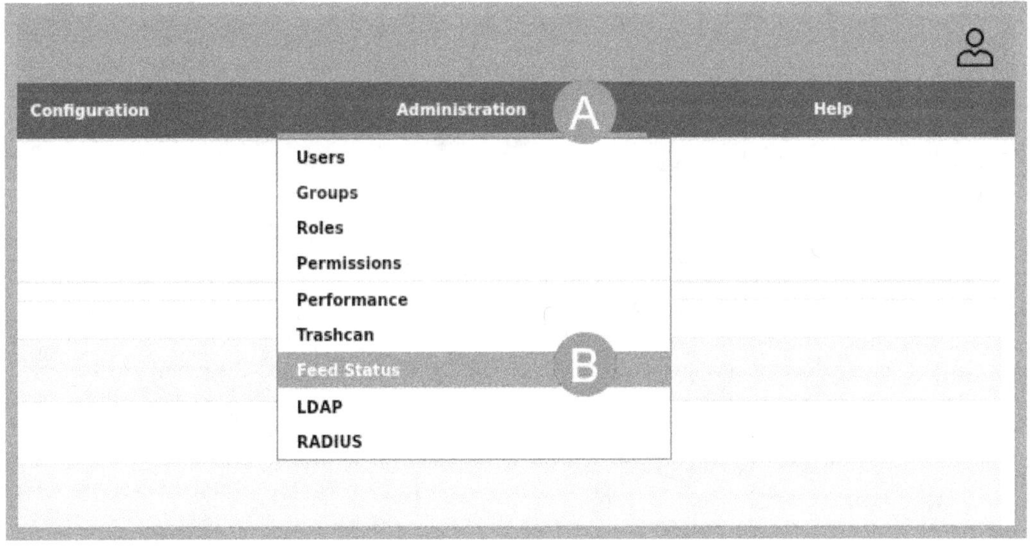

Figure 8.26 – GVM web interface

GVM will continue to download and install additional CTI from various data sources to ensure GVM is able to identify security vulnerabilities on targeted systems using the latest CTI data available:

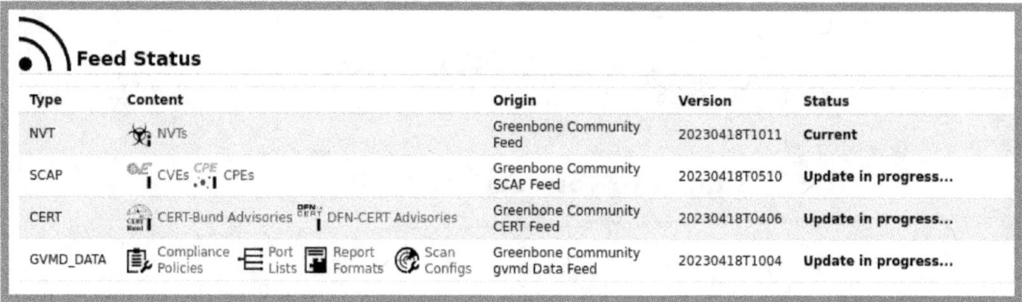

Feed Status

Type	Content	Origin	Version	Status
NVT	NVTs	Greenbone Community Feed	20230418T1011	Current
SCAP	CVEs CPEs	Greenbone Community SCAP Feed	20230418T0510	Update in progress...
CERT	CERT-Bund Advisories DFN-CERT Advisories	Greenbone Community CERT Feed	20230418T0406	Update in progress...
GVMD_DATA	Compliance Policies Port Lists Report Formats Scan Configs	Greenbone Community gvmd Data Feed	20230418T1004	Update in progress...

Figure 8.27 – Feed Status

The download process usually takes quite some time to be completed. Ensure all feed statuses are current and up to date before proceeding to perform any vulnerability scans:

Feed Status

Type	Content	Origin	Version	Status
NVT	NVTs	Greenbone Community Feed	20230418T1011	Current
SCAP	CVEs CPEs	Greenbone Community SCAP Feed	20230418T0510	Current
CERT	CERT-Bund Advisories DFN-CERT Advisories	Greenbone Community CERT Feed	20230418T0406	Current
GVMD_DATA	Compliance Policies Port Lists Report Formats Scan Configs	Greenbone Community gvmd Data Feed	20230418T1004	Current

Figure 8.28 – Current feed status

Part 2 – scanning with GVM

1. To perform a vulnerability scan using GVM, first click on **Configuration** | **Targets** to add a target:

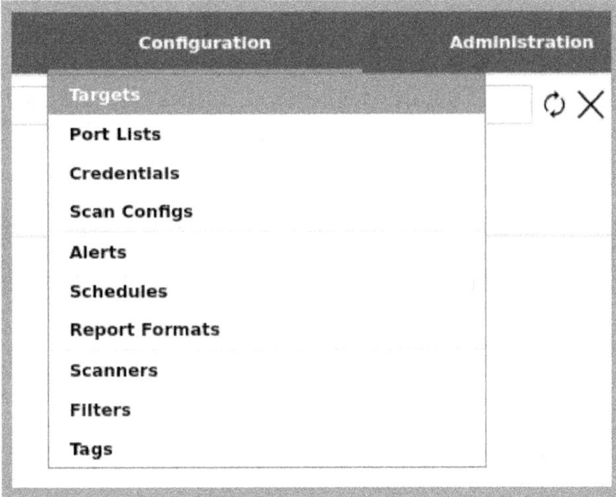

Figure 8.29 – Targets option

2. Next, click on the **New Target** icon in the top-left corner.
3. The **New Target** window will appear. Ensure you set a name for the target and insert the IP address of Metasploitable 3 within the **Hosts** field. Then, click on **Save**:

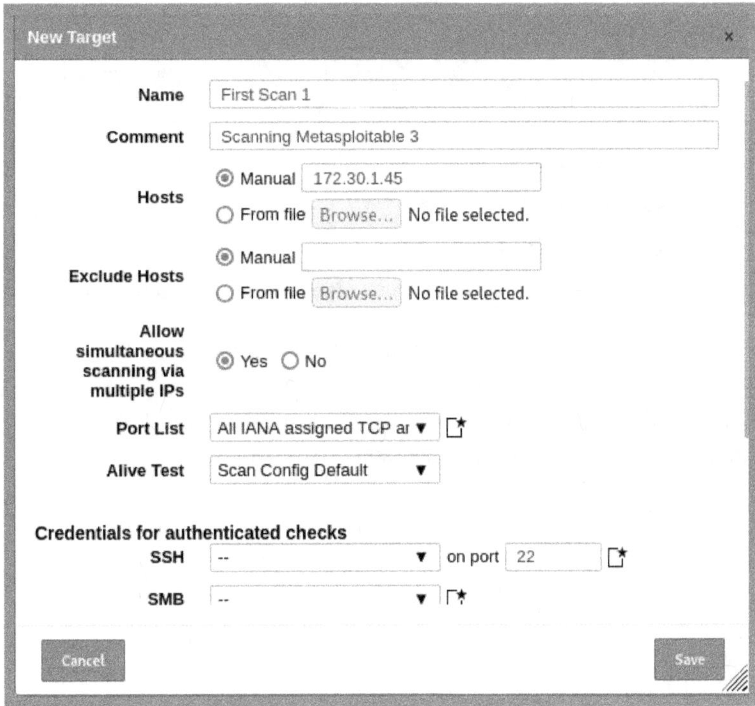

Figure 8.30 – Adding a new target

As shown in the preceding figure, the **New Target** menu provides additional options to scan a list of IP addresses from a text file, exclude either a single or multiple hosts from a targeted list, scan targeted service ports, and choose whether to perform an authenticated or unauthenticated scan on the target.

4. Next, to create a scan task within GVM, click on **Scans** | **Tasks**, as shown:

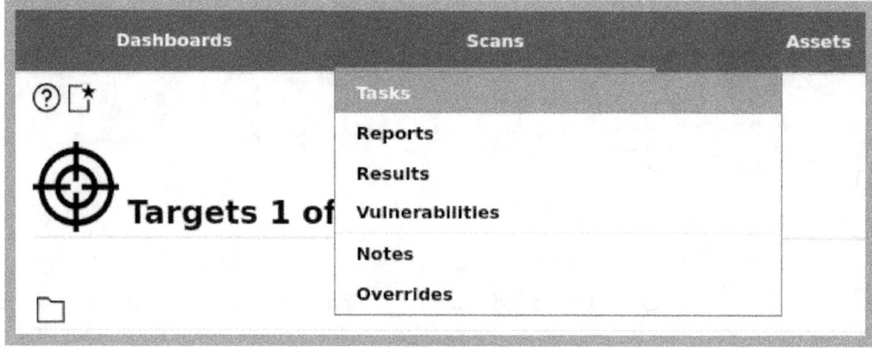

Figure 8.31 – Accessing the task menu

5. Next, click on the **Magic Paper** icon (top-left corner) and then **New Task**, as shown:

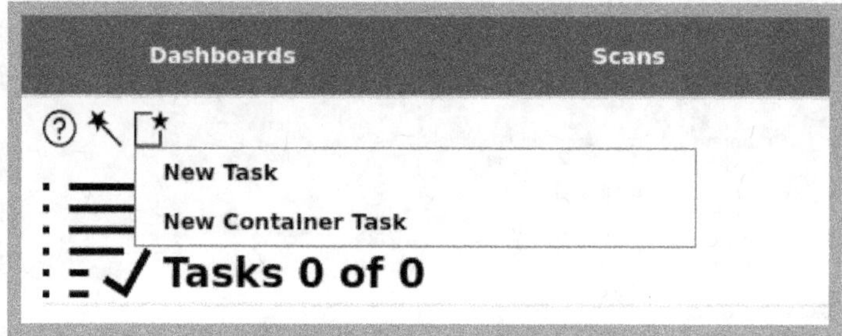

Figure 8.32 – Creating a new task

6. The **New Task** window will appear. Ensure you set a name for the task and select **Scan Targets** from the drop-down menu. Then, click on **Save**, as shown:

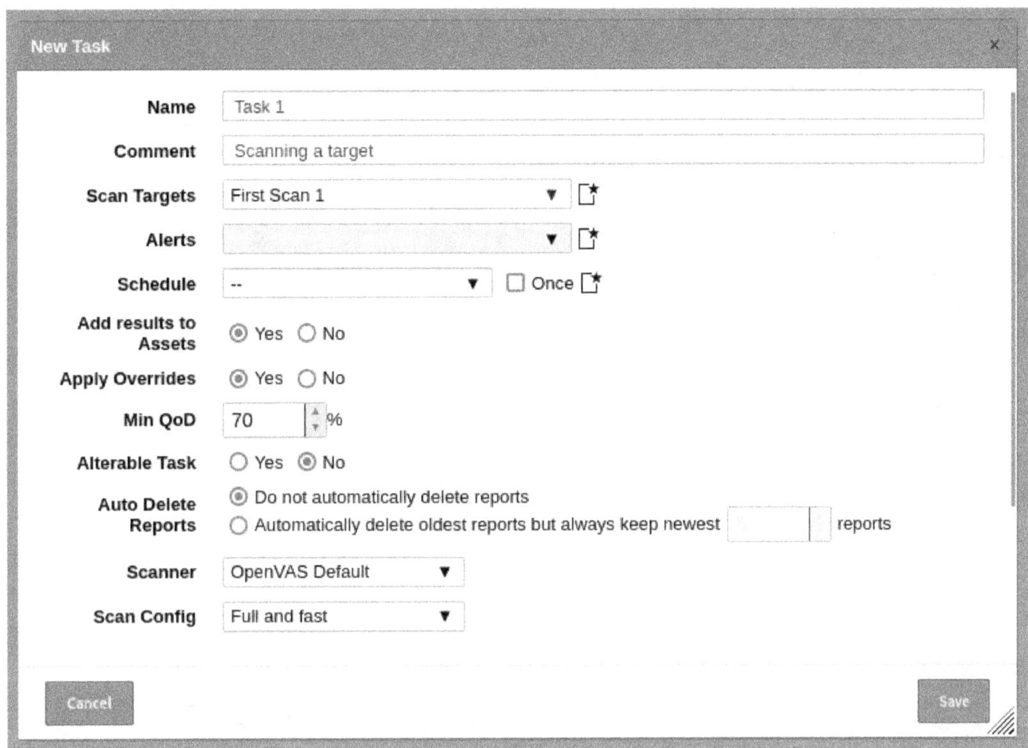

Figure 8.33 – Creating a new task

7. The newly created scan task will appear on the lower section of the same window. Click on the **Play** button to launch the task:

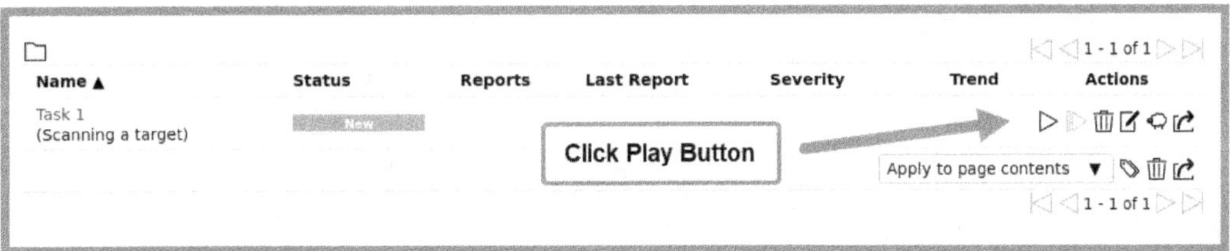

Figure 8.34 – Starting a scan

The scan status will change as the scan progresses. Once the scan is complete, the status will change to **Done** and will display the overall vulnerability severity, as shown:

Figure 8.35 – Scan complete

Part 3 – vulnerability analysis

1. To view the vulnerability analysis from GVM, click on **Scans** | **Reports**, as shown:

Figure 8.36 – Vulnerability summary

As shown in the preceding figure, GVM categorizes the number of security vulnerabilities into **High**, **Medium**, **Low**, **Log**, and even **False Positives**. The severity ratings help cybersecurity professionals to prioritize resources for more critical issues.

2. Next, click on the task name to view its details, such as all the security vulnerabilities that were found:

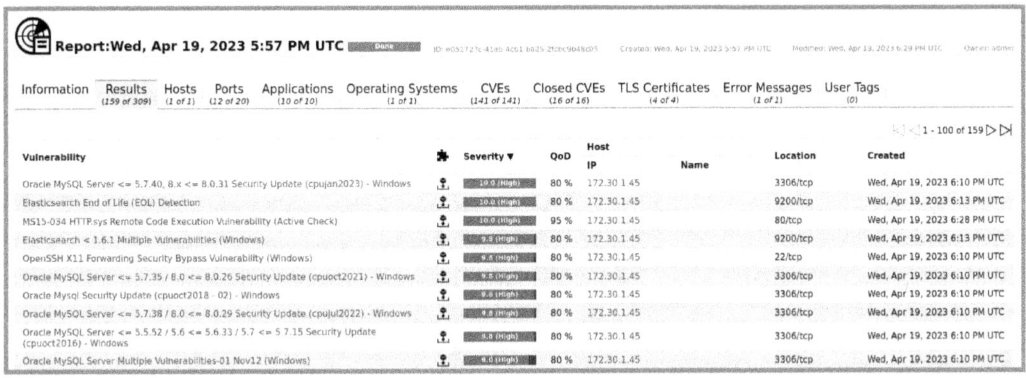

Figure 8.37 – Viewing security vulnerabilities

As shown in the preceding figure, GVM provides the name of the security vulnerability, its severity rating and **Quality of Detection** (**QoD**), and location (port number). The QoD value is used by Greenbone to display the reliability of the vulnerability detection by GVM.

3. To view the details of a vulnerability, simply click on one to view its description, as shown:

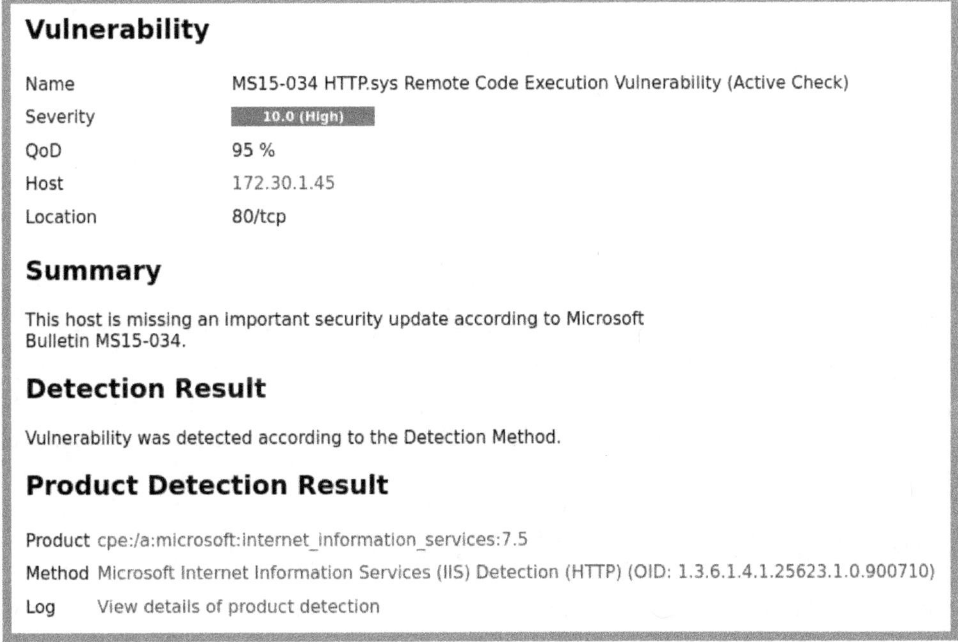

Figure 8.38 – Vulnerability details

Additionally, the details page provides information on how the security vulnerability was detected, the affected software and operating systems, and its impact and solution, as shown:

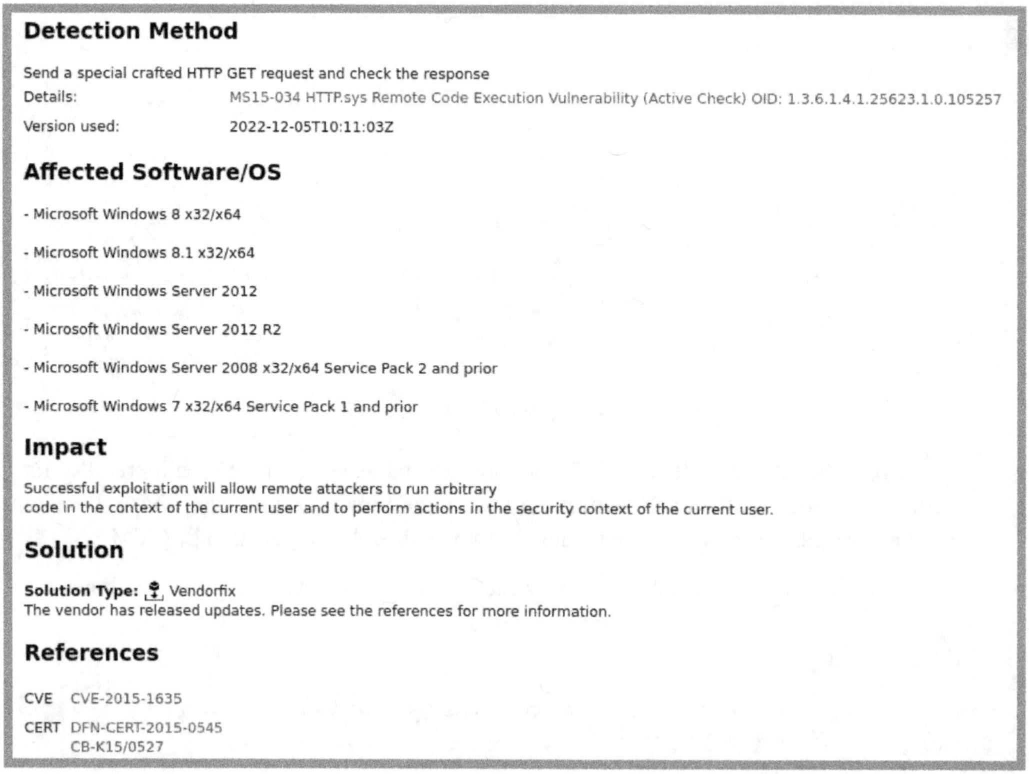

Detection Method

Send a special crafted HTTP GET request and check the response

Details:	MS15-034 HTTP.sys Remote Code Execution Vulnerability (Active Check) OID: 1.3.6.1.4.1.25623.1.0.105257
Version used:	2022-12-05T10:11:03Z

Affected Software/OS

- Microsoft Windows 8 x32/x64
- Microsoft Windows 8.1 x32/x64
- Microsoft Windows Server 2012
- Microsoft Windows Server 2012 R2
- Microsoft Windows Server 2008 x32/x64 Service Pack 2 and prior
- Microsoft Windows 7 x32/x64 Service Pack 1 and prior

Impact

Successful exploitation will allow remote attackers to run arbitrary
code in the context of the current user and to perform actions in the security context of the current user.

Solution

Solution Type: Vendorfix
The vendor has released updates. Please see the references for more information.

References

CVE	CVE-2015-1635
CERT	DFN-CERT-2015-0545 CB-K15/0527

Figure 8.39 – Solutions and recommendations

> **Tip**
>
> To learn more about GVM, please see `https://github.com/greenbone/gvmd`.

In this section, you have learned how to install, set up, and perform a vulnerability scan using GVM. In the next section, you will learn how to use Nmap to identify common security vulnerabilities on a target system.

Vulnerability discovery with Nmap

As you learned in the previous chapter, Nmap is one of the most popular network scanners that's commonly used in the networking and cybersecurity industries to discover host systems, identify open ports, detect service versions, and profile the operating system of a target. Furthermore, there's the **Nmap Scripting Engine** (**NSE**), which is integrated within Nmap and provides advanced scanning capabilities using custom Nmap scripts to detect common security vulnerabilities on targeted systems. However, the Nmap scripts that are used by the NSE component can be aggressive and have the potential to crash a system and cause data loss. Therefore, it's important to consider the potential risk and impact of using NSE during your security assessments as an ethical hacker.

There are a lot of pre-built scripts for NSE that are already pre-installed on Kali Linux. The following are the various categories of NSE scripts:

- **Auth**: This category of NSE scripts is useful for detecting whether the authentication mechanism on a targeted system can be bypassed. This information is useful to ethical hackers who are looking for techniques to gain a foothold on a target.
- **Broadcast**: This category of scripts is useful for discovering host systems on a network.
- **Brute**: This category of scripts helps ethical hackers to determine whether a target is vulnerable to various types of brute-force attacks.
- **Default**: This category contains the default scripts that are used with NSE during network scans.
- **Discover**: The Discover category contains scripts that are commonly used during active reconnaissance to collect sensitive information from live systems on a network.
- **DoS**: The **Denial of Service** (**DoS**) category contains special scripts that are designed to check whether a targeted system is susceptible to DoS-based attacks from a hacker.
- **Exploit**: This category of scripts is commonly used to determine whether there's a known security vulnerability on a targeted system.
- **External**: This category of scripts is used to send data that's collected by Nmap to an external resource for additional and further data processing.
- **Fuzzer**: This category contains scripts that are designed to send malformed data into an application to determine whether there are software-based vulnerabilities and bugs.
- **Intrusive**: The scripts within this category are considered to be *high risk* as they have the potential to crash systems.
- **Malware**: Malware-based scripts are used to detect malware-infected hosts on a network.
- **Safe**: These scripts are considered to be non-intrusive and safe for scanning a target.
- **Version**: This category of scripts is used to collect service versions of running applications on a target.
- **Vuln**: These scripts are used to check whether a target contains a known security vulnerability.

> Tip
>
> To view a listing and description of each NSE script, please refer to `https://nmap.org/nsedoc/scripts/`.

To get started using NSE on Kali Linux, please use the following instructions:

1. Firstly, power on the Kali Linux and Metasploitable 3 virtual machines.
2. Open Terminal on Kali Linux and use the following commands to view the directory and list of Nmap scripts:

   ```
   kali@kali:~$ ls -l /usr/share/nmap/scripts
   ```

   The following figure shows the execution of the preceding command and the listing of NSE scripts:

   ```
   kali@kali:~$ ls -l /usr/share/nmap/scripts
   total 4936
   -rw-r--r-- 1 root root  3901 Oct  6  2022 acarsd-info.nse
   -rw-r--r-- 1 root root  8749 Oct  6  2022 address-info.nse
   -rw-r--r-- 1 root root  3345 Oct  6  2022 afp-brute.nse
   -rw-r--r-- 1 root root  6463 Oct  6  2022 afp-ls.nse
   -rw-r--r-- 1 root root  7001 Oct  6  2022 afp-path-vuln.nse
   -rw-r--r-- 1 root root  5600 Oct  6  2022 afp-serverinfo.nse
   -rw-r--r-- 1 root root  2621 Oct  6  2022 afp-showmount.nse
   -rw-r--r-- 1 root root  2262 Oct  6  2022 ajp-auth.nse
   -rw-r--r-- 1 root root  2983 Oct  6  2022 ajp-brute.nse
   ```

   Figure 8.40 – Viewing NSE scripts

3. Next, use the following command with the wildcard (*) to view a list of HTTP vulnerability detection scripts:

   ```
   kali@kali:~$ ls -l /usr/share/nmap/scripts/http-vuln*
   ```

   As shown in the following figure, the preceding command enables you to filter HTTP vulnerability NSE scripts:

   ```
   kali@kali:~$ ls -l /usr/share/nmap/scripts/http-vuln*
   -rw-r--r-- 1 root root  3273 Oct  6  2022 /usr/share/nmap/scripts/http-vuln-cve2006-3392.nse
   -rw-r--r-- 1 root root  6610 Oct  6  2022 /usr/share/nmap/scripts/http-vuln-cve2009-3960.nse
   -rw-r--r-- 1 root root  2957 Oct  6  2022 /usr/share/nmap/scripts/http-vuln-cve2010-0738.nse
   -rw-r--r-- 1 root root  5607 Oct  6  2022 /usr/share/nmap/scripts/http-vuln-cve2010-2861.nse
   -rw-r--r-- 1 root root  4527 Oct  6  2022 /usr/share/nmap/scripts/http-vuln-cve2011-3192.nse
   -rw-r--r-- 1 root root  5851 Oct  6  2022 /usr/share/nmap/scripts/http-vuln-cve2011-3368.nse
   -rw-r--r-- 1 root root  4403 Oct  6  2022 /usr/share/nmap/scripts/http-vuln-cve2012-1823.nse
   -rw-r--r-- 1 root root  4831 Oct  6  2022 /usr/share/nmap/scripts/http-vuln-cve2013-0156.nse
   ```

   Figure 8.41 – Filtering NSE scripts

4. Next, use the following commands to filter all NSE **Server Message Block** (**SMB**) vulnerability scripts:

   ```
   kali@kali:~$ ls -l /usr/share/nmap/scripts/smb-vuln*
   ```

   The following figure shows the list of SMB vulnerability scripts:

```
kali@kali:~$ ls -l /usr/share/nmap/scripts/smb-vuln*
-rw-r--r-- 1 root root  7524 Oct  6  2022 /usr/share/nmap/scripts/smb-vuln-conficker.nse
-rw-r--r-- 1 root root  6402 Oct  6  2022 /usr/share/nmap/scripts/smb-vuln-cve2009-3103.nse
-rw-r--r-- 1 root root 23154 Oct  6  2022 /usr/share/nmap/scripts/smb-vuln-cve-2017-7494.nse
-rw-r--r-- 1 root root  6545 Oct  6  2022 /usr/share/nmap/scripts/smb-vuln-ms06-025.nse
-rw-r--r-- 1 root root  5386 Oct  6  2022 /usr/share/nmap/scripts/smb-vuln-ms07-029.nse
-rw-r--r-- 1 root root  5688 Oct  6  2022 /usr/share/nmap/scripts/smb-vuln-ms08-067.nse
-rw-r--r-- 1 root root  5647 Oct  6  2022 /usr/share/nmap/scripts/smb-vuln-ms10-054.nse
-rw-r--r-- 1 root root  7214 Oct  6  2022 /usr/share/nmap/scripts/smb-vuln-ms10-061.nse
-rw-r--r-- 1 root root  7344 Oct  6  2022 /usr/share/nmap/scripts/smb-vuln-ms17-010.nse
-rw-r--r-- 1 root root  4400 Oct  6  2022 /usr/share/nmap/scripts/smb-vuln-regsvc-dos.nse
-rw-r--r-- 1 root root  6586 Oct  6  2022 /usr/share/nmap/scripts/smb-vuln-webexec.nse
```

Figure 8.42 – SMB scripts

5. Next, let's use the following command to execute all the SMB vulnerability scripts to scan the Metasploitable 3 virtual machine:

   ```
   kali@kali:~$ sudo nmap --script smb-vuln* 172.30.1.45
   ```

   The following figure shows the `smb-vuln-ms17-010` script has found a known security vulnerability on the target:

```
Host script results:
| smb-vuln-ms17-010:
|   VULNERABLE:
|   Remote Code Execution vulnerability in Microsoft SMBv1 servers (ms17-010)
|     State: VULNERABLE
|     IDs:  CVE:CVE-2017-0143
|     Risk factor: HIGH
|       A critical remote code execution vulnerability exists in Microsoft SMBv1
|        servers (ms17-010).
|
|     Disclosure date: 2017-03-14
|     References:
|       https://blogs.technet.microsoft.com/msrc/2017/05/12/customer-guidance-for-wannacrypt-attacks/
|       https://technet.microsoft.com/en-us/library/security/ms17-010.aspx
|_      https://cve.mitre.org/cgi-bin/cvename.cgi?name=CVE-2017-0143
```

Figure 8.43 – Discovering security vulnerabilities

   As shown in the preceding figure, the NSE script was able to identify the presence of the EternalBlue vulnerability on the targeted host. In addition, the script provides references to better understand the severity and impact if a real attacker were to compromise this security flaw. As an ethical hacker, such information can be used to develop or acquire an exploit that can take advantage of this security weakness to gain a foothold on the target.

There are many Nmap scripts to help you identify vulnerabilities in a system. Depending on your security assessment, you can use keywords to help you filter suitable Nmap scripts to identify specific security vulnerabilities on a target. However, it's important to remember to never depend only on one vulnerability scanner during a security assessment as they can provide false negatives at times.

In this section, you learned how to use the NSE component within Nmap to identify security vulnerabilities in a system.

Summary

During the course of this chapter, you learned about the importance of vulnerability management and how it helps organizations to identify and resolve security weaknesses in their assets. Furthermore, you learned how to set up and use Nessus, GVM, and NSE to identify security vulnerabilities on a target.

Once a security vulnerability is found, the ethical hacker should perform additional research to determine how the vulnerability can be exploited and gain unauthorized access to the target. This means the ethical hacker can either develop exploits on their own or acquire exploits from public sources and perform testing within a simulated environment to determine the likelihood the exploit would successfully compromise the vulnerability on the target.

I hope this chapter has been informative for you and helpful in your journey into entering the cybersecurity industry. In the next chapter, *Chapter 9, Delving into Website Reconnaissance*, you will learn how to profile web technologies and applications.

Further reading

- Vulnerability assessment: `https://www.techtarget.com/searchsecurity/definition/vulnerability-assessment-vulnerability-analysis`
- Benefits of vulnerability analysis: `https://www.hackerone.com/knowledge-center/what-vulnerability-assessment-benefits-tools-and-process`

9

Delving into Website Reconnaissance

As the internet continues to grow, there are many websites created almost every day by organizations to help reach new and potential customers beyond traditional borders. Threat actors usually perform a lot of reconnaissance to gather as much information as possible and collect intelligence about their targets' websites and domains. This information is valuable to threat actors in planning future cyber-attacks on a target. As an ethical hacker, it's important to identify the attack surface and determine how web reconnaissance can be leveraged by real attackers in planning a cyber-attack.

During this chapter, you will learn how to use common tools and techniques to efficiently collect information about a target, such as its IP addresses and sub-domains, discover hidden directories, and identify the attack surface. Furthermore, you will learn how to use web vulnerability scanners to identify web application weaknesses and work with web reconnaissance frameworks to automate the collection and analysis of data.

In this chapter, we will cover the following topics:

- Collecting domain information
- Sub-domain enumeration
- Performing directory enumeration
- Web application vulnerabilities
- Web reconnaissance frameworks

Let's dive in!

Technical requirements

To follow along with the exercises in this chapter, please ensure that you have met the following hardware and software requirements:

- Kali Linux installed – `https://www.kali.org/get-kali/`

Collecting domain information

As an ethical hacker, collecting the **Domain Name System** (**DNS**) information and IP addresses and determining the backend infrastructure helps you to better understand the attack surface and attack vectors of a target. For instance, if you're performing an external network penetration test or **Open Source Intelligence** (**OSINT**) penetration test on an organization, finding the target's domain and website are good starting points. A domain name can lead you to discover the website and the sub-domains and IP addresses assigned to servers owned by the target.

This section focuses on using various tactics and techniques to retrieve the IP addresses, discovering any infrastructure details, and running web technologies on a target's web server and domain.

Retrieving IP addresses

By retrieving the IP addresses of a target domain and its sub-domain, ethical hackers are able to map the external network topology and identify potential security vulnerabilities that can be exploited by an attacker to gain unauthorized access. Furthermore, without identifying the IP addresses of a target, ethical hackers won't be able to simulate real-world cyber-attacks to determine the security posture of an organization.

To get started with identifying the IP addresses for a domain name, use the following instructions:

1. Firstly, power on the **Kali Linux** virtual machine and ensure it has internet connectivity.
2. Next, open the Terminal and use the `host <domain-name>` command to retrieve the IPv4 and IPv6 addresses of a target, as shown:

```
kali@kali:~$ host microsoft.com
microsoft.com has address 20.112.52.29
microsoft.com has address 20.103.85.33
microsoft.com has address 20.53.203.50
microsoft.com has address 20.84.181.62
microsoft.com has address 20.81.111.85
```

Figure 9.1 – host command

A domain name is simply a namespace on a network such as the internet; however, the IP addresses that are associated with a domain or host indicate the network location of the target. As shown in the preceding snippet, performing an IP lookup on the addresses can reveal the geo-location of the hosting servers or target.

> **Tip**
>
> To identify the geo-location of the network address, go to `https://whatismyipaddress.com/ip-lookup` to perform an IP lookup on any public IP address.

3. Next, the **nslookup** tool is used to troubleshoot **DNS** issues and retrieve the IP addresses of a hostname. To get the IP addresses of a domain name, use the `nslookup <domain-name> <DNS-server>` command as shown:

```
kali@kali:~$ nslookup microsoft.com 8.8.8.8
Server:         8.8.8.8
Address:        8.8.8.8#53

Non-authoritative answer:
Name:    microsoft.com
Address: 20.81.111.85
Name:    microsoft.com
Address: 20.84.181.62
Name:    microsoft.com
Address: 20.103.85.33
```

Figure 9.2 – The nslookup tool

As shown in the preceding snippet, `nslookup` sent the DNS queries to Google's DNS server at `8.8.8.8` and the response provides the IP addresses shown in the output.

> **Important note**
>
> There are different DNS records, such as `A`, which resolves a hostname to an IPv4 address; `AAAA`, which resolves a hostname to an IPv6 address; `MX`, which specifies the mail exchange server; `CNAME`, which specifies a canonical name or alias for a domain; `PTR`, which resolves an IP address to a hostname; `TXT`, which specifies a text record; `RP`, which specifies the person responsible for the domain; and `NS`, which specifies the authoritative name server for the domain.

4. The **dig** utility on Linux-based systems enables you to query a specific DNS server using the `@` syntax, then followed by the domain name. To retrieve the DNS records for a domain name, use the `dig @<dns-server> <target-domain>` command as shown:

```
kali@kali:~$ dig @8.8.8.8 microsoft.com

; <<>> DiG 9.18.12-1-Debian <<>> @8.8.8.8 microsoft.com
; (1 server found)
;; global options: +cmd
;; Got answer:
;; ->>HEADER<<- opcode: QUERY, status: NOERROR, id: 11397
;; flags: qr rd ra; QUERY: 1, ANSWER: 5, AUTHORITY: 0, ADDITIONAL: 1

;; ANSWER SECTION:
microsoft.com.          526     IN      A       20.81.111.85
microsoft.com.          526     IN      A       20.84.181.62
microsoft.com.          526     IN      A       20.103.85.33
microsoft.com.          526     IN      A       20.53.203.50
microsoft.com.          526     IN      A       20.112.52.29

;; Query time: 91 msec
;; SERVER: 8.8.8.8#53(8.8.8.8) (UDP)
;; WHEN: Fri Apr 28 14:07:52 AST 2023
;; MSG SIZE  rcvd: 122
```

Figure 9.3 – The dig command

To retrieve a specific DNS record type, simply append the `A`, `AAAA` `MX`, `NS`, `PTR`, `CNAME`, or `RP` record type at the end of the command as shown:

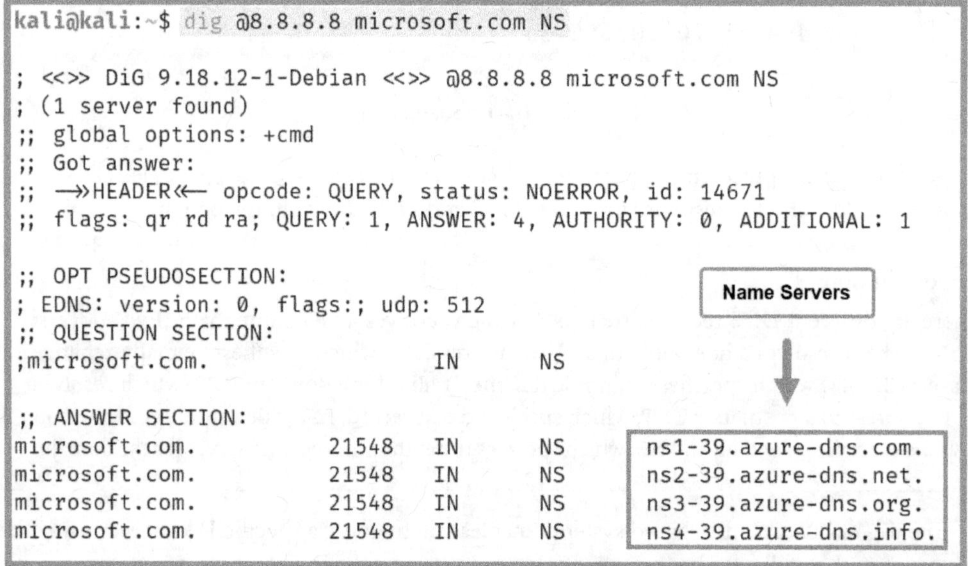

```
kali@kali:~$ dig @8.8.8.8 microsoft.com NS

; <<>> DiG 9.18.12-1-Debian <<>> @8.8.8.8 microsoft.com NS
; (1 server found)
;; global options: +cmd
;; Got answer:
;; ->>HEADER<<- opcode: QUERY, status: NOERROR, id: 14671
;; flags: qr rd ra; QUERY: 1, ANSWER: 4, AUTHORITY: 0, ADDITIONAL: 1

;; OPT PSEUDOSECTION:
; EDNS: version: 0, flags:; udp: 512
;; QUESTION SECTION:
;microsoft.com.                 IN      NS

;; ANSWER SECTION:
microsoft.com.          21548   IN      NS      ns1-39.azure-dns.com.
microsoft.com.          21548   IN      NS      ns2-39.azure-dns.net.
microsoft.com.          21548   IN      NS      ns3-39.azure-dns.org.
microsoft.com.          21548   IN      NS      ns4-39.azure-dns.info.
```


Figure 9.4 – Retrieving specific DNS records

As shown in the preceding snippet, `dig` was able to retrieve the name servers for the domain name.

> **Important note**
> If an attacker is able to compromise the name servers of a domain, the hacker can manipulate the DNS records stored on the server. This enables the attacker to perform various types of DNS-based and spoofing attacks.

5. Next, the **dnsrecon** tool enables you to automate the retrieval of public DNS records for a domain; use the `dnsrecon -d <domain-name> -n <dns-server>` command as shown:

```
kali@kali:~$ dnsrecon -d microsoft.com -n 8.8.8.8
[*] std: Performing General Enumeration against: microsoft.com...
[-] DNSSEC is not configured for microsoft.com
[*]      SOA ns1-39.azure-dns.com 150.171.10.39
[*]      SOA ns1-39.azure-dns.com 2603:1061:0:10::27
[*]      NS ns1-39.azure-dns.com 150.171.10.39
[*]      NS ns1-39.azure-dns.com 2603:1061:0:10::27
[*]      NS ns2-39.azure-dns.net 150.171.16.39
[*]      NS ns2-39.azure-dns.net 2620:1ec:8ec:10::27
[*]      NS ns3-39.azure-dns.org 13.107.222.39
[*]      NS ns3-39.azure-dns.org 2a01:111:4000:10::27
[*]      NS ns4-39.azure-dns.info 13.107.206.39
```

Figure 9.5 – The dnsrecon tool

Having completed this exercise, you have learned how to retrieve the IP addresses and DNS records of a target domain. Next, you'll learn how to identify the target's domain infrastructure.

Identifying domain infrastructure

Identifying the domain infrastructure helps ethical hackers to map the attack surface of a target. This enables the ethical hacker to determine what systems and applications are in use, and this data can be leveraged to identify security vulnerabilities and attack vectors for delivering an exploit to the target.

To get started with identifying domain infrastructure, use the following instructions:

1. It's always recommended to implement a **Web Application Firewall** (**WAF**) in front of a web application server to prevent web application attacks. The **WAFW00F** tool helps ethical hackers to determine whether a target website is behind a WAF, so use the `wafw00f <target>` command as shown:

```
kali@kali:~$ wafw00f cloudflare.com

                   ~ WAFW00F : v2.2.0 ~

[*] Checking https://cloudflare.com
[+] The site https://cloudflare.com is behind Cloudflare (Cloudflare Inc.) WAF.
[~] Number of requests: 2
```

Figure 9.6 – Detecting WAFs

If the target is behind a WAF, you will not be able to retrieve the real public IP address of the target; instead, you'll obtain the address of the WAF application due to the reverse proxy feature. In addition, the WAF will analyze all web traffic going to the targeted website and block any potentially malicious traffic.

2. The **WHOIS** tool enables you to retrieve the public domain registrar records of the target; use the `whois <domain>` command as shown:

```
kali@kali:~$ whois microsoft.com
   Domain Name: MICROSOFT.COM
   Registry Domain ID: 2724960_DOMAIN_COM-VRSN
   Registrar WHOIS Server: whois.markmonitor.com
   Registrar URL: http://www.markmonitor.com
   Updated Date: 2023-04-01T11:51:08Z
   Creation Date: 1991-05-02T04:00:00Z
   Registry Expiry Date: 2024-05-03T04:00:00Z
   Registrar: MarkMonitor Inc.
   Registrar IANA ID: 292
```

Figure 9.7 – Collecting the domain registration details

Sometimes, an organization may not pay for additional privacy protection when registering a domain name, and its sensitive information will be publicly available, which can be leveraged by threat actors in planning future attacks such as social engineering.

3. Next, the site report tool on **Netcraft** helps ethical hackers to profile the infrastructure and technologies of a website on the internet. Go to `https://sitereport.netcraft.com/` and enter a domain as shown:

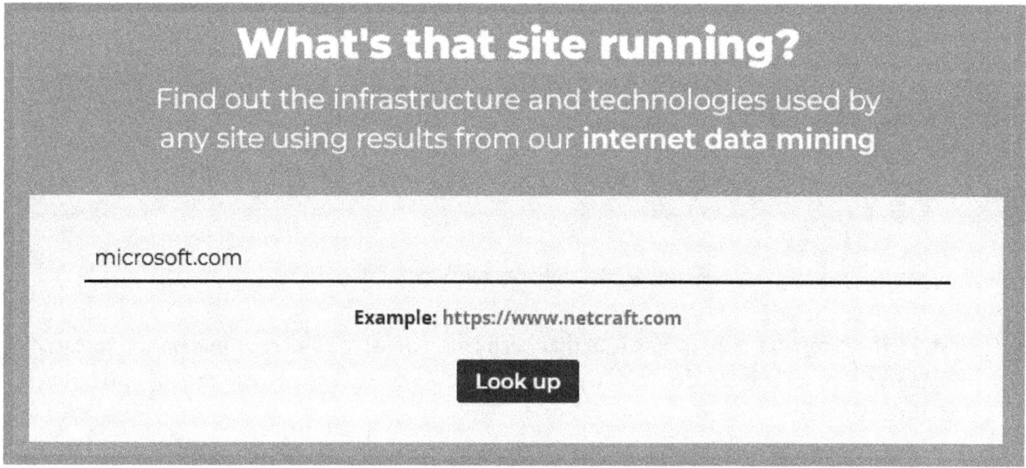

Figure 9.8 – Netcraft

After a few seconds, Netcraft will provide the network and hosting details as shown:

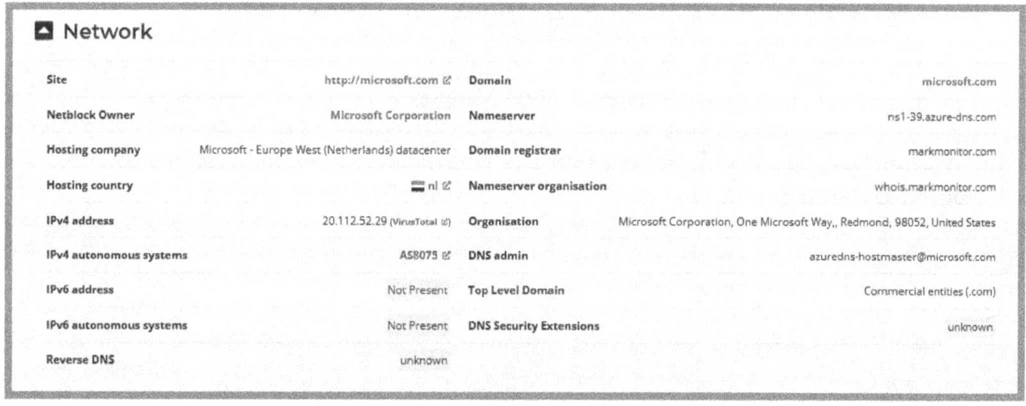

Network

Site	http://microsoft.com	Domain	microsoft.com
Netblock Owner	Microsoft Corporation	Nameserver	ns1-39.azure-dns.com
Hosting company	Microsoft - Europe West (Netherlands) datacenter	Domain registrar	markmonitor.com
Hosting country	nl	Nameserver organisation	whois.markmonitor.com
IPv4 address	20.112.52.29 (VirusTotal)	Organisation	Microsoft Corporation, One Microsoft Way,, Redmond, 98052, United States
IPv4 autonomous systems	AS8075	DNS admin	azuredns-hostmaster@microsoft.com
IPv6 address	Not Present	Top Level Domain	Commercial entities (.com)
IPv6 autonomous systems	Not Present	DNS Security Extensions	unknown
Reverse DNS	unknown		

Figure 9.9 – Network infrastructure details

Additionally, Netcraft provides the IP block information that can be used to geo-locate the physical server that's hosting the domain and website as shown:

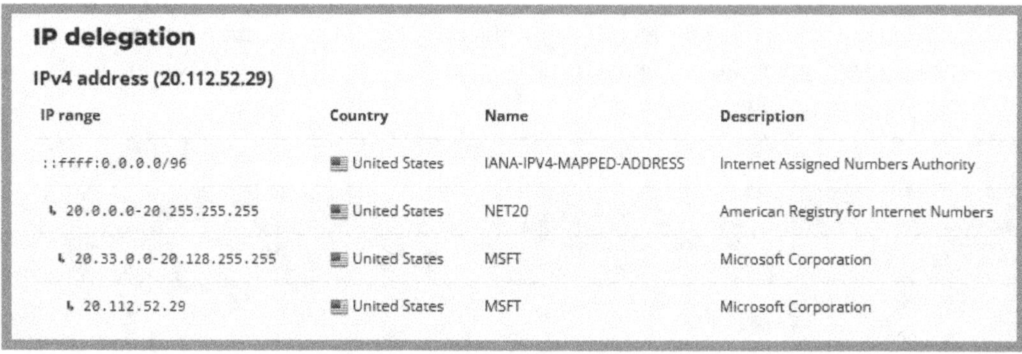

IP delegation

IPv4 address (20.112.52.29)

IP range	Country	Name	Description
::ffff:0.0.0.0/96	United States	IANA-IPV4-MAPPED-ADDRESS	Internet Assigned Numbers Authority
↳ 20.0.0.0-20.255.255.255	United States	NET20	American Registry for Internet Numbers
↳ 20.33.0.0-20.128.255.255	United States	MSFT	Microsoft Corporation
↳ 20.112.52.29	United States	MSFT	Microsoft Corporation

Figure 9.10 – Network block details

4. Next, **DNS Dumpster** is an online tool that performs extensive DNS reconnaissance, research, and analysis on a targeted domain. Go to `https://dnsdumpster.com/` and enter a domain as shown:

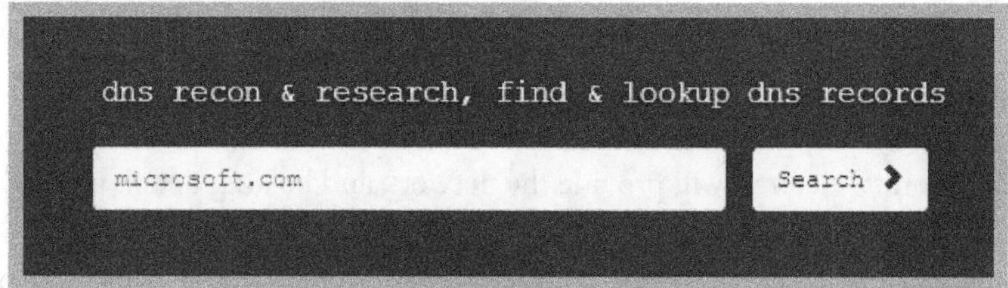

Figure 9.11 – DNS Dumpster

After a few seconds, DNS Dumpster provides all the DNS records and name servers for the domain as shown:

DNS Servers

ns1-39.azure-dns.com.	150.171.10.39 ns1-39.azure-dns.com	MICROSOFT-CORP-MSN-AS-BLOCK United States
ns2-39.azure-dns.net.	150.171.16.39 ns2-39.azure-dns.net	MICROSOFT-CORP-MSN-AS-BLOCK United States
ns3-39.azure-dns.org.	13.107.222.39 ns3-39.azure-dns.org	MICROSOFT-CORP-MSN-AS-BLOCK United States
ns4-39.azure-dns.info.	13.107.206.39 ns4-39.azure-dns.info	MICROSOFT-CORP-MSN-AS-BLOCK United States

MX Records ** This is where email for the domain goes...

10 microsoft-com.mail.protection.outlook.com.	52.101.40.29	MICROSOFT-CORP-MSN-AS-BLOCK United States

Figure 9.12 – DNS information

Additionally, DNS Dumpster automatically creates a map of the domain to provide insights on the available DNS records, IP addresses, and name servers as shown:

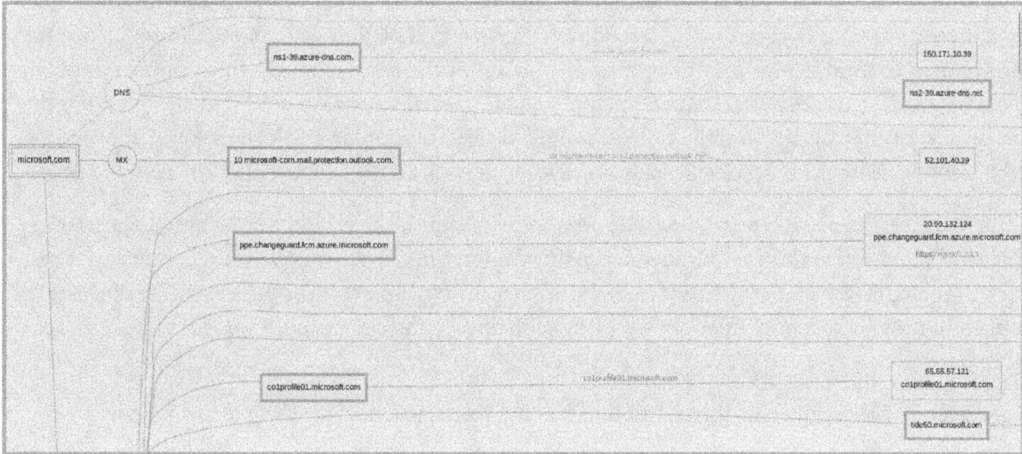

Figure 9.13 – Mapping of a domain

Furthermore, DNS Dumpster provides a graphical representation showing the backend infrastructure and how the records are mapped to each other in the domain:

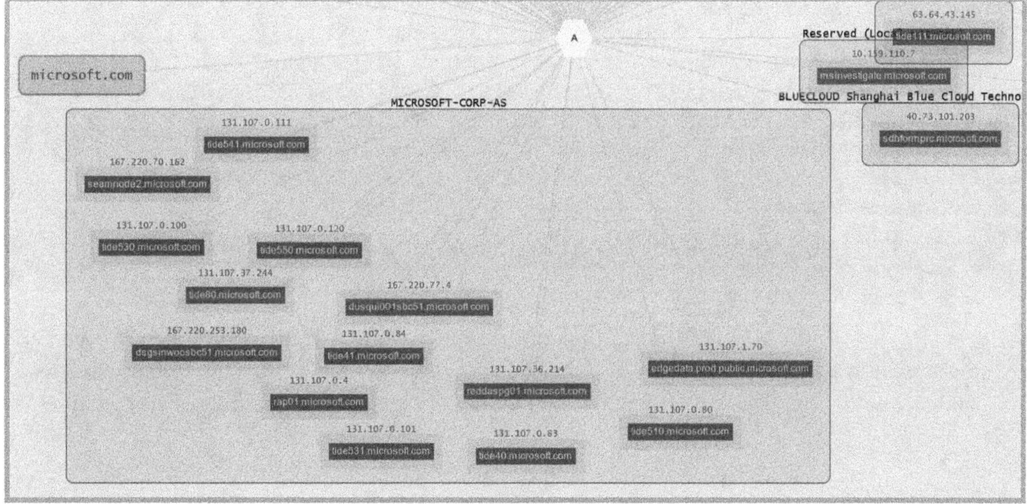

Figure 9.14 – DNS infrastructure graph

As shown in the preceding snippet, the visual map helps ethical hackers to better understand how the DNS records of a domain are associated with IP addresses and whether there are any unintentionally exposed systems on the internet that are owned by the target.

Identifying web technologies

As an ethical hacker, it's important to identify whether a website has any security vulnerabilities that can be exploited by a real attacker. Sometimes, an underlying web application or **Content Management System** (**CMS**) may be outdated or contain vulnerable plugins, which can be exploited to gain unauthorized access, steal data or manipulate the backend database. Additionally, this helps the ethical hacker better understand the web technologies that are running on the target and improve the attack plan to ensure the best suitable tactics and techniques are used to deliver the attack to the target.

While there are many web technologies in the industry, ethical hackers can leverage knowledge, experience, and research to find known security vulnerabilities in a web application just as a real attacker would. However, the intention is to discover security vulnerabilities before a real cyber-attack occurs and provide recommendations on mitigation and countermeasures to prevent future attacks and threats.

To get started with identifying the technologies of a website, use the following instructions:

1. Firstly, power on the **Kali Linux** virtual machine.
2. The **WhatWeb** tools help ethical hackers to profile the web technologies that are found on a target website. To identify web technologies, use the `whatweb <domain>` command within the Terminal as shown:

```
kali@kali:~$ whatweb https://github.com
https://github.com [200 OK] Content-Language[en-US], Cookies[_gh_sess,_octo
,logged_in], Country[UNITED STATES][US], HTML5, HTTPServer[GitHub.com], Htt
pOnly[_gh_sess,logged_in], IP[140.82.114.4], Open-Graph-Protocol[object][14
01488693436528], OpenSearch[/opensearch.xml], Script[application/javascript
], Strict-Transport-Security[max-age=31536000; includeSubdomains; preload],
 Title[GitHub: Let's build from here · GitHub], UncommonHeaders[x-content-t
ype-options,referrer-policy,content-security-policy,x-github-request-id], X
-Frame-Options[deny], X-XSS-Protection[0]
```

Figure 9.15 – Identifying web technologies

As shown in the preceding snippet, WhatWeb was able to retrieve specific web application and technology versions, which can be used to identify security vulnerabilities on the target.

3. Next, **Wappalyzer** is a browser add-on for identifying web technologies on a website. Open the **Mozilla Firefox** web browser, go to `https://addons.mozilla.org/en-US/firefox/addon/wappalyzer/`, and click on **Add to Firefox** to install the plugin as shown:

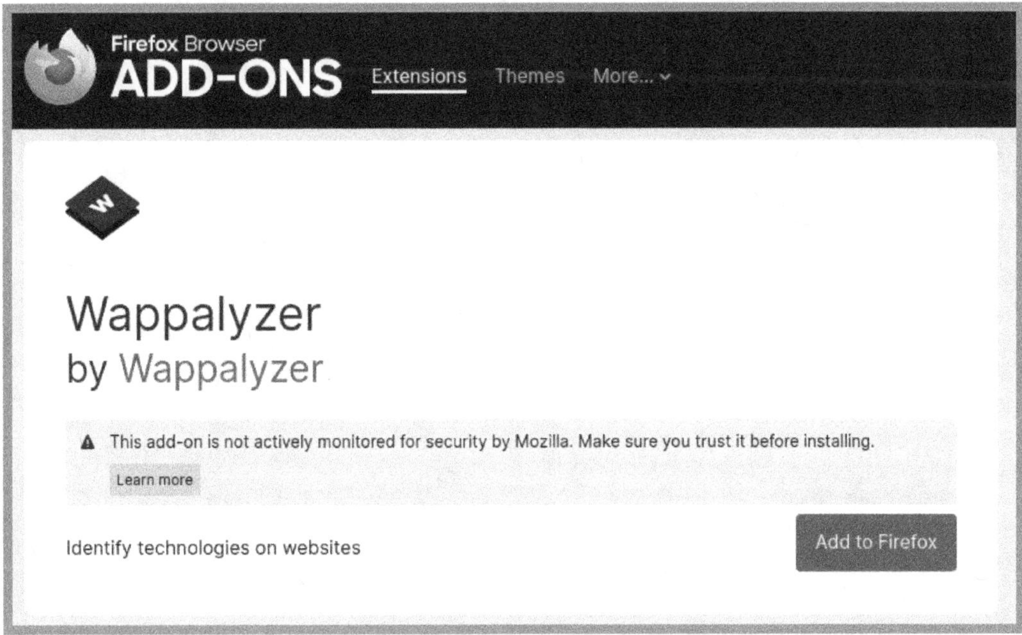

Figure 9.16 – Wappalyzer

After the add-on is installed on Mozilla Firefox, the Wappalyzer icon will automatically appear on the browser's toolbar.

Wappalyzer will analyze the web applications on any website that you visit to identify the web technologies; simply click on the Wappalyzer icon on the browser's toolbar to view the website details as shown:

Figure 9.17 – Wappalyzer details

4. Next, **BuiltWith** is another Mozilla Firefox add-on for profiling the web technologies of a website. Go to `https://addons.mozilla.org/en-US/firefox/addon/builtwith/` and click on **Add to Firefox** as shown:

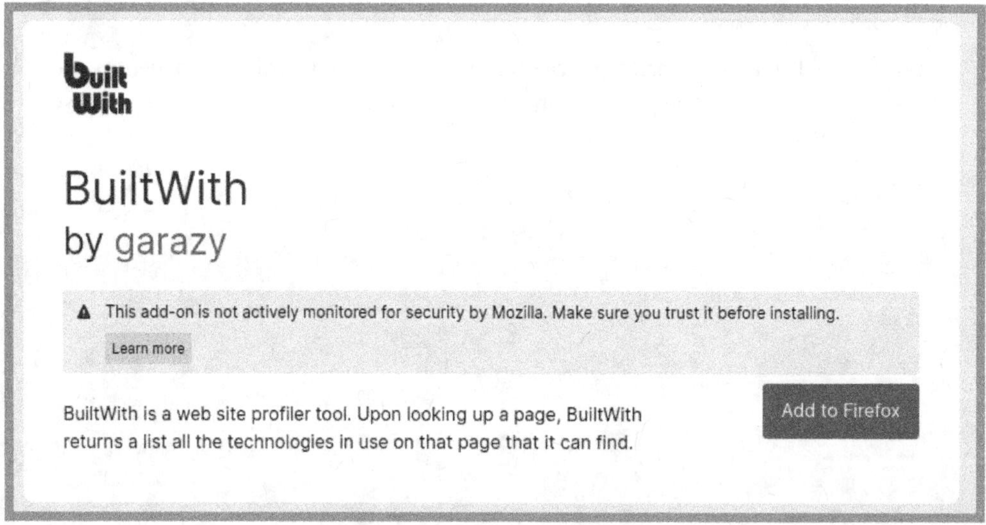

Figure 9.18 – BuiltWith add-on page

After the BuiltWith add-on is installed on Firefox, its icon will automatically appear on the browser's toolbar. Simply visit any website and click on the BuiltWith icon to view the web technologies of the website as shown:

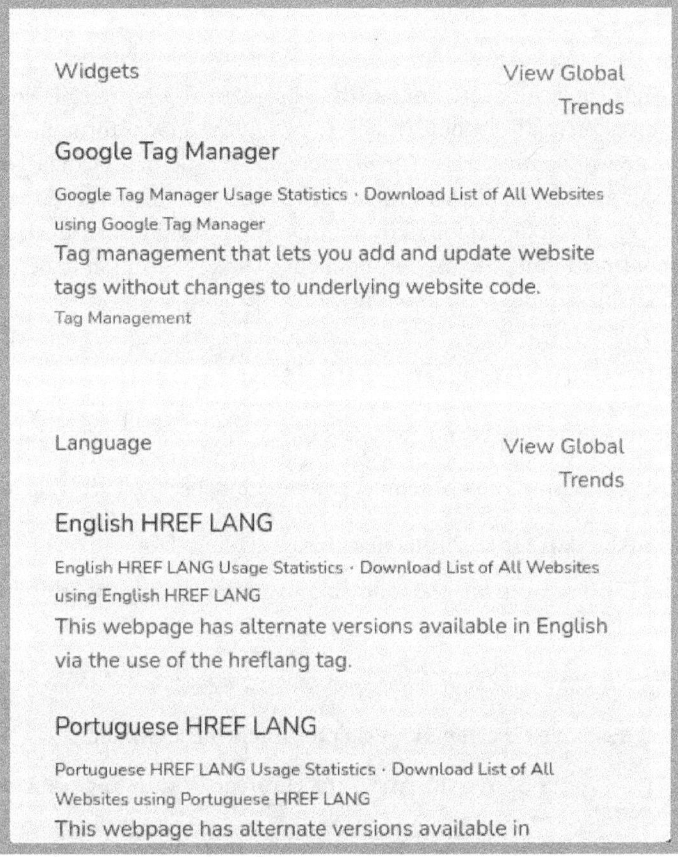

Figure 9.19 – BuiltWith details

These add-ons are very useful and conveniently and quickly identify the technologies that are running on any web application or server that you visit and should be within your arsenal of tools.

Having completed this section, you have learned how to use various tools and techniques to collect information about a targeted domain name and website. Next, you will learn how to enumerate the sub-domains of a target.

Sub-domain enumeration

Threat actors such as hackers use both passive and active reconnaissance techniques to identify the sub-domains of a target. Usually, an organization will register a domain name (parent domain) and create additional sub-domains, where each sub-domain points to a different server that's owned by the target. For instance, `domain.local` points to the IP address of the web server and `mail.domain.local` points to the IP address of the email server. Therefore, enumerating the sub-domains and resolving their IP addresses helps attackers to identify security vulnerabilities and the attack surface of additional systems owned by the target.

Sometimes, a sub-domain is used to host a test environment for users that are misconfigured, running a vulnerable or less secure application, and connected to the internal corporate network. If an attacker were to compromise this system, they would be able to pivot their attacks through the compromised system to the internal network of the target. Furthermore, the sub-domains of a target are not always configured with the same security policies and posture, which can further lead to having different security vulnerabilities that can be exploited in multiple ways to gain unauthorized access. As an ethical hacker, it's important to identify the sub-domains of an organization to better determine its security posture and identify hidden vulnerabilities.

Discovering sub-domains using Sublist3r

Sublist3r is a popular tool that collects and analyzes OSINT to identify the sub-domains of a target.

To get started with using Sublist3r for sub-domain enumeration, use the following instructions:

1. Firstly, power on the **Kali Linux** virtual machine.
2. Next, open the Terminal and use the following commands to update the software package repository list:

   ```
   kali@kali:~$ sudo apt update
   ```

3. Next, download the Sublist3r setup files using the following command:

   ```
   kali@kali:~$ git clone https://github.com/huntergregal/
   Sublist3r.git
   ```

   The preceding GitHub repository is a working branch from the official repository at `https://github.com/aboul3la/Sublist3r`. Unfortunately, the packages from the official GitHub repository no longer work at the time of writing.

4. Next, use the following commands to change the present working directory to the `Sublist3r` folder and install its requirements:

   ```
   kali@kali:~$ cd Sublist3r
   kali@kali:~/Sublist3r$ sudo pip install -r requirements.txt
   ```

The following snippet shows the execution of the preceding commands:

```
kali@kali:~$ cd Sublist3r

kali@kali:~/Sublist3r$ sudo pip install -r requirements.txt
Collecting argparse
  Using cached argparse-1.4.0-py2.py3-none-any.whl (23 kB)
Requirement already satisfied: dnspython in /usr/lib/python3/dist-packages
Requirement already satisfied: requests in /usr/lib/python3/dist-packages
Installing collected packages: argparse
Successfully installed argparse-1.4.0
```

Figure 9.20 – Installing the requirements

5. Next, use the following commands to perform sub-domain enumeration on a domain:

```
kali@kali:~/Sublist3r$ python ./sublist3r.py -d <domain-name>
```

The following snippet shows Sublist3r has searched through multiple data sources and discovered the sub-domains of the target:

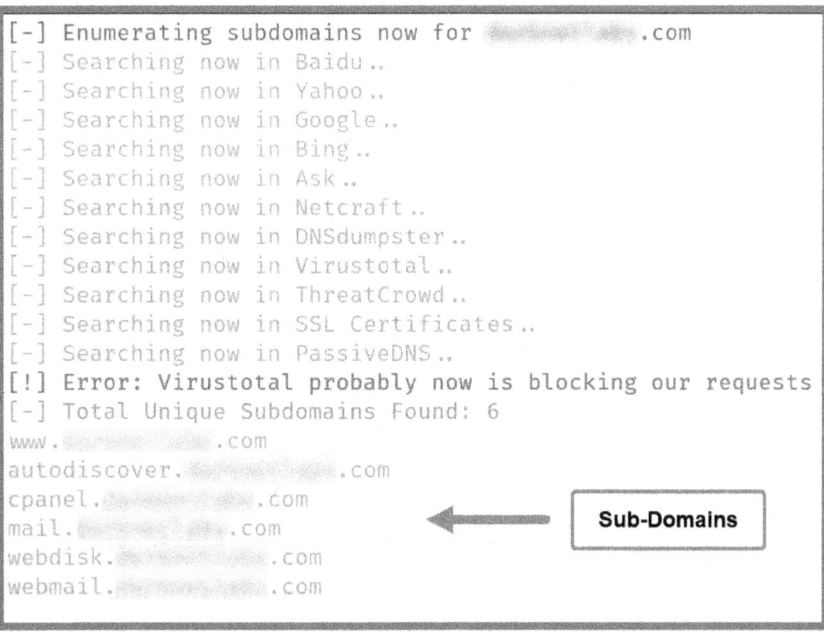

Figure 9.21 – Discovering sub-domains

Having completed this exercise, you have learned how to set up and work with Sublist3r to enumerate the sub-domains of a target. Next, you will learn how to use theHarvester to find additional sub-domains.

Finding sub-domains with theHarvester

theHarvester is a popular tool used by ethical hackers during passive reconnaissance to collect email addresses, hostnames, IP addresses, and sub-domains of a target from multiple online data sources.

To get started with using theHarvester to find sub-domains, use the following instructions:

1. Firstly, power on the **Kali Linux** virtual machine.
2. Next, open the Terminal and use the following commands to view available options and syntax for theHarvester:

   ```
   kali@kali:~$ theHarvester -h
   ```

   The following snippet shows the various forms of syntax and their usage:

```
options:
  -h, --help            show this help message and exit
  -d DOMAIN, --domain DOMAIN
                        Company name or domain to search.
  -l LIMIT, --limit LIMIT
                        Limit the number of search results, default=500.
  -S START, --start START
                        Start with result number X, default=0.
  -p, --proxies         Use proxies for requests, enter proxies in proxies.yaml.
  -s, --shodan          Use Shodan to query discovered hosts.
  --screenshot SCREENSHOT
                        Take screenshots of resolved domains specify output directory: --screenshot output_directory
  -v, --virtual-host    Verify host name via DNS resolution and search for virtual hosts.
  -e DNS_SERVER, --dns-server DNS_SERVER
                        DNS server to use for lookup.
  -r, --take-over       Check for takeovers.
  -n, --dns-lookup      Enable DNS server lookup, default False.
  -c, --dns-brute       Perform a DNS brute force on the domain.
  -f FILENAME, --filename FILENAME
                        Save the results to an XML and JSON file.
  -b SOURCE, --source SOURCE
                        anubis, baidu, bevigil, binaryedge, bing, bingapi, bufferoverun, censys, certspotter, crtsh,
                        dnsdumpster, duckduckgo, fullhunt, github-code, hackertarget, hunter, intelx, omnisint, otx,
                        pentesttools, projectdiscovery, qwant, rapiddns, rocketreach, securityTrails, sublist3r,
                        threatcrowd, threatminer, urlscan, virustotal, yahoo, zoomeye
```

Figure 9.22 – theHarvester options

3. To gather the sub-domains, use the `theHarvester -d <target> -b <source>` command as follows:

   ```
   kali@kali:~$ theHarvester -d microsoft.com -b dnsdumpster
   ```

   The `source` specifies the data source to use when collecting information about the target. The following snippet shows that theHarvester was able to collect sub-domains and their IP addresses for the target:

```
[*] Hosts found: 109
---------------------
assessment.changeguard.fcm.azure.microsoft.com:20.69.174.99
auditboard-ppe.microsoft.com:52.143.78.92
awsuiu1.microsoft.com:13.77.211.251
bayprofile10.microsoft.com:64.4.17.21
bayprofile11.microsoft.com:64.4.17.22
bn1vlsctest01.microsoft.com:134.170.22.43
changeguard.fcm.azure.microsoft.com:20.69.174.99
changemanager.fcm.azure.microsoft.com:20.69.145.220
cmsbn1test01.microsoft.com:134.170.22.53
cmsco2test60.microsoft.com:134.170.184.45
co1profile01.microsoft.com:65.55.57.121
comm-image.microsoft.com:13.88.11.232
cus.dlsppe.microsoft.com:52.154.223.20
dsgsinwoosbc51.microsoft.com:167.220.253.180
```

Figure 9.23 – Sub-domains

As an ethical hacker, ensure you use different data sources when collecting intelligence on a target, as some data sources will provide more information than others. Having completed this exercise, you have learned how to use theHarvester to collect sub-domains and their IP addresses. Next, you will learn how to use Knockpy to aggressively enumerate sub-domains.

Collecting sub-domains using Knockpy

Knockpy is a passive, sub-domain enumeration tool that leverages OSINT and performs dictionary scanning to identify sub-domains of a target.

To get started with using Knockpy for sub-domain enumeration, use the following instructions:

1. Firstly, power on the **Kali Linux** virtual machine.
2. Next, open the Terminal and use the following command to update the software package repository list and install **Knockpy**:

```
kali@kali:~$ sudo apt update
kali@kali:~$ sudo apt-get install knockpy
```

3. Next, to perform sub-domain enumeration on a target, use the following command:

```
kali@kali:~$ knockpy <domain>
```

The following snippet shows the Knockpy results when tested with a public domain name:

```
Ip address       Code Subdomain                                          Server
                                                    Real hostname
_______________ ____ ____________________________________________________ ________
_______________________________________________________ __________________________

_______________
(ctrl+c) | 1.15% | 6d75338fc909107.twitter.com
104.244.42.131  400  0.twitter.com                                        tsa_b
                                                    s.twitter.com
104.244.42.195  200  2012.twitter.com                                     tsa_b
                                                    s.twitter.com
104.244.42.67   200  2013.twitter.com                                     tsa_b
                                                    s.twitter.com
104.244.42.131  200  2010.twitter.com                                     tsa_b
                                                    s.twitter.com
104.244.42.67   200  2011.twitter.com                                     tsa_b
                                                    s.twitter.com
104.244.42.67   200  2014.twitter.com                                     tsa_b
```

Figure 9.24 – Knockpy results

As shown in the preceding snippet, Knockpy was able to retrieve IP addresses, hostnames, and server information about each sub-domain. Additionally, the HTTP status code indicates whether the resource was found (`200`) or not.

Having completed this section, you have learned how to set up and use various tools to efficiently identify the sub-domains of a target. In the next section, you will learn how to discover hidden directories on a web application.

Performing directory enumeration

Sometimes, web administrators and IT professionals unintentionally expose sensitive and restricted directories and files on their web applications and servers on the internet. If a threat actor were to find confidential data within a hidden directory on a target's web server, it can be leveraged to plan and perform future attacks on the target.

This section focuses on using various tools and techniques to discover hidden directories and files as an ethical hacker.

Using GoBuster to find hidden directories

GoBuster is a brute-force tool used to identify the sub-domains, directories, files, and hostnames of a target.

To get started with using GoBuster to find hidden directories and files of a domain, use the following instructions:

1. Firstly, power on the **Kali Linux** virtual machine.
2. Next, open the Terminal and use the following commands to update the software package repository list and install **GoBuster**:

```
kali@kali:~$ sudo apt update
kali@kali:~$ sudo apt install gobuster
```

3. To perform DNS sub-domain enumeration on a target, use the following commands:

```
kali@kali:~$ gobuster dns -d microsoft.com -w /usr/share/
wordlists/dirb/common.txt
```

`dns` is a syntax-specific DNS brute-force mode, and `-w` specifies an offline wordlist.

> **Tip**
>
> Within Kali Linux, there are many offline wordlists within the `/usr/share/wordlists/` directory that are commonly used for brute-force attacks and enumerating directories. Furthermore, you can download additional wordlist files from `https://github.com/danielmiessler/SecLists`.

The following snippet shows GoBuster enumerating the sub-domains of a target:

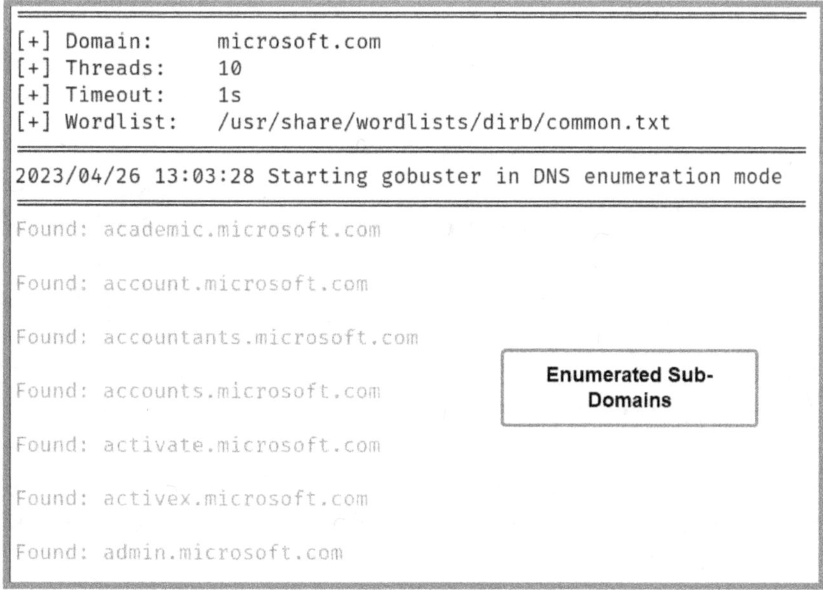

Figure 9.25 – Sub-domain enumeration

4. Next, to discover the hidden directories of a domain/website, use the following commands:

```
kali@kali:~$ gobuster dir -u https://github.com/ -w /usr/share/
wordlists/dirb/common.txt
```

The following snippet shows various directories that were found and a relevant HTTP status code:

```
/~administrator (Status: 301)
/~amanda (Status: 301)
/~ftp (Status: 301)
/~apache (Status: 301)
/~bin (Status: 301)
/~guest (Status: 301)
/~httpd (Status: 301)
/~www (Status: 301)
/08 (Status: 200)
/04 (Status: 200)
/02 (Status: 200)
/01 (Status: 200)
```

Figure 9.26 – Directories

HTTP status code 301 means the resource has permanently changed its location and a new **Uniform Resource Locator** (**URL**) is given in the response from the web server. Additionally, the HTTP status code `200` means the resource was found, which is good for ethical hackers.

> **Tip**
>
> To learn more about HTTP status codes and their meaning, please see `https://developer.mozilla.org/en-US/docs/Web/HTTP/Status`.

As you saw in the preceding snippet, Gobuster performs an aggressive directory and file enumeration on a target. This tool can help ethical hackers identify sensitive and unintentionally exposed resources on a web server.

> **Important note**
>
> To learn more about GoBuster, please see `https://github.com/OJ/gobuster`.

Having completed this exercise, you have learned how to use GoBuster to discover the sub-domains and hidden directories of a target. Next, you will learn how to perform directory enumeration using Dirb.

Directory enumeration with DIRB

DIRB is a popular web application scanner that looks for hidden directories and files on a web server. DIRB uses a wordlist to perform a dictionary-based attack on a target, querying each word from the wordlist to identify web resources.

To get started with using DIRB to identify hidden web directories and files, use the following instructions:

1. Firstly, power on the **Kali Linux** virtual machine.
2. Next, open the Terminal and run the **OWASP JuiceShop** Docker container using the following command:

```
kali@kali:~$ sudo docker run --rm -p 3000:3000 bkimminich/juice-shop
```

3. Next, open a new Terminal window and use the following commands to perform a directory-based brute-force on the **Open Web Application Security Project** (**OWASP**) JuiceShop web application:

```
kali@kali:~$ dirb http://127.0.0.1:3000/
```

The following snippet shows hidden directories that were found on the target using DIRB:

```
---- Scanning URL: http://127.0.0.1:3000/ ----
+ http://127.0.0.1:3000/assets (CODE:301|SIZE:179)
+ http://127.0.0.1:3000/ftp (CODE:200|SIZE:11062)
+ http://127.0.0.1:3000/profile (CODE:500|SIZE:1243)
+ http://127.0.0.1:3000/promotion (CODE:200|SIZE:6586)
+ http://127.0.0.1:3000/redirect (CODE:500|SIZE:3119)
+ http://127.0.0.1:3000/robots.txt (CODE:200|SIZE:28)
+ http://127.0.0.1:3000/snippets (CODE:200|SIZE:683)
```

Figure 9.27 – Directory enumeration

It's important to manually visit every URL to determine whether the hidden resource is valuable and can be leveraged when planning future attacks on the target. For instance, the `/assets` directory may contain sensitive information such as the IP addresses of additional systems owned by the target, and the `/robots.txt` resource may contain additional sub-directories.

> **Tip**
>
> To learn more about DIRB, please see `https://www.kali.org/tools/dirb/`.

Having completed this section, you have learned how to use various tools to assist when discovering hidden directories and files on a web server of a target. In the next section, you will learn how to get started with discovering web vulnerabilities.

Web application vulnerability

As an ethical hacker, it's essential to understand web application security and the OWASP **Top 10** web application security risks. Web applications are special software that runs on a server to host websites and are easily accessible using a standard web browser. For instance, when you access your favorite search engine, your web browser creates a **Hypertext Transfer Protocol** (**HTTP**) or **HTTP Secure** (**HTTPS**) `GET` message to request the resource on the destination web server (the internet search engine). Once the web application receives the HTTP `GET` message (the request), it processes and provides a response with the requested data back to the web browser:

The following diagram shows the communication between the browser and a web application:

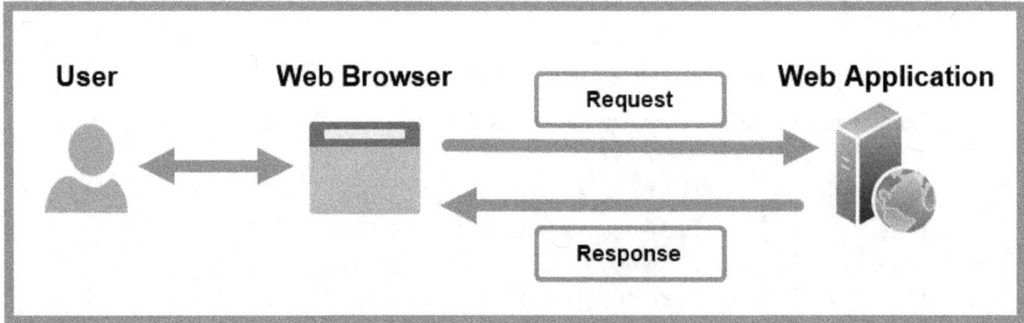

Figure 9.28 – Web communication

Web applications are created with lots of code by developers. However, many web servers on the internet run outdated and insecure web applications that can be exploited by potential hackers. As with any software, web applications need to be regularly tested by developers and ethical hackers to identify any hidden security vulnerabilities and implement secure coding practices to resolve any issues. OWASP regularly updates its Top 10 web application security risks to help cybersecurity professionals, researchers, and software developers to better understand how to find security weaknesses using thorough security testing and recommends how to resolve these security flaws before a real hacker is able to discover and exploit them.

> **Important note**
>
> This book focuses on reconnaissance for ethical hackers and does not cover exploitation. If you're interested in learning more about web application security risks, please visit the OWASP website at `https://owasp.org/www-project-top-ten/`.

There are many commercial web application scanners in the industry that work well. However, **Nikto** is a free web application scanner that's designed to perform fast security checks on a target web application or server, which we will use here.

To get started with using Nikto to discover web application vulnerabilities, please use the following instructions:

1. Firstly, power on the **Kali Linux** virtual machine.
2. Next, open the Terminal and execute the following commands to run the **OWASP JuiceShop** Docker container:

   ```
   kali@kali:~$ sudo docker run --rm -p 3000:3000 bkimminich/juice-shop
   ```

3. Next, open another Terminal and perform a web vulnerability scan on the OWASP JuiceShop web application using the following command:

   ```
   kali@kali:~$ nikto -h http://127.0.0.1:3000
   ```

   As shown in the following snippet, Nikto is able to identify common web application security vulnerabilities in the target:

```
kali@kali:~$ nikto -h http://127.0.0.1:3000
- Nikto v2.1.6
---------------------------------------------------------------------------
+ Target IP:          127.0.0.1
+ Target Hostname:    127.0.0.1
+ Target Port:        3000
+ Start Time:         2023-04-26 13:15:50 (GMT-4)
---------------------------------------------------------------------------
+ Server: No banner retrieved
+ Retrieved access-control-allow-origin header: *
+ The X-XSS-Protection header is not defined. This header can hint to the user agent
+ Uncommon header 'x-recruiting' found, with contents: /#/jobs
+ Uncommon header 'feature-policy' found, with contents: payment 'self'
+ No CGI Directories found (use '-C all' to force check all possible dirs)
+ Entry '/ftp/' in robots.txt returned a non-forbidden or redirect HTTP code (200)
+ "robots.txt" contains 1 entry which should be manually viewed.
+ /site.jks: Potentially interesting archive/cert file found.
```

Figure 9.29 – Nikto

It's important to thoroughly research each bullet point (+) listed in the Nikto results as it will provide brief details on a potential security issue. Additionally, Nikto provides **Open Sourced Vulnerability Database** (**OSVDB**) references for known security vulnerabilities.

> **Tip**
>
> For a list of web vulnerability scanning tools, please see `https://owasp.org/www-community/Vulnerability_Scanning_Tools`.

Having completed this section, you have learned how to use Nikto, a free web application vulnerability scanner, to identify common security flaws on a web application. Next, you will learn how to automate your web reconnaissance techniques using reconnaissance frameworks.

Web reconnaissance frameworks

Web reconnaissance frameworks help ethical hackers and penetration testers to simply automate many manual tasks, such as running multiple tools one after another, then collecting and consolidating the results to improve the analysis phase. With a reconnaissance framework, an ethical hacker can set the target domain and let the framework take care of thoroughly collecting information using passive and active techniques from multiple data sources and generating human-readable reports for post-analysis.

In this section, you will learn how to set up and use the following web reconnaissance frameworks:

- Sn1per
- Amass

I hope you're excited – let's dive in.

Automating reconnaissance with Sn1per

Sn1per is designed to be an all-in-one web reconnaissance framework that's built to perform extensive information gathering on discovering hidden assets and security vulnerabilities on a target, hence helping cybersecurity professionals and organizations with **Attack Surface Management** (**ASM**). Rather than running multiple reconnaissance tools for collecting and analyzing data from various sources, the Sn1per framework helps automate the entire process of data collection for a target, from analysis to providing a human-readable report at the end. Hence, it helps ethical hackers to automate time-consuming and mundane tasks to improve their reconnaissance phase and analyze data efficiently to create intelligence about a target.

To get started with using Sn1per for reconnaissance, use the following instructions:

1. Firstly, power on the **Kali Linux** virtual machine.
2. Next, open the Terminal and use the following command to update the software packages repository list:

```
kali@kali:~$ sudo apt update
```

3. Next, download the Sn1per files from their official GitHub repository:

```
kali@kali:~$ git clone https://github.com/1N3/Sn1per
```

4. Next, use the following command to change the present working directory to the `Sn1per` folder and execute the installation script:

```
kali@kali:~$ cd Sn1per
kali@kali:~/Sn1per$ sudo bash install.sh
```

The installation process usually takes a few minutes to complete.

5. To perform active reconnaissance (a normal scan) using Sn1per, use the following command:

```
kali@kali:~$ sudo sniper -t <domain-name>
```

During a normal scan, Sn1per will attempt to perform DNS enumeration, carry out sub-domain discovery, perform multiple Nmap scans, scan HTTP/DNS scripts, grab HTTP banners, check for the presence of a WAF, and discover hidden files and directories on the target.

> **Important note**
>
> During active reconnaissance (normal scanning) using Sn1per, it will attempt to directly connect to the target to collect sensitive information. Keep in mind that active reconnaissance scanning can trigger security alerts and notify the target.

The following snippet shows an Nmap scan that was performed during a normal scan with Sn1per:

```
RUNNING TCP PORT SCAN

Starting Nmap 7.93 ( https://nmap.org ) at 2023-04-30 11:20 AST
Nmap scan report for [illegible].com (172.67.168.43)
Host is up (0.065s latency).
Other addresses for [illegible].com (not scanned): [illegible]
Not shown: 59 filtered tcp ports (no-response)
Some closed ports may be reported as filtered due to --defeat-rst-ratelimit
PORT     STATE SERVICE
80/tcp   open  http
443/tcp  open  https
8080/tcp open  http-proxy
8443/tcp open  https-alt

Nmap done: 1 IP address (1 host up) scanned in 2.89 seconds
```

Figure 9.30 – Port scanning

Additionally, the following snippet shows that Sn1per was able to identify the hidden files and directories on a target:

```
RUNNING COMMON FILE/DIRECTORY BRUTE FORCE

v0.4.2

Target: http://[illegible].com:80/
[11:21:42] Starting:
[11:21:44] 200 -    3KB - /1.txt
[11:21:45] 200 -    3KB - /123.txt
[11:21:45] 200 -    3KB - /2.txt
[11:21:46] 200 -    3KB - /access.txt
[11:21:46] 200 -    3KB - /accounts.txt
[11:21:46] 200 -    3KB - /admin/access.txt
[11:21:47] 200 -    3KB - /admin/error.txt
[11:21:49] 200 -    3KB - /admins/log.txt
[11:21:49] 200 -    3KB - /admin/_logs/login.txt
[11:21:49] 200 -    3KB - /admin/logs/login.txt
```

Figure 9.31 – Discovering hidden files and directories

6. To perform passive reconnaissance using Sn1per, use the following commands:

```
kali@kali:~$ sudo sniper -t <domain-name> -m stealth -o -re
```

The -o syntax specifies to use OSINT and -re specifies reconnaissance.

Important note

Passive reconnaissance techniques and tools do not directly connect to the target but rather collect and analyze OSINT to create a profile about the target. Passive reconnaissance is recommended if you're interested in reducing your threat level in triggering alerts on the target.

7. There are many types of modes and scans available in Sn1per; use the following command to view the entire listing:

```
kali@kali:~$ sudo sniper --help
```

Sn1per scans can take a few minutes to complete each task as it performs extensive information gathering on the target. Once a scan is completed, the loot (collected data) is stored within the /usr/share/sn1per/loot/ directory and a summary report in HTML format is generated and stored in the /usr/share/sn1per/loot/workspace/<target-name>/reports/ directory.

Important note

To learn more about Sn1per, please see the following link: `https://github.com/1N3/Sn1per`.

Next, you will learn how to perform web reconnaissance using Amass.

Using Amass for web reconnaissance

Amass is a web reconnaissance framework developed by OWASP to perform advanced DNS enumeration and help cybersecurity professionals and organizations with attack surface mapping. Amass works by identifying the external assets of an organization by using both passive and active reconnaissance techniques and tools through automation. Hence, it helps ethical hackers to efficiently collect and analyze data from multiple sources to create a profile of a target.

Amass can collect information from various online databases using **Application Programming Interfaces** (**APIs**), digital certificates, DNS information, public routing databases, WHOIS records, and web-archiving databases.

To get started using Amass for web reconnaissance, please use the following instructions:

1. Firstly, power on the **Kali Linux** virtual machine.
2. Next, open the Terminal and use the following command to update the software package repository list and install Amass:

   ```
   kali@kali:~$ sudo apt update
   kali@kali:~$ sudo apt install amass
   ```

   The installation process usually takes a few minutes to complete.

3. To perform sub-domain enumeration using passive reconnaissance on a target, use the following commands:

   ```
   kali@kali:~$ amass enum -d <domain-name>
   ```

The following snippet shows that Amass was able to identify the sub-domains of the target:

```
kali@kali:~$ amass enum -d microsoft.com
tide510.microsoft.com
mail-db9lp0239.outbound.messaging.microsoft.com
wus2-stagingretrieval-dcatbackend-int.staging.bigcatalog-int.commerce.microsoft.com
tide530.microsoft.com
52-114-125-40.relay.teams.microsoft.com
52-114-124-131.relay.teams.microsoft.com
order.rest.store.internal.co1c.microsoft.com
52-114-124-248.relay.teams.microsoft.com
52-114-125-41.relay.teams.microsoft.com
52-114-125-25.relay.teams.microsoft.com
52-114-124-41.relay.teams.microsoft.com
spteam2010.microsoft.com
```

Figure 9.32 – Sub-domain enumeration

Once the task is completed, Amass provides the network block information showing where the sub-domains and their IP addresses are associated:

```
38 names discovered - scrape: 29, api: 6, archive: 1, cert: 2

ASN: 262589 - INTERNEXA BRASIL OPERADORA DE TELECOMUNICACOES S.A
        23.219.152.0/22         2     Subdomain Name(s)
ASN: 20940 - AKAMAI-ASN1
        23.33.40.0/21           2     Subdomain Name(s)
        23.56.5.0/24            4     Subdomain Name(s)
        23.50.8.0/24            2     Subdomain Name(s)
ASN: 0 - Not routed
        64.5.32.0/19            1     Subdomain Name(s)
ASN: 3598 - MICROSOFT-CORP-AS - Microsoft Corporation
        131.107.0.0/17          5     Subdomain Name(s)
ASN: 8075 - MICROSOFT-CORP-MSN-AS-BLOCK - Microsoft Corporation
        213.199.128.0/18        3     Subdomain Name(s)
        52.112.0.0/14           11    Subdomain Name(s)
        70.37.128.0/18          1     Subdomain Name(s)
        20.64.0.0/10            9     Subdomain Name(s)
```

Figure 9.33 – Network block information

The network block data can be used to identify the geo-location, hosting provider, country, **Autonomous System** (**AS**), and **Internet Service Provider** (**ISP**) details.

4. To perform an intel scan to discover targets for enumeration, use the following command:

```
kali@kali:~$ amass intel -whois -d microsoft.com -dir /home/
kali/amass-target1
```

The `-dir` syntax specifies where to store the output in a local folder on Kali Linux. The intel scan helps ethical hackers to collect OSINT about the target and identify additional root domain names that are associated with the target as shown:

```
kali@kali:~$ amass intel -whois -d microsoft.com -dir /home/kali/amass-target1
0fficeteam.com
1-800-microsoft.com
121hotmailhelp.com
123hotmail.com
123hotmail.net
123lovehotmail.com
123xbox.com
1drive.com
1drive.net
1drv.com
1hotmail.com
1xbox.com
```

Figure 9.34 – Intel scan

5. Next, to perform passive DNS enumeration and map the attack surface of exposed assets to a target, use the following command:

```
kali@kali:~$ amass enum -passive -d microsoft.com -src
```

The following snippet shows Amass was able to identify the sub-domains of a target by collecting information from various online sources:

```
kali@kali:~$ amass enum -passive -d microsoft.com -src
[RapidDNS]        wus2-frontdoor-displaycatalog-int.bigcatalog.microsoft.com
[AnubisDB]        000dco2o40pl1.redmond.corp.microsoft.com
[AnubisDB]        accountservices.microsoft.com
[AnubisDB]        globalmigration.partners.extranet.microsoft.com
[AnubisDB]        servicedeliveryws.microsoft.com
[RapidDNS]        36-usce.noam.prd.audience.teams.microsoft.com
[AnubisDB]        dns11.one.microsoft.com
[RapidDNS]        reportingapi.bingads.microsoft.com
```

Figure 9.35 – DNS and attack surface mapping

6. To perform active reconnaissance on a target, use the following commands:

```
kali@kali:~$ amass enum -d <domain-name> -src -ip -brute -dir /
home/kali/amass-target1
```

The following snippet shows Amass was able to identify the associated IP addresses for each sub-domain of the target:

```
kali@kali:~$ amass enum -d microsoft.com -src -ip -brute -dir /home/kali/amass-target1
[Brute Forcing]    broadcast.microsoft.com 209.240.199.60
[DNS]              microsoft.com 20.112.52.29,20.103.85.33,20.84.181.62,20.81.111.85
[Brute Forcing]    demo.microsoft.com 20.112.52.29,20.84.181.62,20.103.85.33,20.81.111.85
[Brute Forcing]    mail2.microsoft.com 131.107.115.215
[DNSSpy]           mail6.microsoft.com 205.248.106.32
[Brute Forcing]    mail.microsoft.com 167.220.71.19
[Brute Forcing]    shop.microsoft.com 20.84.181.62,20.81.111.85,20.103.85.33,20.112.52.29
[Brute Forcing]    mi.microsoft.com 20.84.181.62,20.81.111.85,20.112.52.29,20.103.85.33
[DNS]              epgdata.microsoft.com 209.240.199.60
```

Figure 9.36 – Mapping IP addresses to sub-domains

7. Next, to list the enumeration data that was collected on a target using Amass, use the following commands:

```
kali@kali:~$ amass db -dir /home/kali/amass-target1 -list
```

The following snippet shows the reports that were created by Amass during the enumeration process:

```
kali@kali:~$ amass db -dir /home/kali/amass-target1 -list
1) 04/30 17:22:19 2023 UTC → 04/30 17:25:21 2023 UTC: adobe.com, windows.n
et, microsoft.com, azure.com, akadns.net, azure-dns.org, mktoweb.com, azure
-dns.com, office.com, 1eslivesecrets.azurewebsites.net, azure-dns.info, a-m
sedge.net, wpeproxy.com, azurefd.net, dispositionjournalfd-eastus2-commerci
al.azurewebsites.net, azure-dns.net, sharepoint.com, lync.com, b-msedge.net
, nsatc.net, ost-thirdpartycode-prod.azurewebsites.net, audience-prd-noam-u
sce-14.cloudapp.net, trafficmanager.net, edgekey.net, dynamitecdn.com, best
practice-prod.azurewebsites.net, outlook.com, office.net, p.azurewebsites.n
et
```

Figure 9.37 – Listing reports

8. To generate a visual graph of the collected data, please use the following commands:

```
kali@kali:~$ amass viz -d3 -dir /home/kali/amass-target1
```

9. To open the visual graph using the web browser, use the following commands:

```
kali@kali:~$ firefox /home/kali/amass-target1/amass.html
```

The following snippet shows the visual graph from Amass:

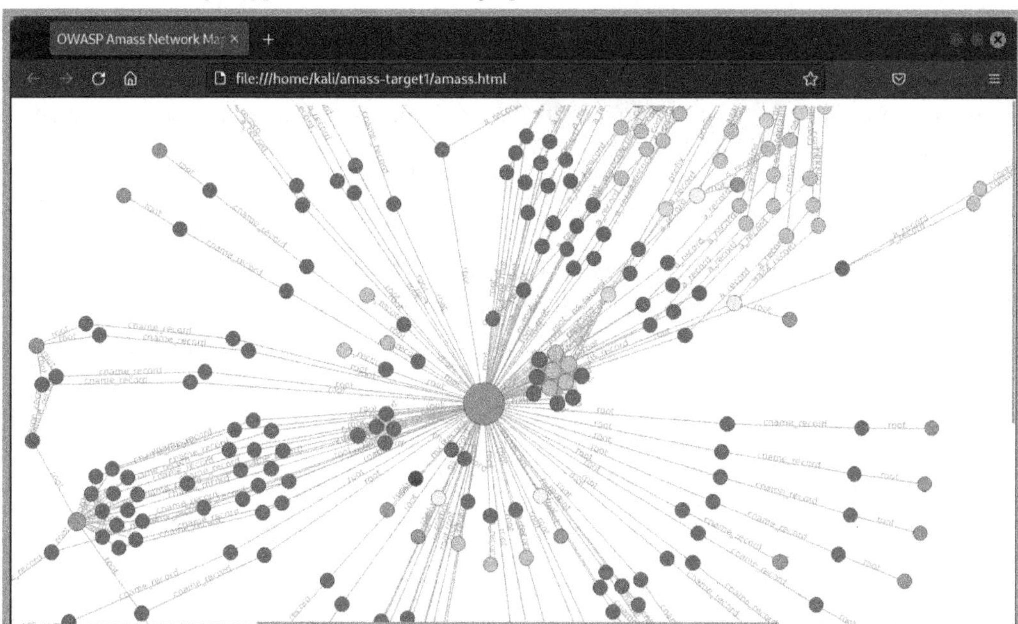

Figure 9.38 – Data visualization graph

The visual graph helps ethical hackers to easily map sub-domains and IP addresses and identify the attack surface of a target.

> **Important note**
>
> To learn more about Amass and its usage, please see `https://github.com/owasp-amass/amass/wiki/User-Guide`.

Having completed this section, you have learned how to automate web reconnaissance using both the Sn1per and Amass frameworks.

Summary

Over the course of this chapter, you learned how to perform web and domain reconnaissance using both passive and active information-gathering techniques. Additionally, you gained the hands-on skills that are needed by ethical hackers to identify the attack surface of a target and its sub-domains and discover hidden files and directories. Lastly, you explored how web reconnaissance frameworks can be used by ethical hackers to automate the discovery of an attack surface for a target.

I hope this chapter has been informative for you and helpful in your journey in the cybersecurity industry. In the next chapter, *Implementing Reconnaissance Monitoring and Detection Systems*, you will learn how to implement various open source threat monitoring tools to identify network intrusions and determine whether reconnaissance-based attacks are happening within your organization.

Further reading

- OWASP Top 10 – `https://owasp.org/www-project-top-ten/`
- OWASP web reconnaissance – `https://owasp.org/www-project-web-security-testing-guide/stable/4-Web_Application_Security_Testing/01-Information_Gathering/`

10

Implementing Recon Monitoring and Detection Systems

The longer an organization takes to discover that its network infrastructure and systems are compromised, the more time adversaries spend in their network compromising additional systems to expand their foothold within the victim's network. Implementing and working with network security tools helps **Security Operation Centers** (**SOCs**) and **Digital Forensic and Incident Response** (**DFIR**) teams to effectively monitor network traffic, detect potential threats, and provide real-time alerts for faster response times, containment, and eradication of threats.

In this chapter, you will learn the fundamentals of using Wireshark to perform packet analysis as an ethical hacker. Additionally, you will learn how to deploy and use Security Onion to monitor a network and detect suspicious activities to further determine whether it's a security event or an incident within your organization.

In this chapter, we will cover the following topics:

- Wireshark for ethical hackers
- Monitoring and detection systems

Let's dive in!

Technical requirements

To follow along with the exercises in this chapter, please ensure that you have met the following hardware and software requirements:

- Wireshark - `https://www.wireshark.org/`
- Security Onion – `https://securityonionsolutions.com/`
- VMware Workstation Pro – `https://www.vmware.com/products/workstation-pro.html`

Wireshark for ethical hackers

Wireshark is a popular and very powerful network protocol analyzer that's commonly used by both networking and cybersecurity professionals in the industry to identify networking issues and threats within an organization's network. Wireshark enables professionals to capture network packets and perform analysis to gain better insights into what's happening in the network by viewing the contents of each packet or frame.

As you may know, the **Network Interface Card** (**NIC**) of a sender device is responsible for converting data into a signal that can be transported over a certain form of media, such as electrical signals for copper wires, light signals for fiber optics, and radio frequency for wireless communication. Wireshark can capture these signal types and convert them into human-readable language, helping us to better analyze network traffic.

Wireshark helps ethical hackers to detect security vulnerabilities within network-based applications, services, and protocols that can be exploited by threat actors. The following are common reasons why ethical hackers use Wireshark during security assessments:

- **Packet sniffing** – Ethical hackers can set up a network implant on the network to capture sensitive and confidential data sent over an organization's network, which can be leveraged to perform future operations. In addition, ethical hackers can identify whether the organization is using any insecure network protocols that transmit messages in plaintext.
- **Protocol analysis** – Protocol analysis enables ethical hackers to better understand the behavior of the network, employees, and devices by observing network protocols and identifying services and applications that are running on critical systems such as servers. By identifying network protocols, ethical hackers can research whether there are any known security vulnerabilities within a protocol and how they can be exploited by an adversary.
- **Malware analysis** – Cybersecurity professionals such as threat hunters and incident response teams use Wireshark to identify whether there's malware on their network, the type of malware, and whether it has established a **Command and Control** (**C2**) communication channel to an online server.

- **Exploit development** – An important stage when developing an exploit is testing to ensure it's working as expected and determining its threat level. For instance, if a threat actor creates and delivers an exploit that establishes a remote connection to the attacker's machine, the network connection may be detected as suspicious activity and blocked by the organization's security team. Using Wireshark helps you better understand what is seen by networking and cybersecurity professionals.

Wireshark can be a bit daunting for beginners as it provides a lot of details of everything it has captured on the network, and its filters enable you to fully leverage the capabilities of this tool to analyze network traffic.

The following are Wireshark operators and their descriptions:

- `==` – This operator enables you to specify an exact match. For instance, if you want to filter packets with TCP port `80`, the filter will be `tcp.port == 80`.
- `!=` – This operator means *not equal to* and it's commonly used to exclude something. For instance, if you want to filter traffic that does not include TCP port `22`, the filter will be `tcp.port != 22`.
- < or > – The *less than* or *greater than* operators are used to specify a value either less than or greater than within a packet. For instance, if you want to filter TCP packets that are greater than 150 bytes, the filter will be `tcp.len > 150`.
- `&&` – The *and* operator is used to combine multiple criteria that are all true. For instance, if you want to filter traffic from a specific sender to a destination address, the filter will be `(ip.src == 172.16.254.128) && (ip.dst == 8.8.8.8)`.
- `||` – The *or* operator is used to specify multiple criteria where only one must be true. For instance, if you want to filter either HTTP (port `80`) or HTTPS (port `443`) packets, the filter will be `(tcp.port == 80 ) || (tcp.port == 443)`.

The following are common Wireshark display filters:

- To filter traffic based on a specific device with the **Media Access Control** (**MAC**) address `00:0c:29:b6:b5:48`:

  ```
  eth.src == 00:0c:29:b6:b5:48
  ```

- To filter **Address Resolution Protocol** (**ARP**) messages from a device with a source MAC address of `00:0c:29:b6:b5:48`:

  ```
  arp.src.hw_mac == 00:0c:29:b6:b5:48
  ```

- To filter traffic based on a specific device with an IP address of `172.16.254.128`:

  ```
  ip.src == 172.16.254.128
  ```

- To filter ARP messages from a device with the IP address `172.16.254.128`:

```
arp.src.proto_ipv4 == 172.16.254.128
```

- The following common operators to filter traffic types:
    - `eth` – Filters Ethernet messages
    - `ip` – Filters IPv4 packets
    - `ipv6` – Filters IPv6 packets
    - `icmp` – Filters **Internet Control Message Protocol** (**ICMP**) v4 messages
    - `icmpv6` – Filters ICMPv6 messages
    - `tcp` – Filters **Transmission Control Protocol** (**TCP**) messages
    - `udp` – Filters **User Datagram Protocol** (**UDP**) messages
    - `http`, `http2`, and `http3` – Filters HTTP, HTTP/2 and HTTP/3 protocols
    - Common application layer protocols such as **Domain Name System** (**DNS**), **Simple Mail Transfer Protocol** (**SMTP**), and **Secure Shell** (**SSH**) are also filters when written in lowercase letters, such as `dns`, `smtp`, `ssh`, and so on.

> **Tip**
> To learn more about Wireshark display filters, please see `https://wiki.wireshark.org/DisplayFilters/`.

To get started with using Wireshark, please use the following instructions:

1. Go to the official Wireshark website at `https://www.wireshark.org/` to download and install Wireshark on your computer, or you can open the Wireshark application within the Kali Linux virtual machine.
2. Let's download a sample packet capture file, go to `https://www.honeynet.org/challenges/forensics-challenge-14-weird-python/` and download the `conference.pcapng` file. Additionally, you can download more Wireshark sample files from `https://wiki.wireshark.org/SampleCaptures`.

3. Next, when you launch the Wireshark application on your computer, you will see a list of available network adapters as shown:

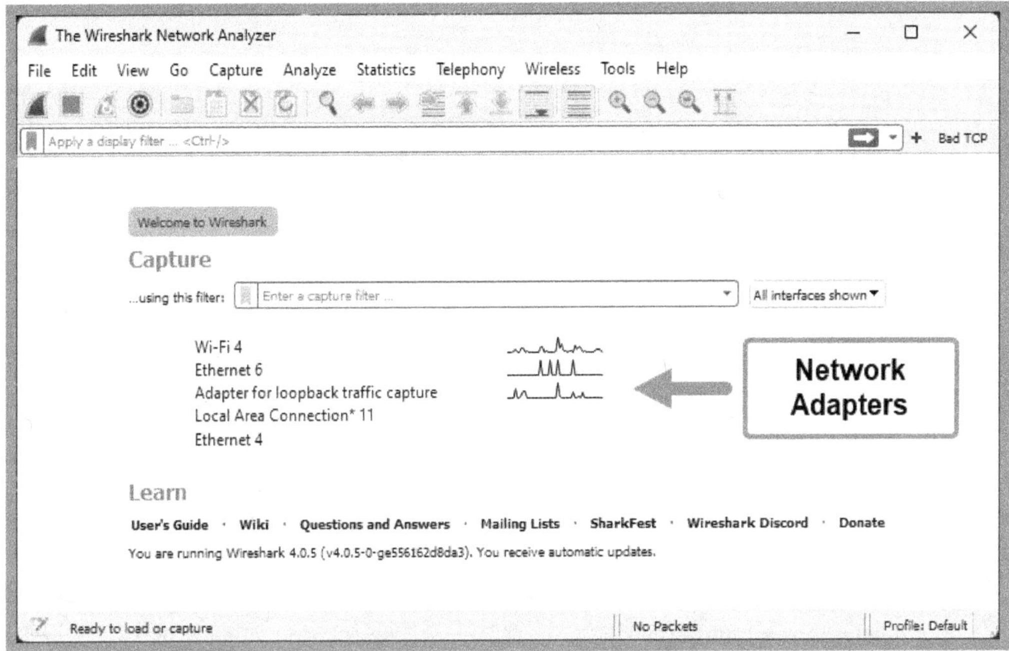

Figure 10.1 – Wireshark interface

If you double-click on an interface, Wireshark will begin to capture packets on that same interface and show you the live traffic.

4. To capture traffic, it's important to set your monitoring/capturing interface to *promiscuous mode* by clicking on **Capture** | **Options**, then enabling **Promiscuous** mode on your preferred interface:

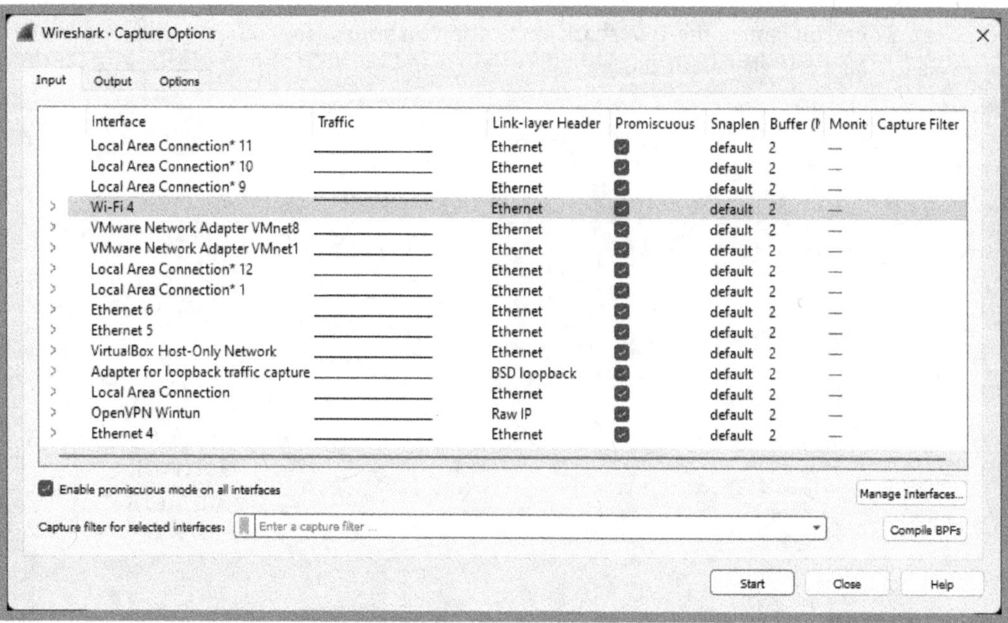

Figure 10.2 – Network adapters

Additionally, the **Output** tab enables you to customize the duration, number of packets, and location when storing the captured data:

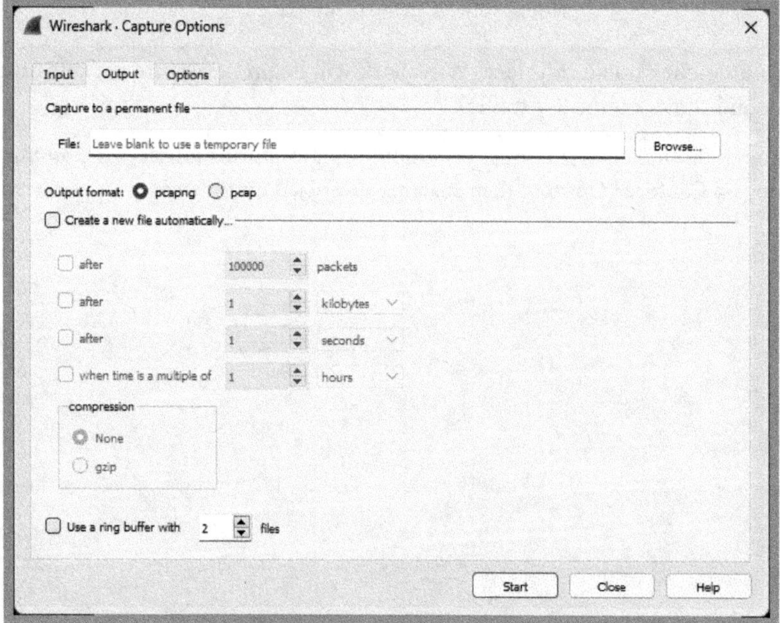

Figure 10.3 – Capture options

Once you click on **Start**, the packet capture will begin on the selected interface. However, do not start the capture; instead, we'll use the `conference.pcapng` file for our analysis.

5. To load the `conference.pcapng` file, click on **File | Open**, and select the sample file; then the packets will populate in Wireshark as shown:

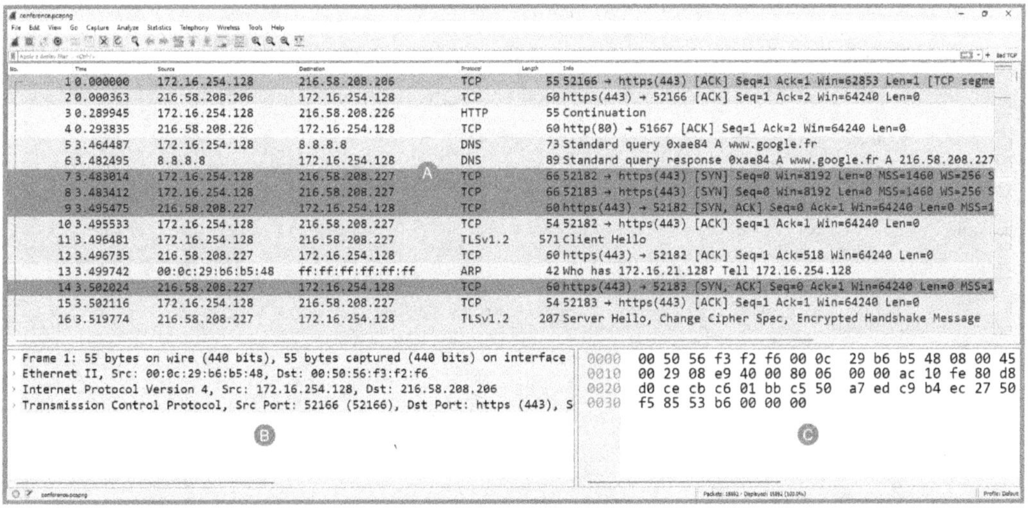

Figure 10.4 – Wireshark

As shown in the preceding snippet, **A** shows the Packet List pane, **B** shows the Packet Details pane and **C** shows the Packet Bytes pane. When a packet is selected in the Packet List, its details are shown in the Packet Details pane.

6. To filter packets with a source IP address of `172.16.254.128`, select packet 5 from the **Packet List** pane, then expand the **Internet Protocol Version 4** header, and right-click on the **Source Address** field as shown:

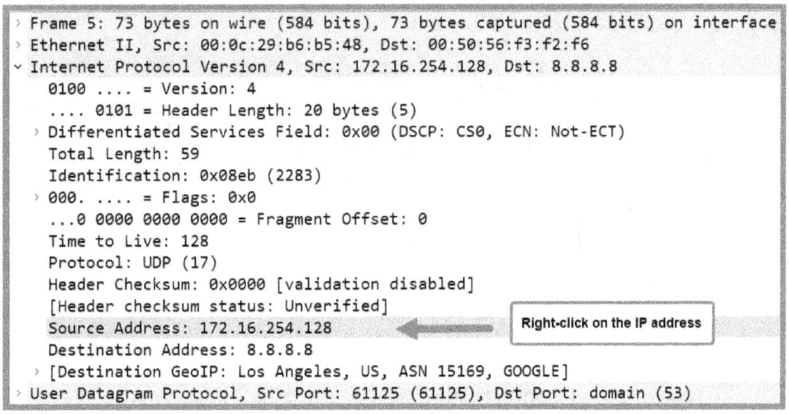

Figure 10.5 – Packet details

Then, click on **Apply as Filter | Selected** as shown:

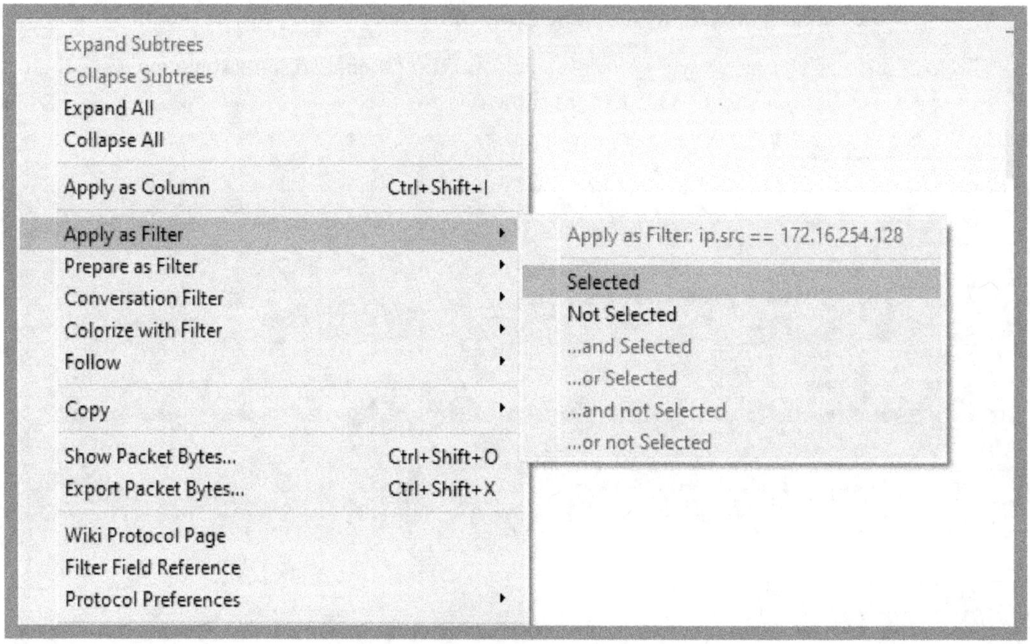

Expand Subtrees
Collapse Subtrees
Expand All
Collapse All
Apply as Column Ctrl+Shift+I
Apply as Filter
Prepare as Filter
Conversation Filter
Colorize with Filter
Follow
Copy
Show Packet Bytes... Ctrl+Shift+O
Export Packet Bytes... Ctrl+Shift+X
Wiki Protocol Page
Filter Field Reference
Protocol Preferences

Apply as Filter: ip.src == 172.16.254.128
Selected
Not Selected
...and Selected
...or Selected
...and not Selected
...or not Selected

Figure 10.6 – Creating filters

As shown in the preceding snippet, the **Apply as Filter** options will immediately apply the display filter. The **Prepare as Filter** options enable you to create filters without applying them immediately.

The following snippet shows the display filter was applied and shows all the packets with a source IP address of `172.16.254.128`:

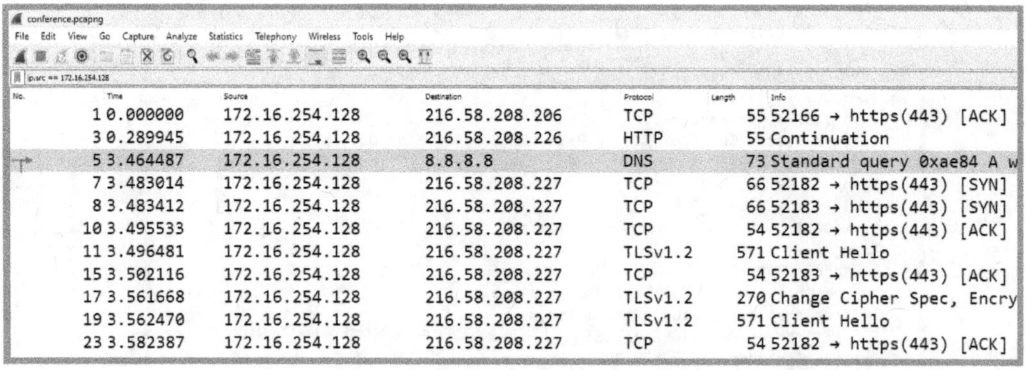

conference.pcapng

File Edit View Go Capture Analyze Statistics Telephony Wireless Tools Help

ip.src == 172.16.254.128

No.	Time	Source	Destination	Protocol	Length	Info
1	0.000000	172.16.254.128	216.58.208.206	TCP	55	52166 → https(443) [ACK]
3	0.289945	172.16.254.128	216.58.208.226	HTTP	55	Continuation
5	3.464487	172.16.254.128	8.8.8.8	DNS	73	Standard query 0xae84 A w
7	3.483014	172.16.254.128	216.58.208.227	TCP	66	52182 → https(443) [SYN]
8	3.483412	172.16.254.128	216.58.208.227	TCP	66	52183 → https(443) [SYN]
10	3.495533	172.16.254.128	216.58.208.227	TCP	54	52182 → https(443) [ACK]
11	3.496481	172.16.254.128	216.58.208.227	TLSv1.2	571	Client Hello
15	3.502116	172.16.254.128	216.58.208.227	TCP	54	52183 → https(443) [ACK]
17	3.561668	172.16.254.128	216.58.208.227	TLSv1.2	270	Change Cipher Spec, Encry
19	3.562470	172.16.254.128	216.58.208.227	TLSv1.2	571	Client Hello
23	3.582387	172.16.254.128	216.58.208.227	TCP	54	52182 → https(443) [ACK]

Figure 10.7 – Filtering traffic based on the source IP address

7. Next, to filter traffic between `172.16.254.128` and `8.8.8.8`, expand Packet Details for packet 5 again, right-click on **Destination Address: 8.8.8.8**, and select **Apply as Filter | Selected**:

No.	Time	Source	Destination	Protocol	Length	Info
5	3.464487	172.16.254.128	8.8.8.8	DNS	73	Standard query 0xae84 A www.google.fr
63	3.932662	172.16.254.128	8.8.8.8	DNS	75	Standard query 0xfec2 A ssl.gstatic.com
204	4.398457	172.16.254.128	8.8.8.8	DNS	74	Standard query 0xc9f8 A www.reddit.com
462	5.853759	172.16.254.128	8.8.8.8	DNS	75	Standard query 0x7ff9 A www.gstatic.com
525	6.026648	172.16.254.128	8.8.8.8	DNS	75	Standard query 0x391b A apis.google.com
970	7.861516	172.16.254.128	8.8.8.8	DNS	84	Standard query 0x1553 A www.google-analytics.com
971	7.861831	172.16.254.128	8.8.8.8	DNS	76	Standard query 0x00d0 A en.wikipedia.org
978	7.881850	172.16.254.128	8.8.8.8	DNS	74	Standard query 0x2b60 A soundcloud.com
1048	8.227485	172.16.254.128	8.8.8.8	DNS	71	Standard query 0x18df A i.imgur.com
1049	8.227771	172.16.254.128	8.8.8.8	DNS	69	Standard query 0x0dd0 A imgur.com

Figure 10.8 – Creating display filters

As shown in the preceding snippet, the amended display filters help us to view conversations between specific source and destination hosts on a network.

8. To perform automatic name resolution of IP addresses, port numbers, and MAC addresses, click on **Edit | Preferences | Name Resolution** to enable these features as shown:

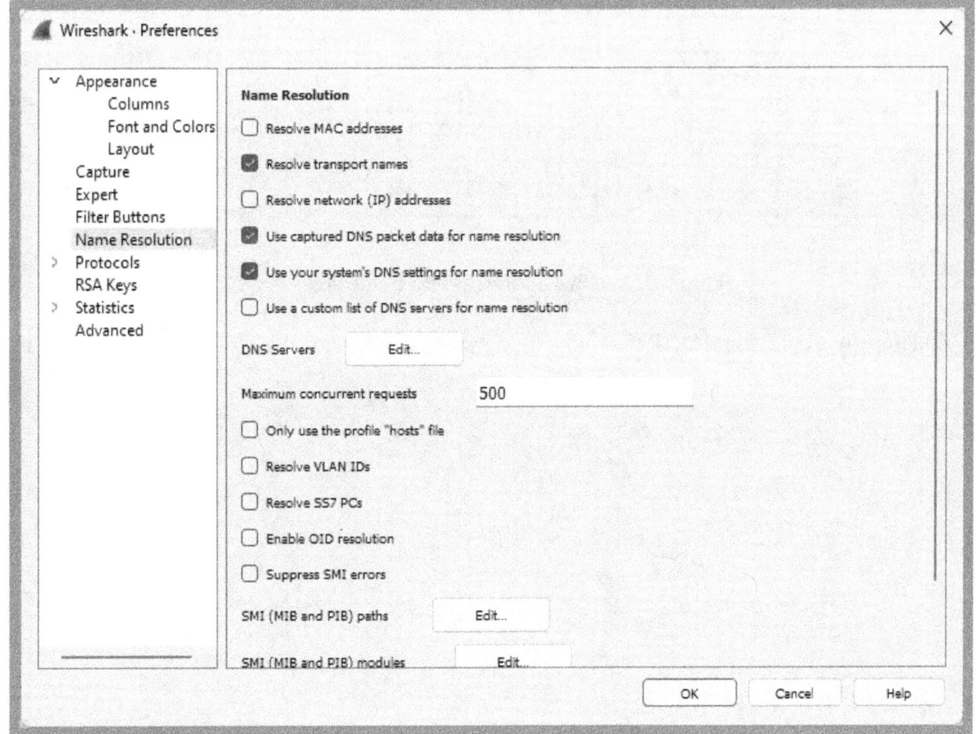

Figure 10.9 – Name resolution

9. To view all network conversations using Wireshark, click on **Statistics** | **Conversations**:

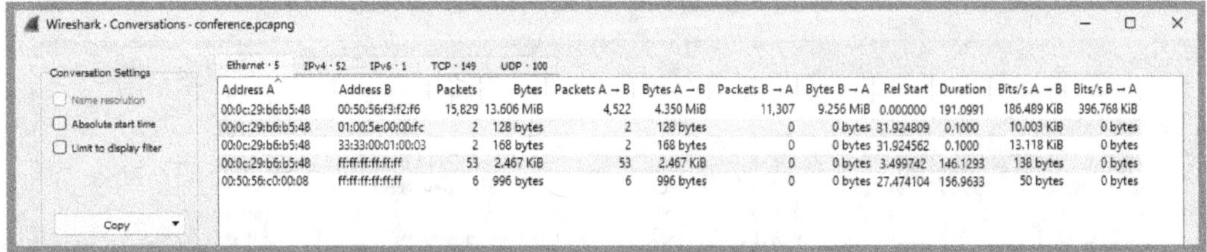

Figure 10.10 – Ethernet conversations

As shown in the preceding snippet, Wireshark shows all the source and destination MAC addresses, the numbers of frames, and the traffic within this sample file.

Next, click on **IPv4** to view all IP addresses that are detected within the sample file and traffic flows:

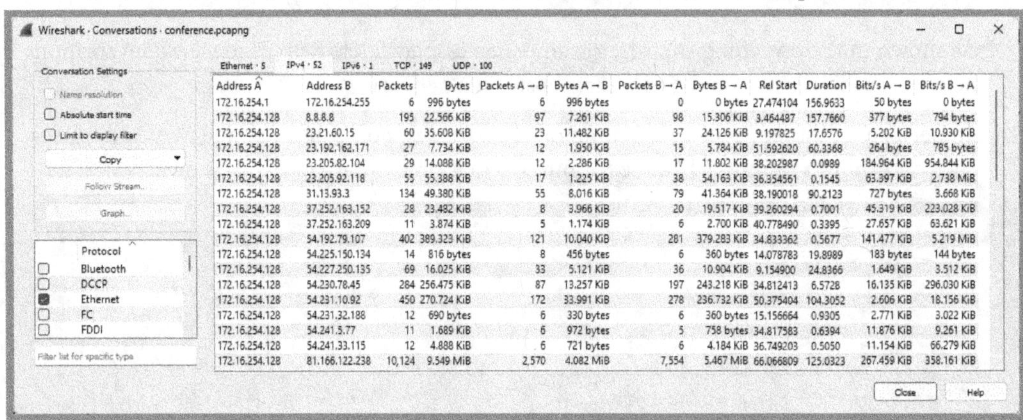

Figure 10.11 – IPv4 addresses and traffic flows

Additionally, the **TCP** and **UDP** tabs show the network services and protocols of the packet capture:

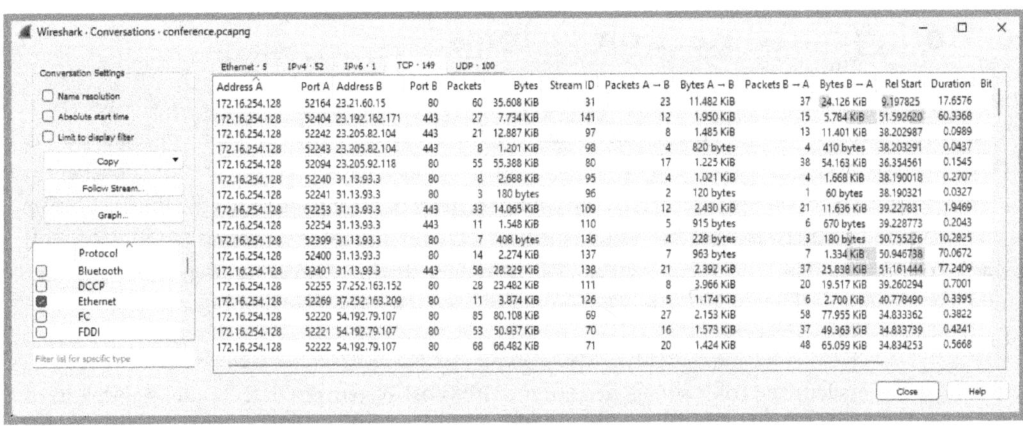

Figure 10.12 – Port numbers

10. Lastly, to view and export files that are found within the capture file, click on **File** | **Export Objects** | **HTTP**:

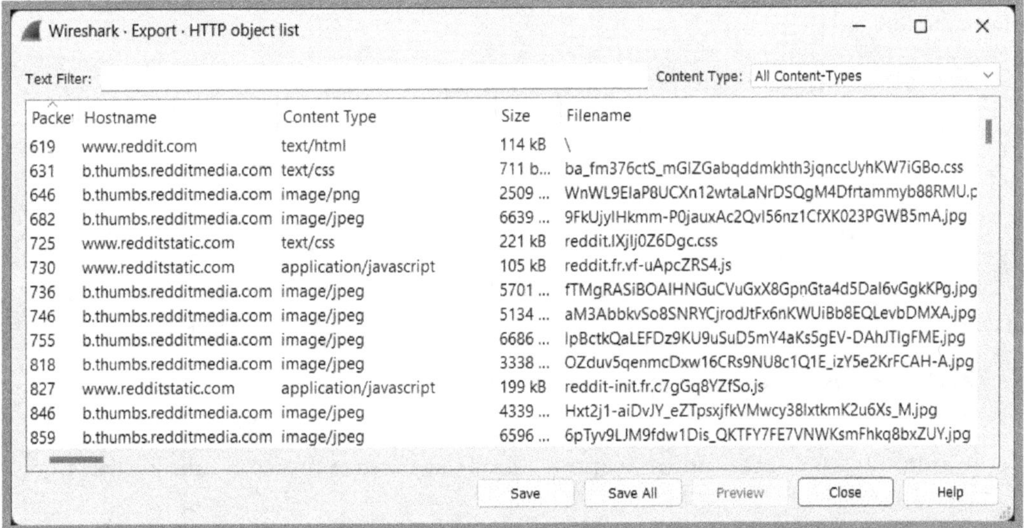

Figure 10.13 – Objects

You have the option to select a file or all files to extract from this sample capture and save on your desktop.

During this section, you have gained an introduction to leveraging Wireshark to capture network traffic as a cybersecurity professional. In the next section, you will learn how to use security tools to detect suspicious activities.

Monitoring and detection systems

Actively monitoring network traffic within an organization helps cybersecurity professionals to detect whether there's a threat on their system and which systems are compromised. For instance, the security team may notice an increase in DNS queries from an internal server that's sending packets to an unknown DNS server on the internet. By analyzing the network traffic, it seems the internal server was compromised by malware that has established a C2 channel to a C2 server on the internet and is using a common network protocol such as DNS to reduce its threat level and detection. Without monitoring network traffic, this security incident may have gone unnoticed and the adversary would continue to expand their attack.

Furthermore, cybersecurity professionals need to monitor host systems to detect potential threats at an early stage. Once a threat is found, the security team can apply immediate actions to contain and eradicate the threat before it can spread and exploit additional systems and cause more damage. The data collected during this security incident can be used to improve incident response and handling, and threat detection on other systems.

Many free and commercial threat monitoring and detection tools are widely available from trusted providers on the internet. However, it's important for organizations to implement the core tools and perform the core tasks:

- **Host-based Intrusion Detection Systems (HIDSes)**
- **Network-based Intrusion Detection Systems (NIDSes)**
- Network traffic and packet analysis
- Data visualization

The following are popular free and open source network security tools that help organizations monitor and detect threats:

- **Zeek** – This is an open source network security tool that's used to monitor and analyze network traffic
- **Wazuh** – This is an open source security platform that's used to collect and analyze data from host devices
- **Suricata** – This is an open source network analysis tool that is used to monitor network traffic and provide alerts on potential threats

> **Note**
>
> You can find links on how to set up Zeek, Wazuh, and Suricata in the *Further reading* section.

Setting up the preceding tools can be quite fun and is a good learning experience to ensure each tool is working as expected and sharing data. However, it can be quite a time-consuming process, especially if there's any troubleshooting that needs to resolve any issues along the way.

Security Onion is an all-in-one security platform that has Zeek, Wazuh, Suricata, and many more open source tools already installed and configured to work together to help cybersecurity professionals and security teams get up and running quickly with threat monitoring, detection, and hunting.

The following host-based tools can be installed on host/client devices that enable Security Onion to collect additional data from assets:

- **Wazuh** – This is a HIDS platform on Security Onion, enabling security professionals to install the Wazuh agent on a host to collect additional data and logs for analysis
- **Osquery** – This is an endpoint agent that's installed on a host device and used for monitoring and collecting analytics
- **Beats** – This is a data shipping component for Elasticsearch that's used to collect logs and metrics from host devices and send them to Security Onion for analysis

The following are network-based tools that enable Security Onion to identify suspicious activities on a network:

- **Google Stenographer** – This tool enables Security Onion to perform full packet capture of network traffic
- **Suricata** – This performs network traffic analysis and generates alerts for potential threats
- **Zeek** – It provides connection logs and metadata on network traffic and extracts files from network data for further analysis
- **Strelka** – This performs file analysis to detect threats

The following are analyst tools that are used by cybersecurity professionals on Security Onion:

- **SOC** –Security Onion Console, which provides the user access to all other tools
- **Hunt** – Provides a dashboard that's used to perform queries for threat hunting
- **Kibana** – Used to provide data visualization on network traffic and threats
- **Cases** – Allows the user to escalate an alert to a case for tracking the investigation during its life cycle
- **CyberChef** – Allows you to decode and analyze artifact data
- **Playbook** – Enables users to create detection playbooks for identifying threats based on rules
- **FleetDM** – Allows you to run live and scheduled queries on endpoints
- **Navigator** – Provides visualization on the MITRE ATT&CK framework

When setting up Security Onion on a network, it's important to ensure there's a dedicated network adapter for management and another for monitoring. The management interface is used to administratively access Security Onion and leverage its pre-installed tools for monitoring and detecting threats. The monitor interface is used to collect data and traffic from the network such as from a network tap or **Switched Port Analyzer** (**SPAN**) port on a switch.

The following is a simplified diagram showing the placement of Security Onion:

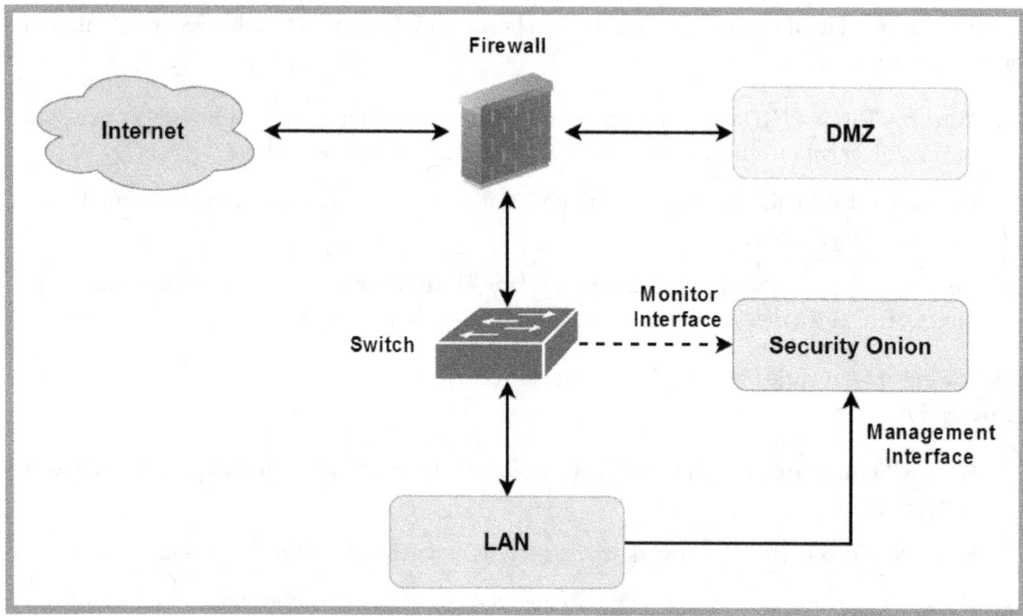

Figure 10.14 – Security Onion

As shown in the preceding diagram, the switch is configured with a mirrored or SPAN interface to send a copy of network traffic between the **Local Area Network** (**LAN**) and the firewall to the monitor interface on Security Onion. This enables Security Onion to monitor and analyze the traffic that's flowing between the internet and the internal network. However, a network tap needs to be implemented between the firewall and the **Demilitarized Zone** (**DMZ**) to enable Security Onion to monitor traffic within that area of the network.

The following are the recommended specifications and considerations for a Security Onion virtual environment:

- **CPU**: 4 cores
- **Memory**: 12 GB RAM
- **Storage**: 200 GB
- 2 NICs

To get started with setting up Security Onion, please use the following instructions:

Part 1 – setting up the environment

1. Firstly, download the **Security Onion** ISO image from `https://securityonionsolutions.com/software/` or `https://github.com/Security-Onion-Solutions/securityonion/blob/master/VERIFY_ISO.md`.
2. Next, open **VMware Workstation Pro** and click on **File** | **New Virtual Machine...** as shown:

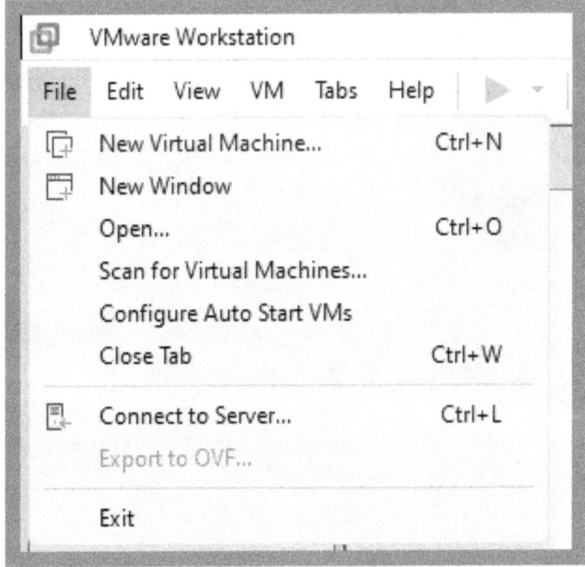

Figure 10.15 – VMware Workstation

3. Next, the **New Virtual Machine Wizard** window will appear; select **Custom (advanced)** and click on **Next**:

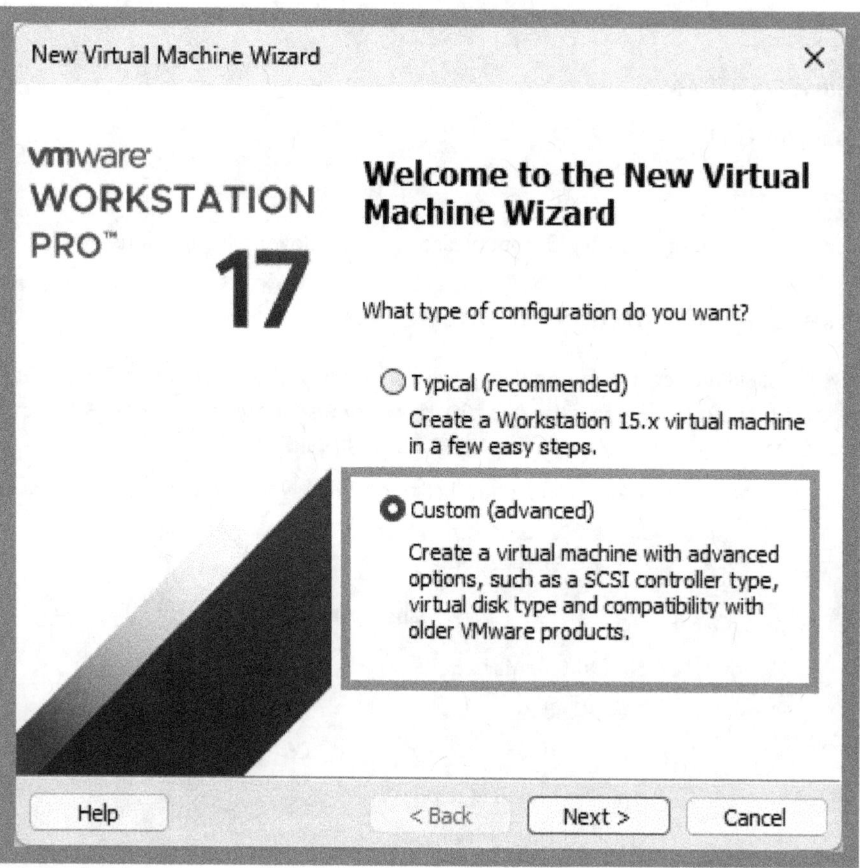

Figure 10.16 – Type of configuration

4. In the **Choose the Virtual Machine Hardware Compatibility** menu, use the default option and click on **Next >**:

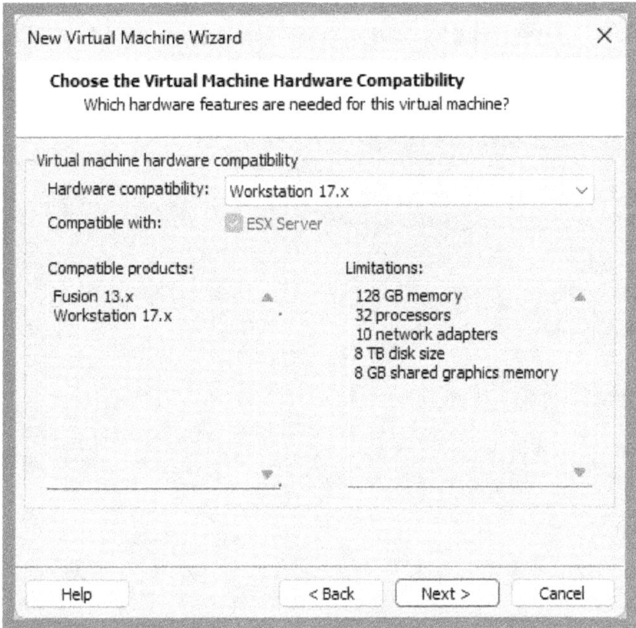

Figure 10.17 – Hardware compatibility

5. In the **Guest Operating System Installation** window, select **Browse…** under **Installer disc image file (iso)** to attach the Security Onion ISO and then click on **Next** >:

Figure 10.18 – Attaching the ISO file

6. In the **Select a Guest Operating System** window, select **Linux** and **CentOS 8 64-bit** and click on **Next >**:

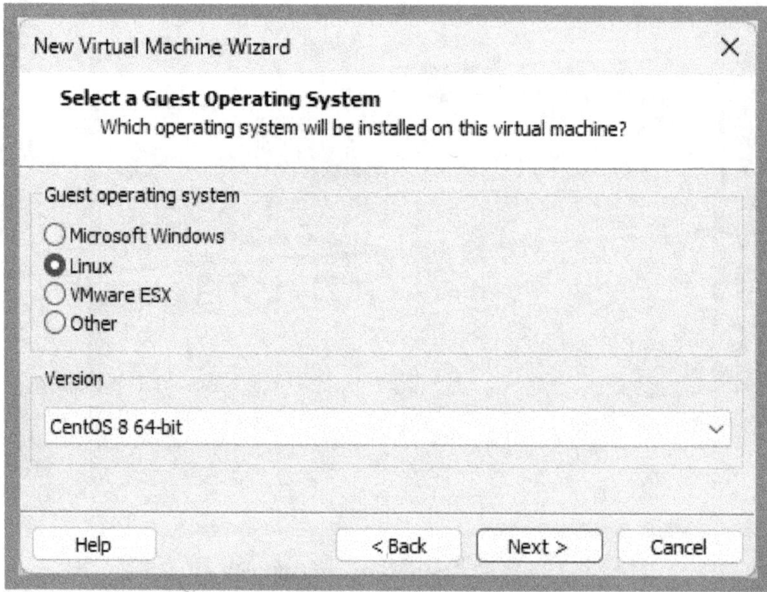

Figure 10.19 – Guest operating system

7. Next, create a name for this new virtual machine and click on **Next >**:

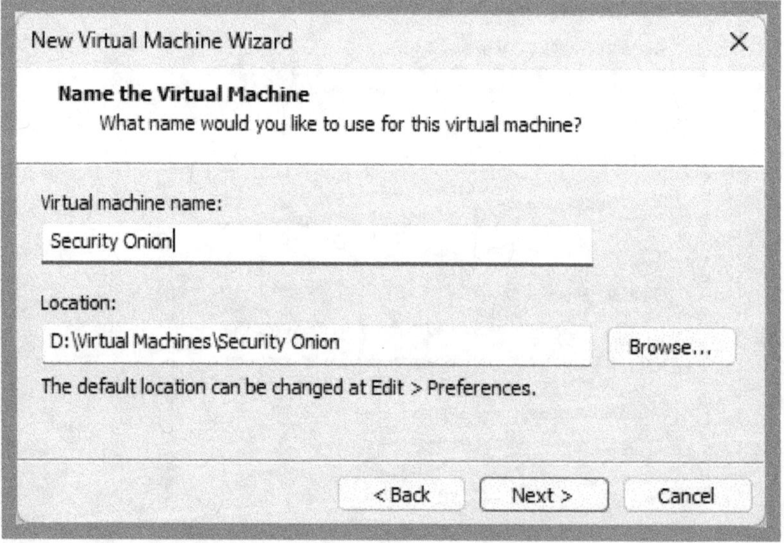

Figure 10.20 – Naming the virtual machine

8. In the **Processor Configuration** window, it's recommended to assign **4** CPU cores to the virtual machine; then, click on **Next** >:

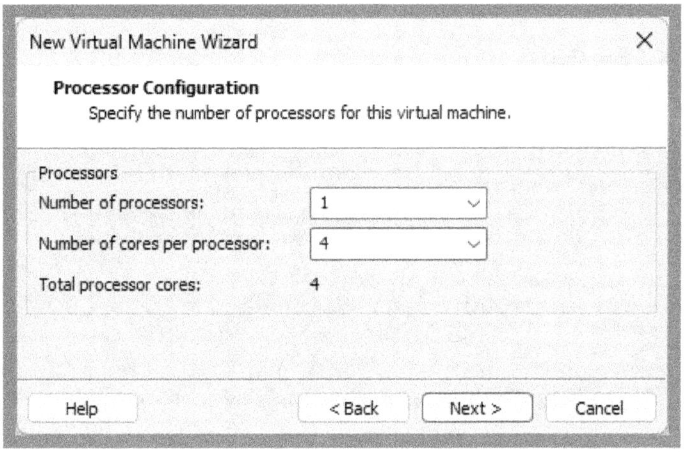

Figure 10.21 – Assigning processor cores

9. In the **Memory for the Virtual Machine** window, it's recommended to assign 12 GB (**12288** MB) of memory to the virtual machine and click on **Next** >:

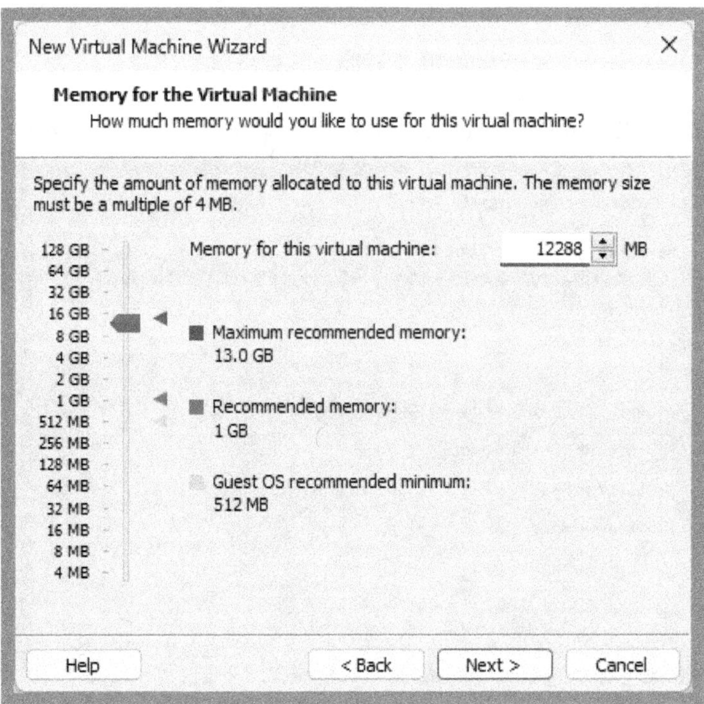

Figure 10.22 – Memory allocation

10. In the **Network Type** window, select **Use bridged networking** and click on **Next** >:

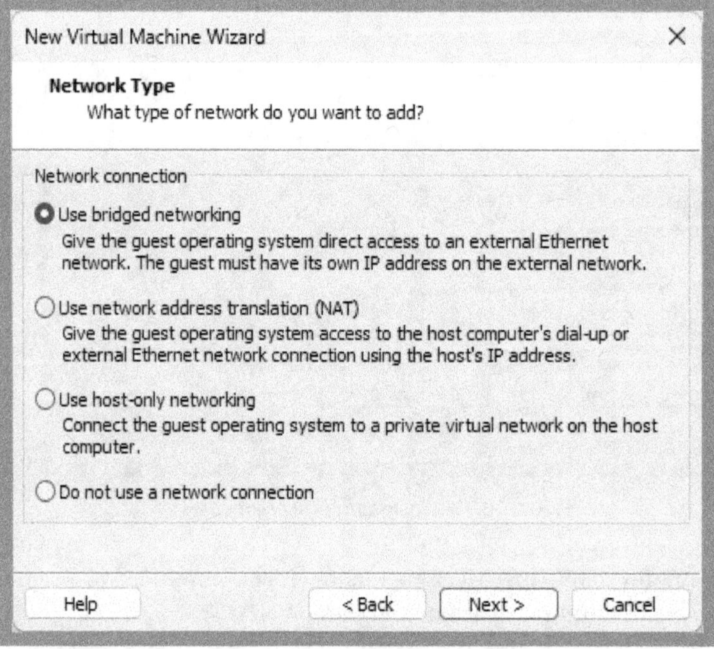

Figure 10.23 – Network type

11. Under **Select I/O Controller Types**, use the default option and click on **Next** >:

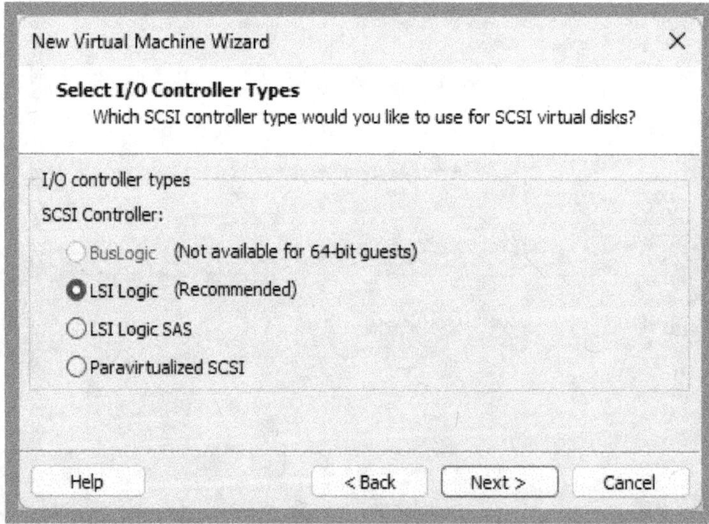

Figure 10.24 – I/O controller type

12. In the **Select a Disk Type** window, use the default option and click on **Next >**:

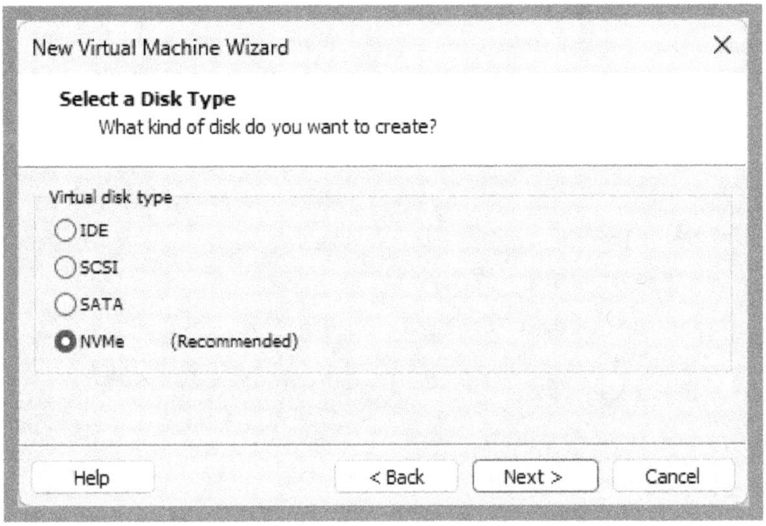

Figure 10.25 – Disk type

13. In the **Select a Disk** window, select **Create a new virtual disk** and click on **Next >**:

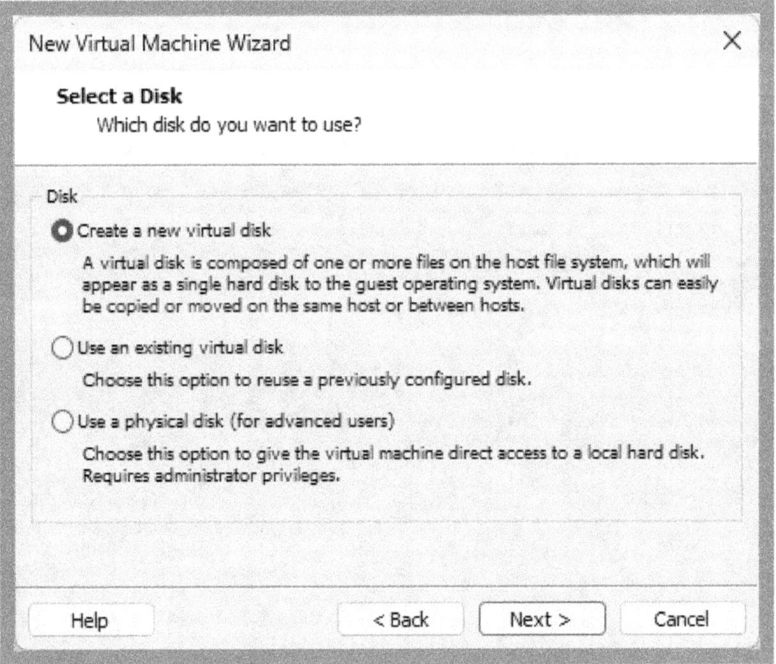

Figure 10.26 – Selecting a disk option

14. In the **Specify Disk Capacity** window, set the maximum disk size to **200.0** GB, select **Split virtual disk into multiple files**, then click on **Next >**:

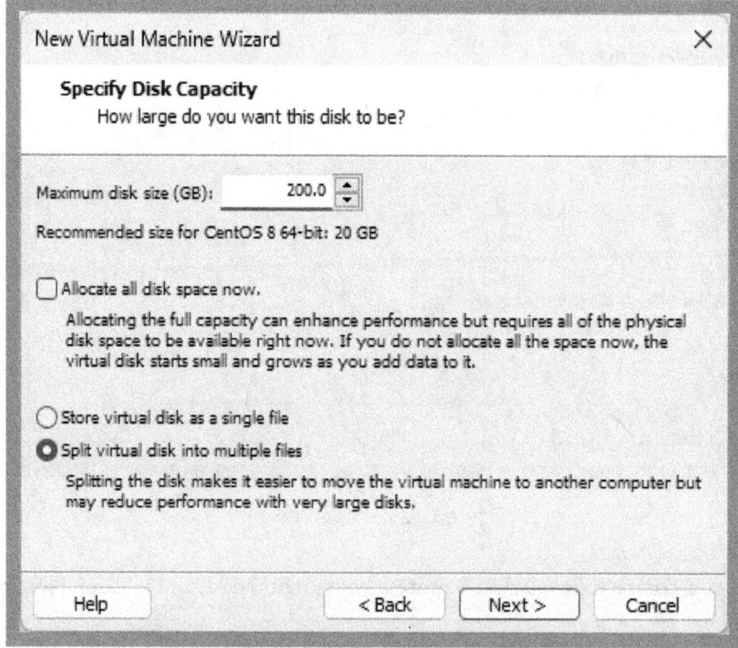

Figure 10.27 – Disk capacity

15. In the **Specify Disk File** window, use the default option and click on **Next >**:

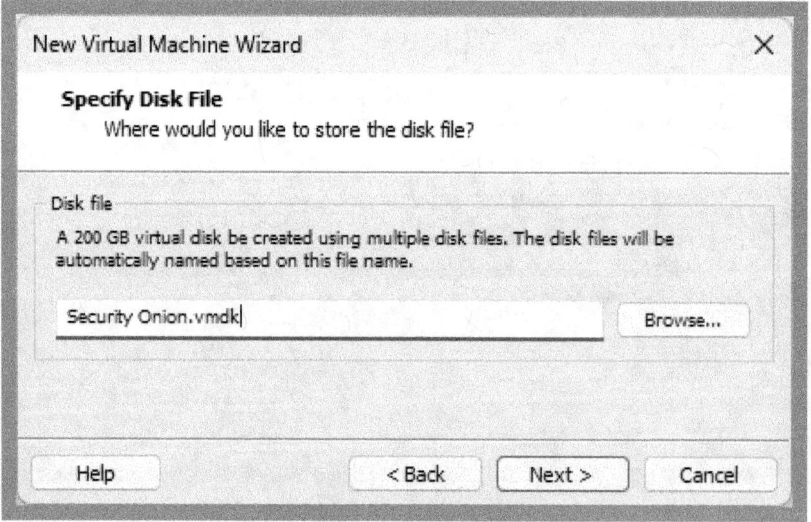

Figure 10.28 – Disk file

16. In the **Ready to Create Virtual Machine** window, click on **Finish**:

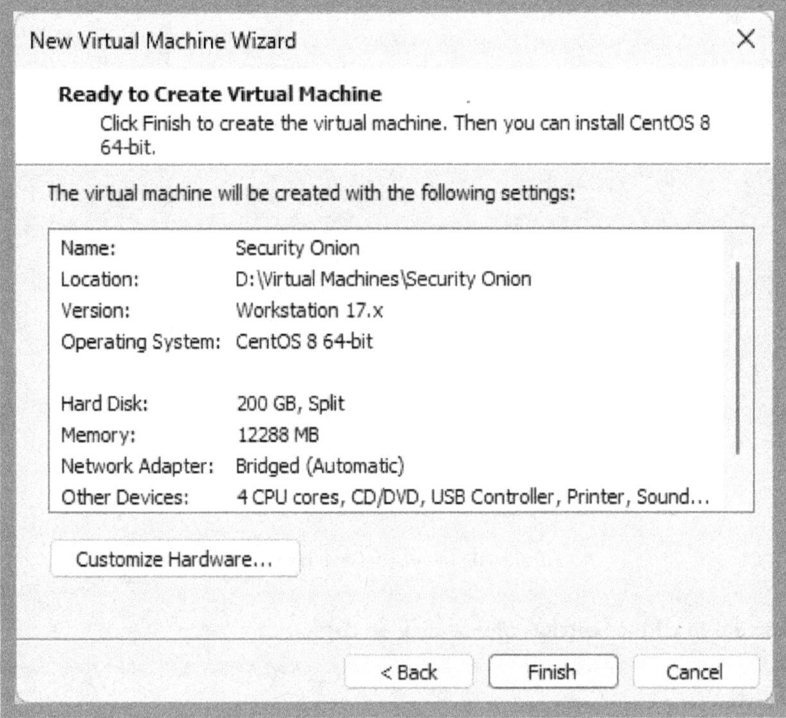

Figure 10.29 – Virtual machine summary

After clicking on **Finish**, the virtual machine environment will automatically be saved within VMware Workstation.

Part 2 – attaching an additional network adapter

1. Within **VMware Workstation**, select the **Security Onion** virtual machine and click on **Edit virtual machine settings** as shown:

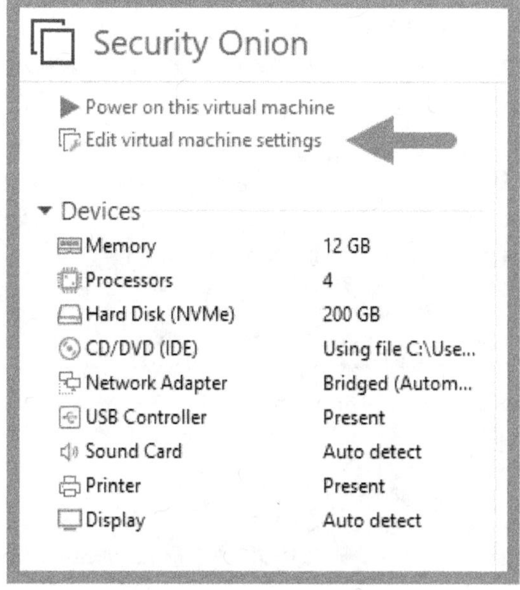

Figure 10.30 – Virtual machine

2. In the **Virtual Machine Settings** menu, click on **Add…**:

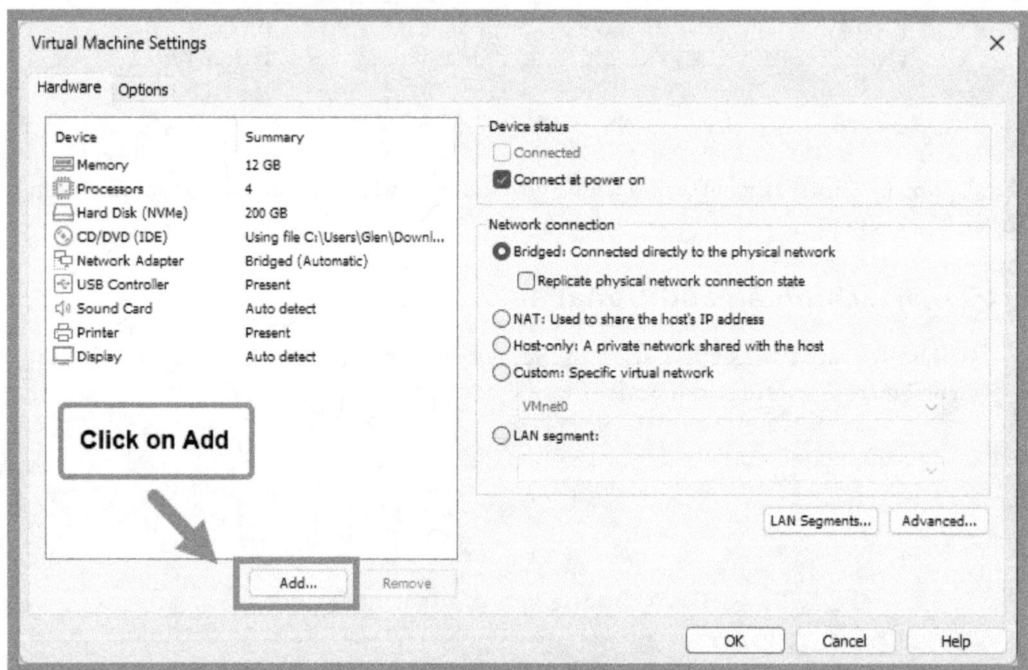

Figure 10.31 – Virtual Machine Settings menu

3. In the **Add Hardware Wizard** menu, select **Network Adapter** and click on **Finish**:

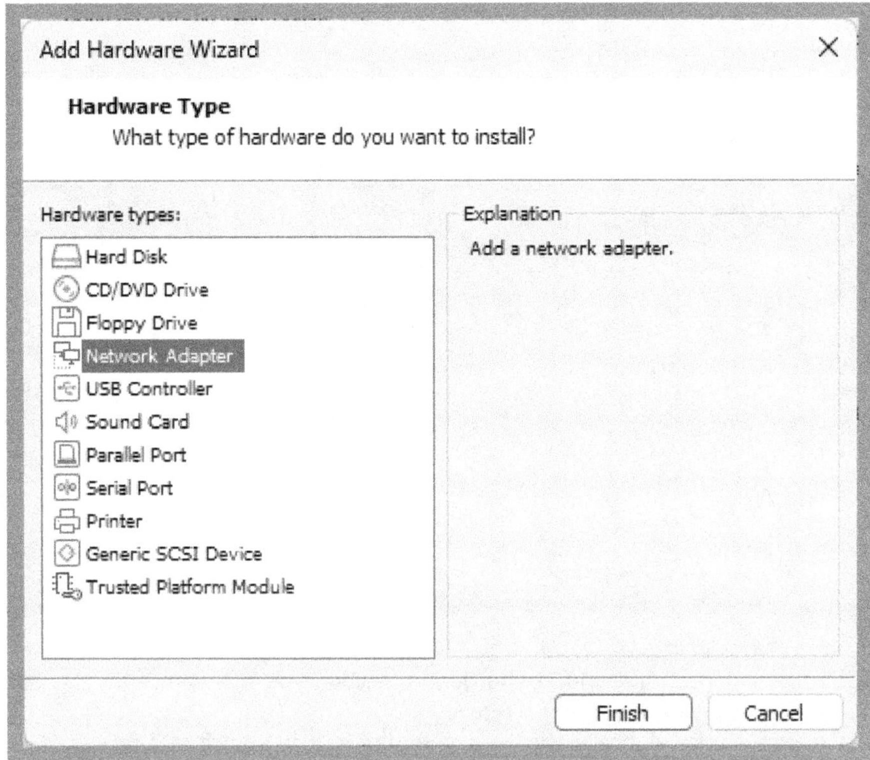

Figure 10.32 – Add Hardware Wizard

4. Next, you'll return to the **Virtual Machine Settings** window, select **Network Adapter 2**, choose **Bridged: Connected directly to the physical network** mode, and click on **OK**:

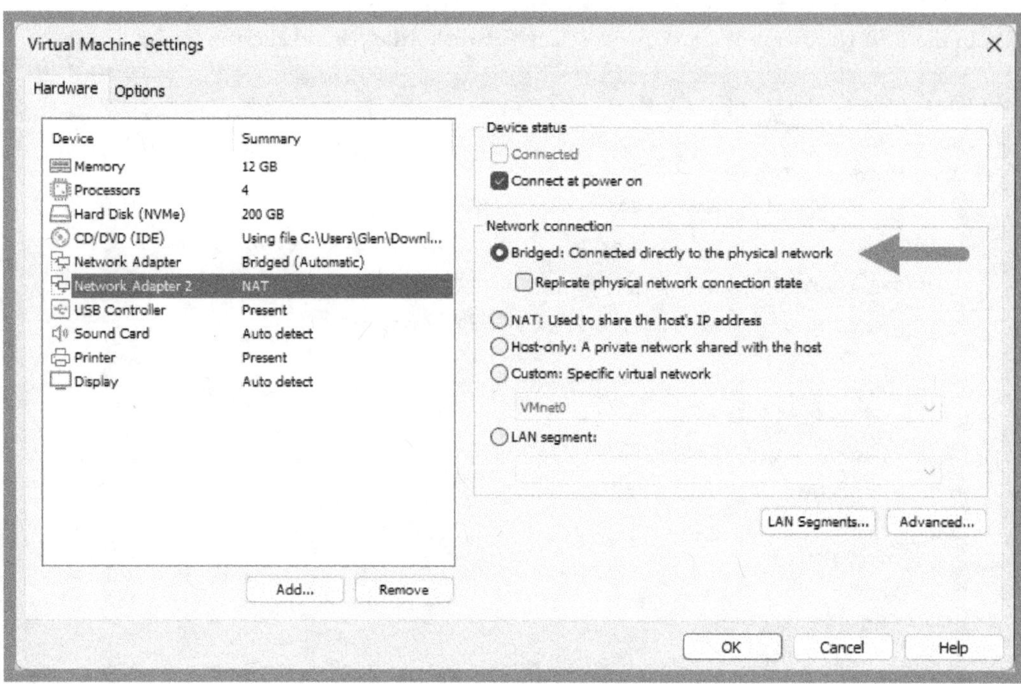

Figure 10.33 – Network connection type

This network adapter will be used as the monitor interface in Security Onion.

5. To ensure your physical network adapter is associated with the bridge connection on VMware Workstation, on the menu bar, click on **Edit** > **Virtual Network Editor** > **Change Settings** as shown:

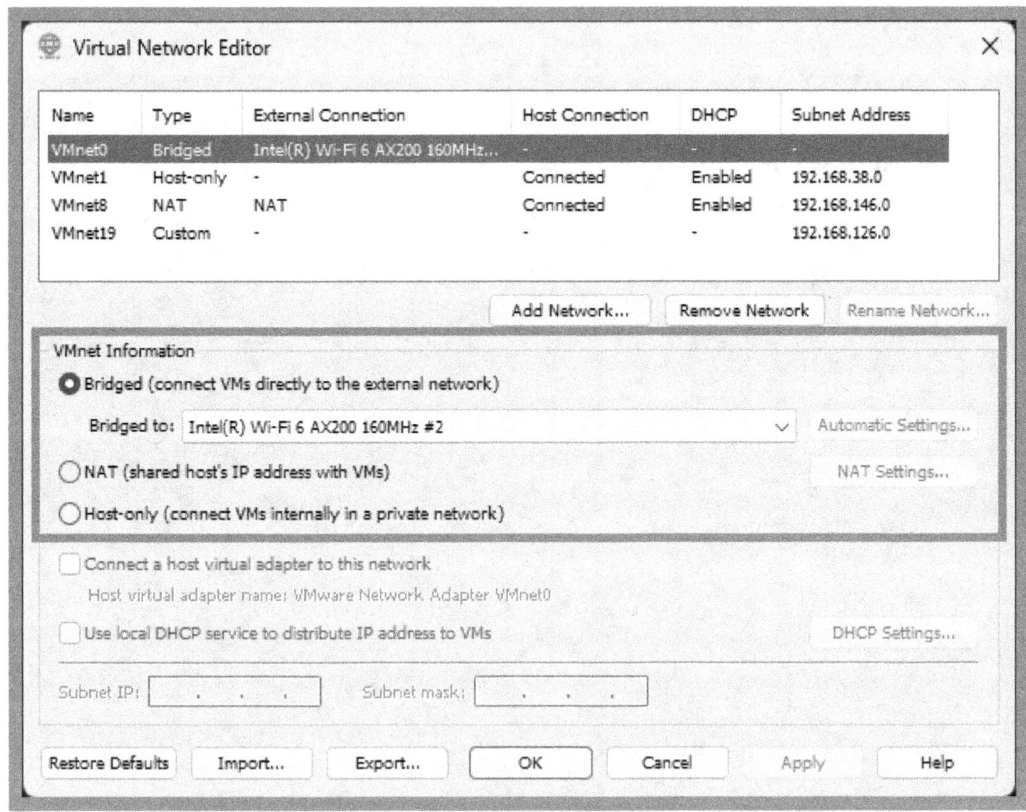

Figure 10.34 – Network editor

As shown in the preceding snippet, **VMnet0** is the **Bridged** network adapter; ensure it's bridged to your physical network adapter on your host machine.

Part 3 – installing Security Onion

1. In VMware Workstation, select the **Security Onion** virtual machine and power on this virtual machine.
2. Next, the Security Onion installation menu will appear; use the default option and hit *Enter*:

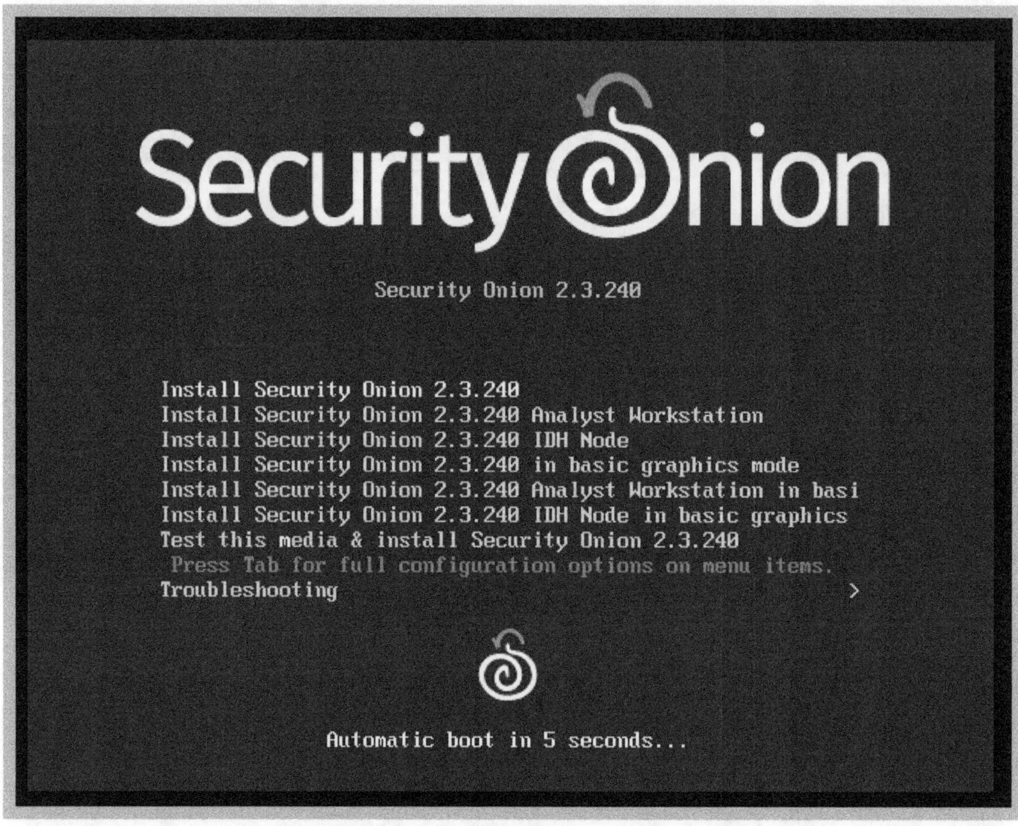

Figure 10.35 – Security Onion installation type

3. Next, the installation wizard will appear; hit *Enter* to begin the initialization process:

```
 - Press the <ENTER> key to begin the installation process.

[    0.000000] Detected CPU family 19h model 80
[    0.000000] Warning: AMD Processor - this hardware has not undergone upstream
 testing. Please consult http://wiki.centos.org/FAQ for more information
[    1.291062] core perfctr but no constraints; unknown hardware!
```

Figure 10.36 – Starting the installation

4. Next, type `yes` to continue and create a user account as shown:

```
###############################################
##          ** W A R N I N G **              ##
##   _________________________________       ##
##                                           ##
##   Installing the Security Onion ISO       ##
## on this device will DESTROY ALL DATA      ##
##           and partitions!                 ##
##                                           ##
##       ** ALL DATA WILL BE LOST **         ##
###############################################
Do you wish to continue? (Type the entire word 'yes' to proceed.) yes

A new administrative user will be created. This user will be used for setting up and administering S
ecurity Onion.

Enter an administrative username: glen

Let's set a password for the glen user:

Enter a password:
Re-enter the password: _
```

Passwords are invisible

Figure 10.37 – Installing Security Onion

The username and password created in this step are used to log in to the console only.

5. Next, the installation of the base operating system and software packages will start. This process will take a few minutes and the virtual machine will automatically reboot when finished.

Part 4 – configuring networking in Security Onion

1. After the Security Onion virtual machine reboots, log in using the username and password you created in the previous steps. The **Security Onion Setup** wizard will appear on the Terminal/Console; using the direction keys (arrows) on your keyboard to select **<Yes>** and hit *Enter*:

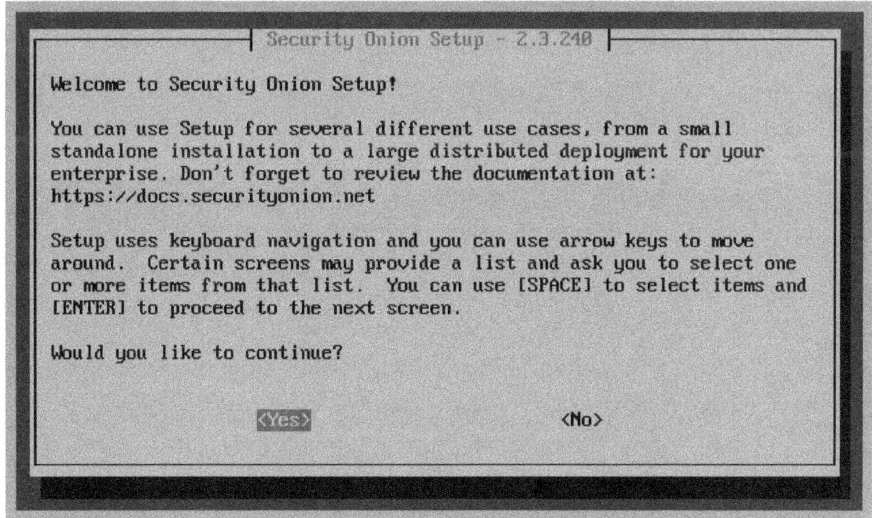

Figure 10.38 – Security Onion Setup

2. Next, click **Install** to start the standard installation of Security Onion:

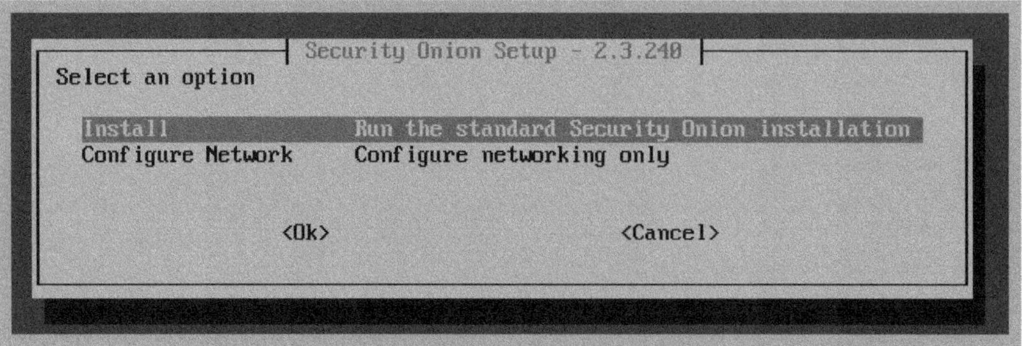

Figure 10.39 – Selecting a setup option

As shown in the preceding figure, you have the choice to proceed with the installation and setup or configure the network addresses and settings in Security Onion.

3. Next, using the down arrow key on your keyboard, select **STANDALONE**, press the spacebar to place an asterisk (`*`), and then hit *Enter*:

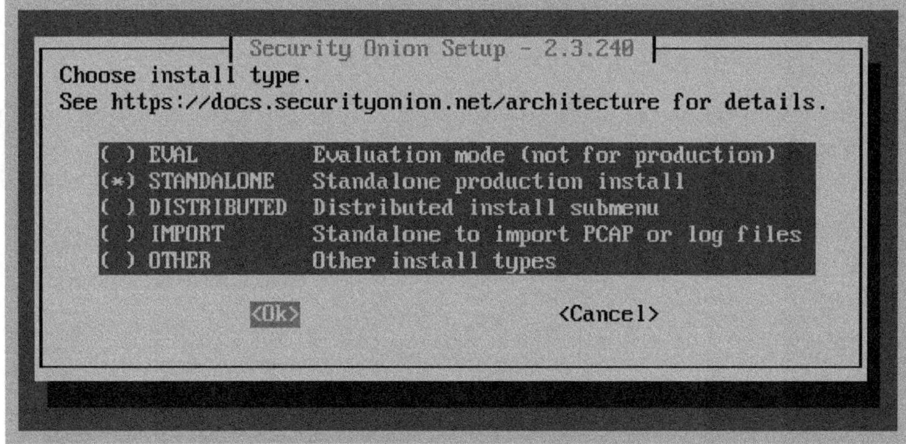

Figure 10.40 – Installation type

4. Next, you will need to type `agree` within the input field to accept the terms and agreement to use Elastic Stack within Security Onion and then hit *Enter*:

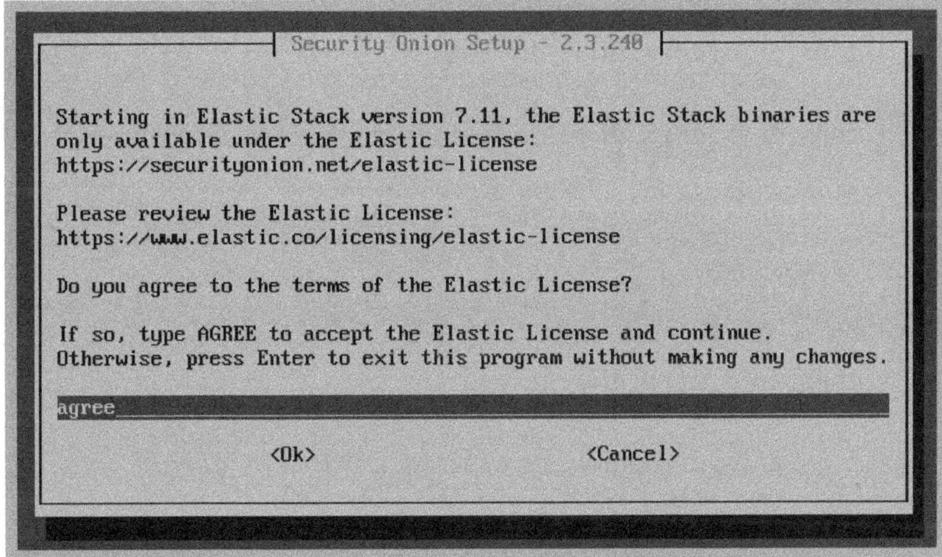

Figure 10.41 – Elastic Stack license agreement

5. Next, insert a hostname for the Security Onion virtual machine and hit *Enter*:

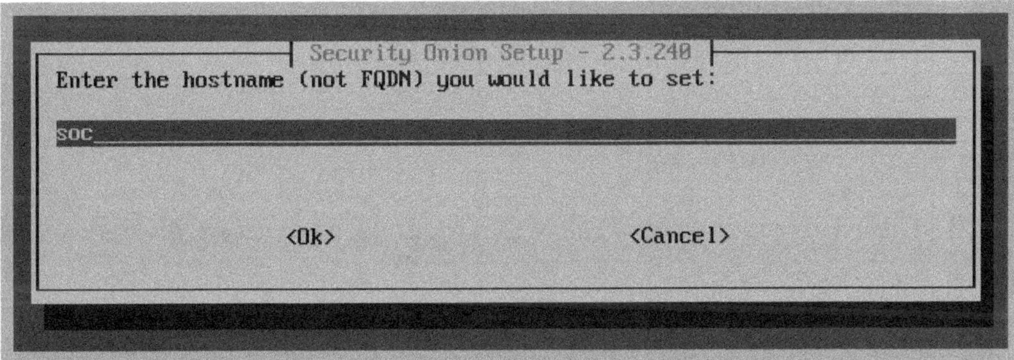

Figure 10.42 – Setting a hostname

6. Next, you'll be provided with an option to insert a description for this instance of Security Onion. This is optional but recommended if you're doing a distributed installation in a network. Since we're doing a standalone setup, leave it as the default and hit *Enter*:

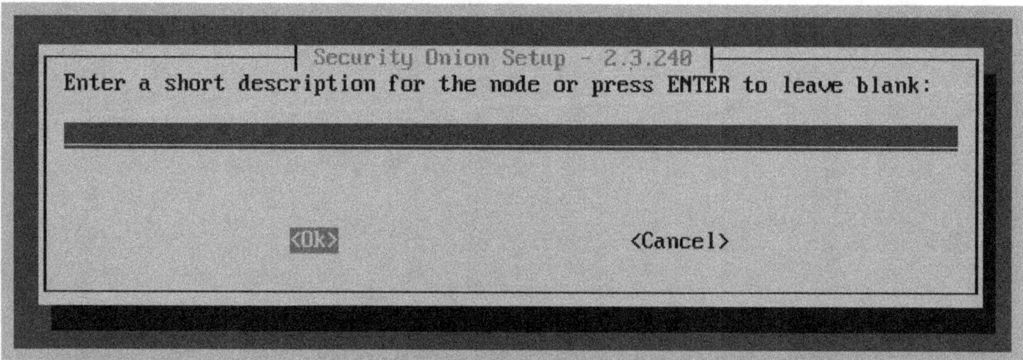

Figure 10.43 – The description window

7. Next, select the first network adapter as the management NIC in Security Onion and hit *Enter*:

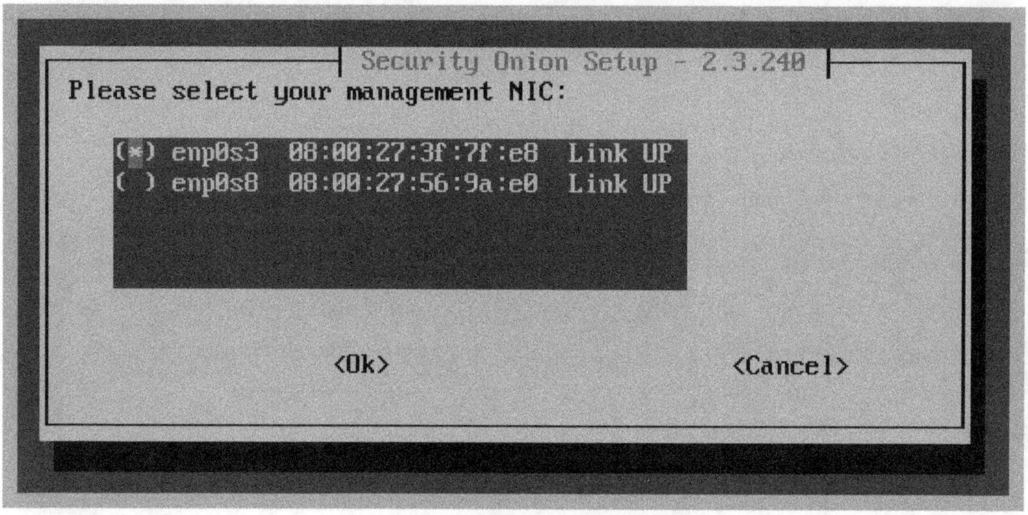

Figure 10.44 – Management interface

8. Next, you are provided with the option to statically assign an IP address to the management interface or dynamically receive an address from the **Dynamic Host Configuration Server** (**DHCP**) server on the network. Select **STATIC** and hit *Enter*:

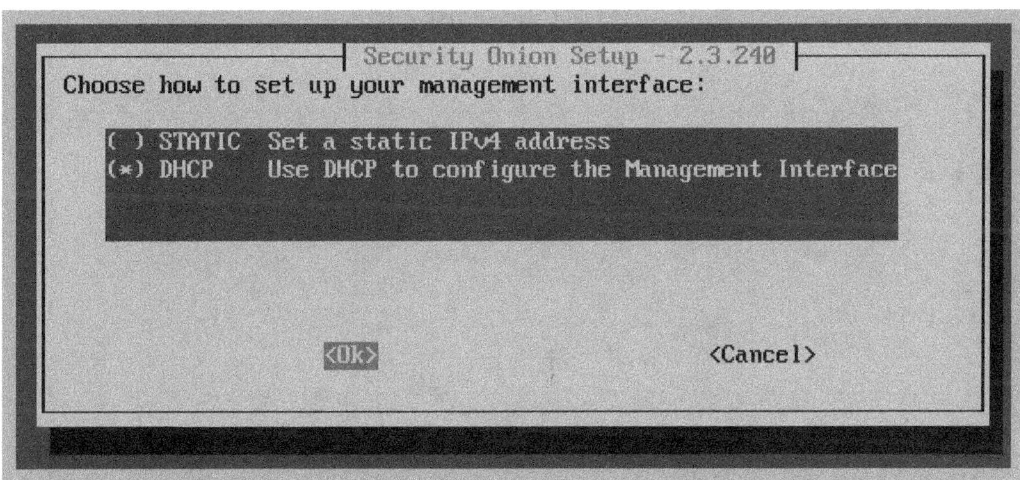

Figure 10.45 – Management interface setup

9. Next, enter an unused/unallocated IP address and network prefix from your network and hit *Enter*:

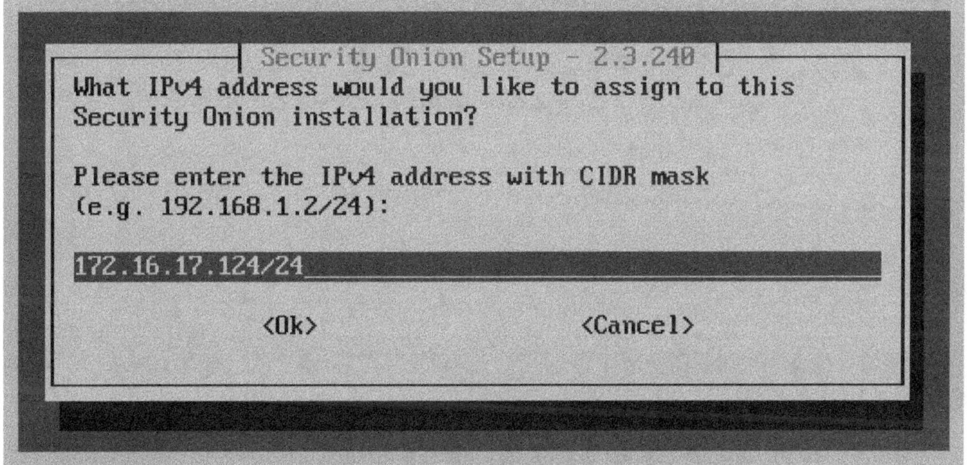

Figure 10.46 – Setting a network address

10. Next, insert your default gateway address and hit *Enter*:

Figure 10.47 – Default gateway address

11. Next, Google's DNS server addresses are automatically set; you can leave them as the default or change them if needed and hit *Enter*:

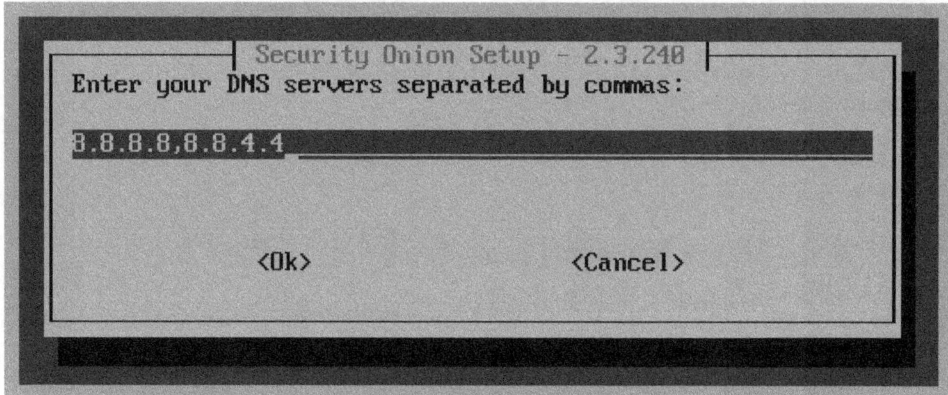

Figure 10.48 – DNS server settings

12. Next, create a DNS search domain and hit *Enter*:

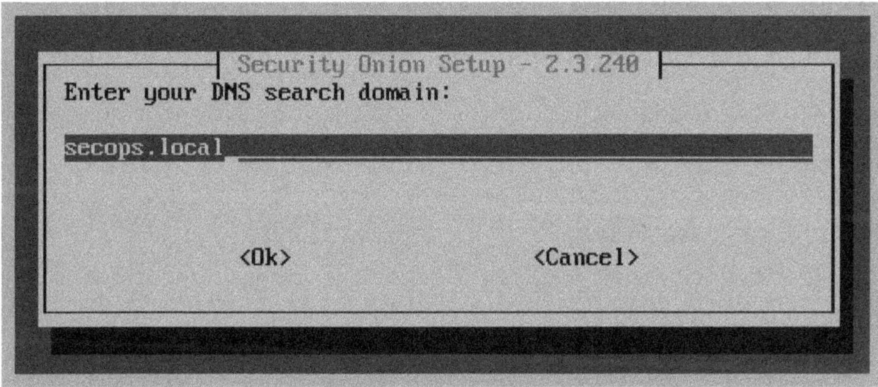

Figure 10.49 – Setting a search domain

13. Next, hit *Enter* to confirm the setup process:

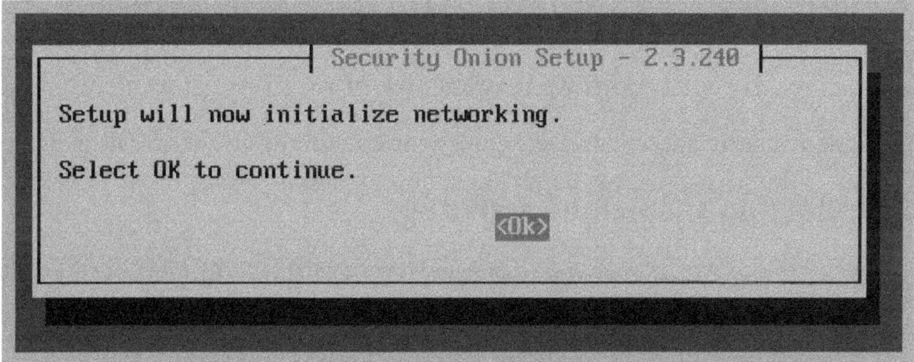

Figure 10.50 – Confirmation

14. Lastly, the setup process will take a few minutes to complete. Once completed, the Security Onion virtual machine will automatically reboot. Then, log in and use the following command to check the status of each process and container:

```
[glen@soc ~]$ sudo so-status
```

After a few minutes, all the processes and containers will be running as shown in the following snippet:

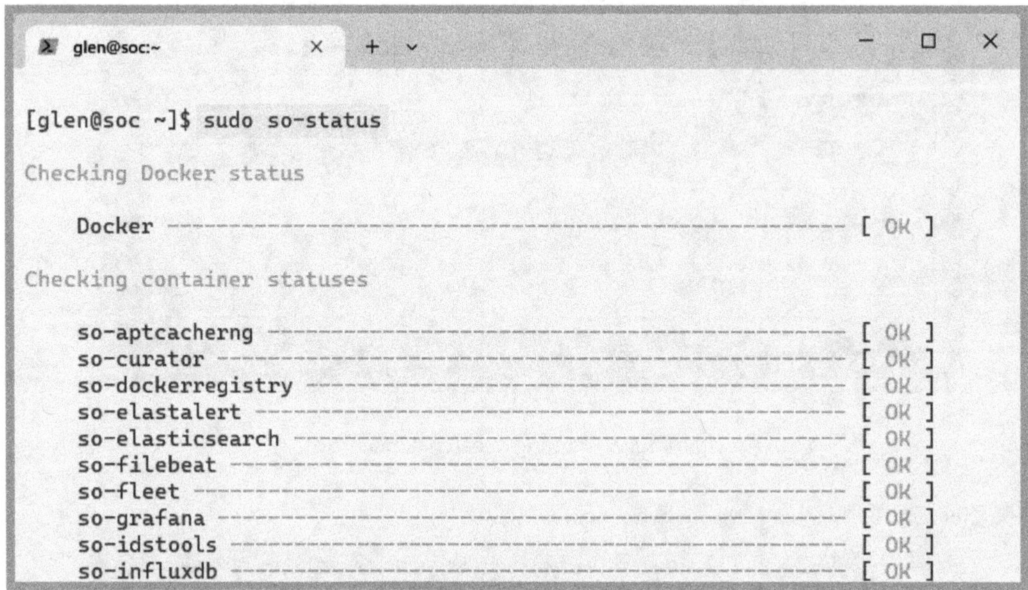

Figure 10.51 – Verifying the status

Ensure all containers and processes are running and OK before proceeding to the next phase.

Part 5 – detecting suspicious activities

1. To access the Security Onion user interface and leverage the security tools, open your web browser and go to https://<ip-address-security-onion>. Here, you will create a user account to log in and access the security tools within Security Onion. This account will be used to access all the tools in Security Onion.

> **Tip**
>
> To find the IP address of the Security Onion virtual machine, login to its console and execute the `ifconfig` or `ip address` command to view the IP address on the management interface.

2. Once you've logged in, you'll see the main dashboard. Click on **Alerts** to see any security events and potential threats that have been detected:

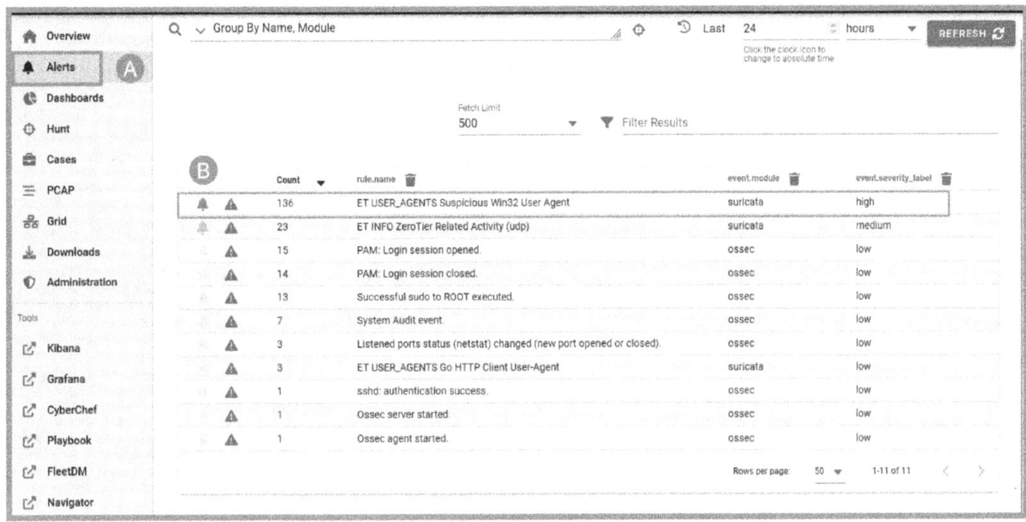

Figure 10.52 – Security alerts

As shown in the preceding snippet, Suricata identified a potential threat with high severity and provided the name of the rule that was used for detection. Furthermore, Suricata recorded 136 counts (events) for this alert.

3. To view all related events for an alert, click on the alert name to open the drop-down menu and select **Drilldown** as shown:

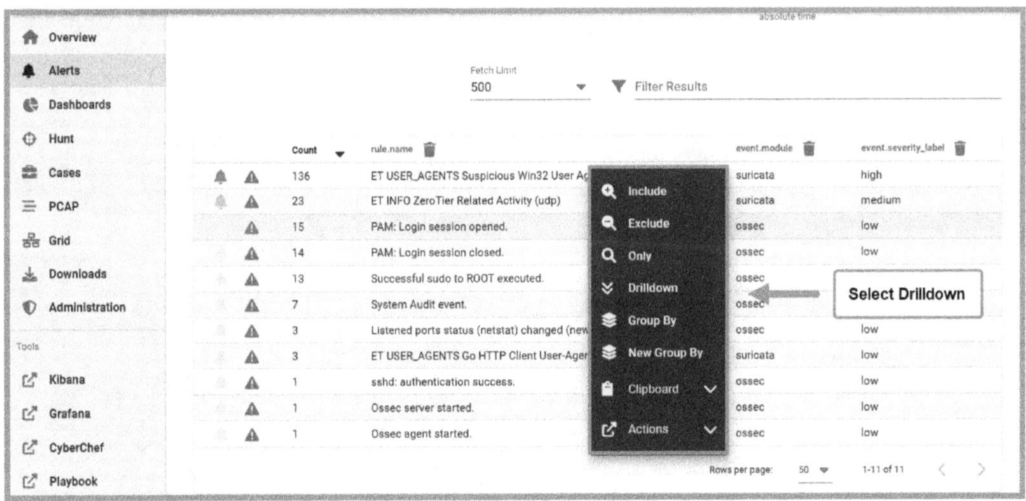

Figure 10.53 – Alert options

Next, Security Onion will show a list of each count (event) for the alert while providing the timestamps, detection rules, severity levels, source IP addresses, and source port numbers as shown:

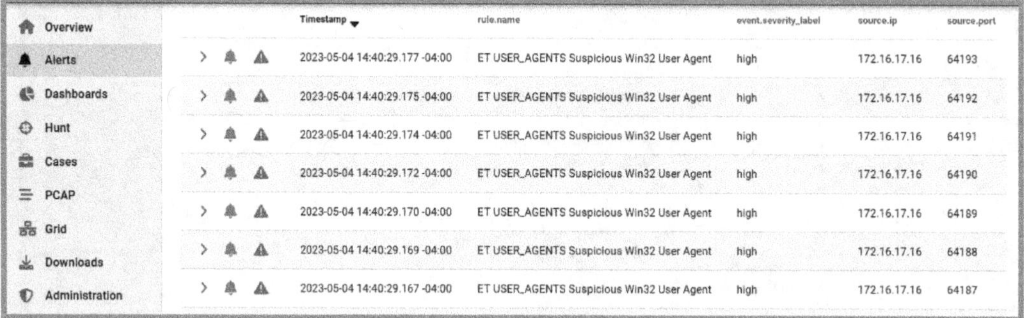

Figure 10.54 – Viewing alerts

The information shown in the preceding snippet helps cybersecurity professionals to identify the source of the threat and the sequence of events that occurred using the timestamps.

4. Next, click on one of the events to view event data such as its source and destination IP addresses and port numbers, the security sensor that triggered the alert, the event severity, and a message that contains specific details about the threat, as shown:

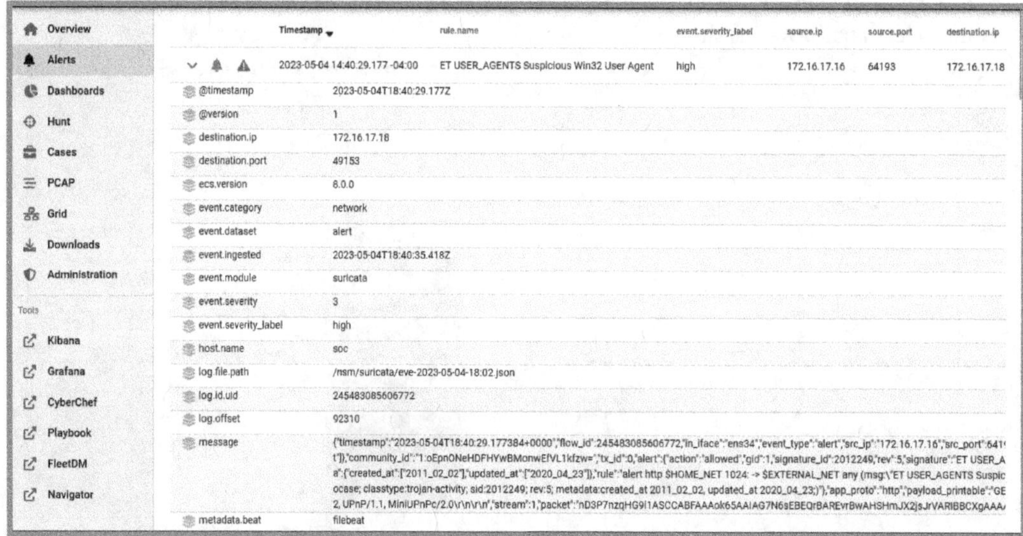

Figure 10.55 – Viewing alert data

The **message** field contains the following information, which can be used by cybersecurity professionals to better understand the event:

- The time and date of the event
- The monitoring interface in Security Onion which detected the event
- The source and destination IP addresses
- The source and destination port numbers
- The signature that was found to trigger the event
- A description of the potential threat

5. To determine the trajectory of the threat and correlate all related events, click on an event and select **Correlate** as shown:

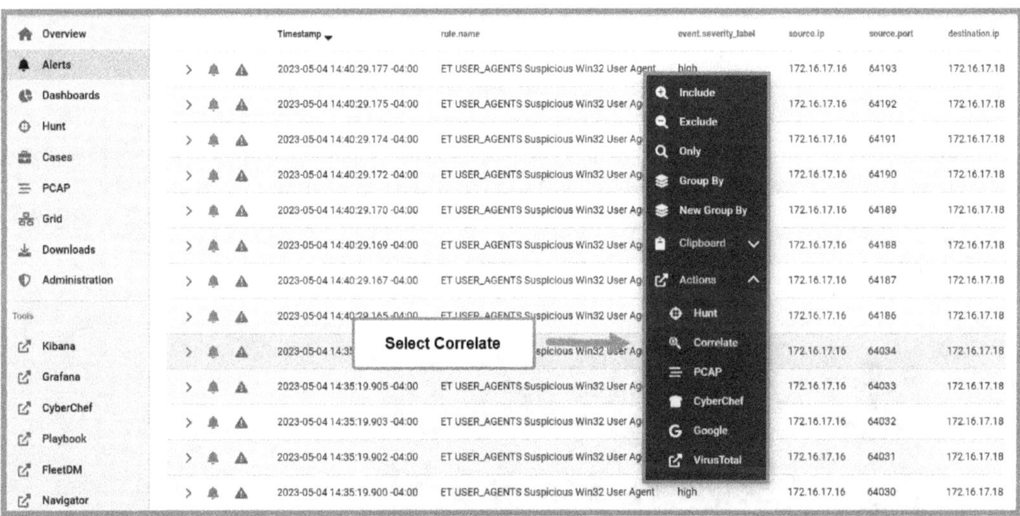

Figure 10.56 – Correlating events

Next, Security Onion will provide a correlation of all events that belong to the same alert as shown:

	Count	event.module	event.dataset
▲	1	suricata	alert
▲	1	zeek	http
▲	1	zeek	conn

Events

Timestamp	source.ip	source.port	destination.ip	destination.port	rule.name
2023-05-04 14:40:29.177 -04:00	172.16.17.16	64193	172.16.17.18	49153	ET USER_AGENTS Suspicious Win32 User Agent
2023-05-04 14:40:29.176 -04:00	172.16.17.16	64193	172.16.17.18	49153	
2023-05-04 14:40:29.175 -04:00	172.16.17.16	64193	172.16.17.18	49153	

Figure 10.57 – Correlation of events

Additionally, clicking on an alert will provide the option to escalate an alert to a case. This case can be used by a security analyst to track the event while performing further analysis:

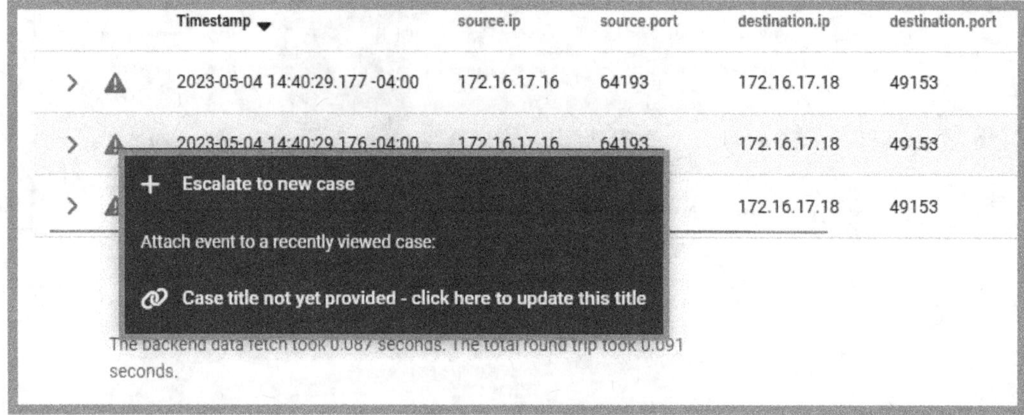

Figure 10.58 – Creating a case

6. Next, click on the **Hunt** category to view all network traffic that's associated with the threat. Then, click on a traffic flow from the list to open the additional menu and select **PCAP** to view the packet contents:

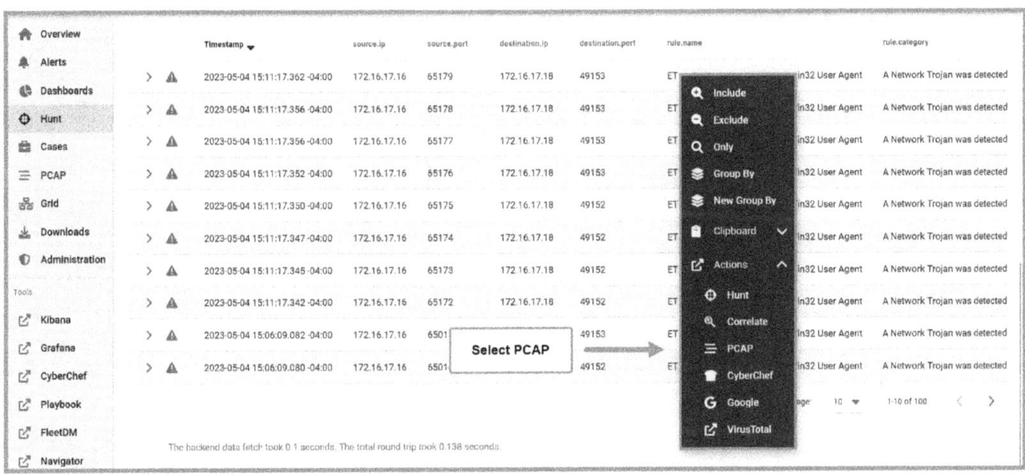

Figure 10.59 – Traffic flow

Next, Security Onion will provide a list of network packets for the selected flow as shown:

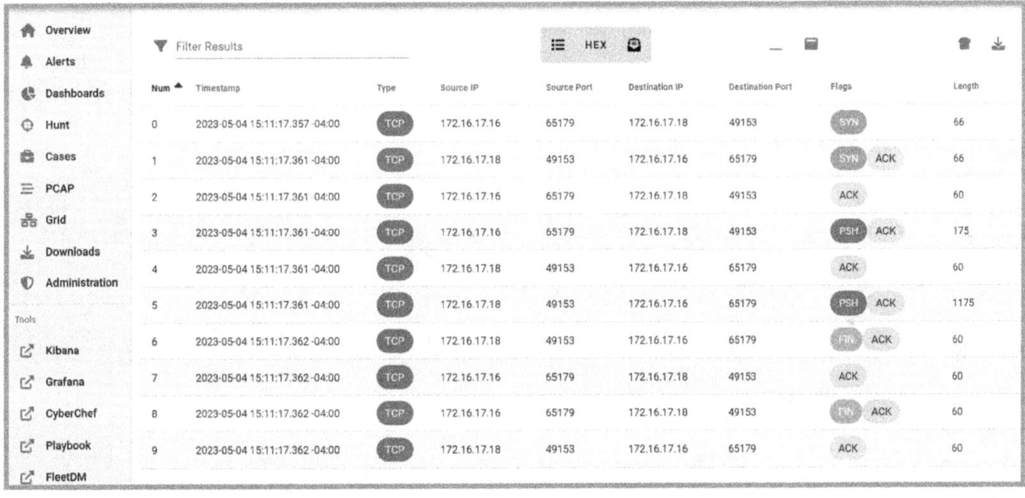

Figure 10.60 – Network packets per flow

Clicking on the **Expand and Detailed** icon will display the details within each packet as shown:

Num	Timestamp	Type	Source IP	Source Port	Destination IP	Destination Port	Flags	Length
0	2023-05-04 15:11:17.357 -04:00	TCP	172.16.17.16	65179	172.16.17.18	49153	SYN	66
1	2023-05-04 15:11:17.361 -04:00	TCP	172.16.17.18	49153	172.16.17.16	65179	SYN ACK	66
2	2023-05-04 15:11:17.361 -04:00	TCP	172.16.17.16	65179	172.16.17.18	49153	ACK	60
3	2023-05-04 15:11:17.361 -04:00	TCP	172.16.17.16	65179	172.16.17.18	49153	PSH ACK	175

```
0000 9C 3D CF EE 7C EA 1C 6F  65 D4 04 82 08 00 45 00
0016 00 34 A2 88 40 00 80 06  DD F8 AC 10 11 10 AC 10
0032 11 12 FE 9B C0 01 80 4A  FE D8 00 00 00 00 80 02
0048 FA F0 BC 1B 00 00 02 04  05 B4 01 03 03 08 01 01
0064 04 02
```

Figure 10.61 – View packet details

To export the packet capture file for offline analysis, click on the download icon.

7. The **Dashboards** category shows a summary of all events and alerts that were detected using Security Onion as shown:

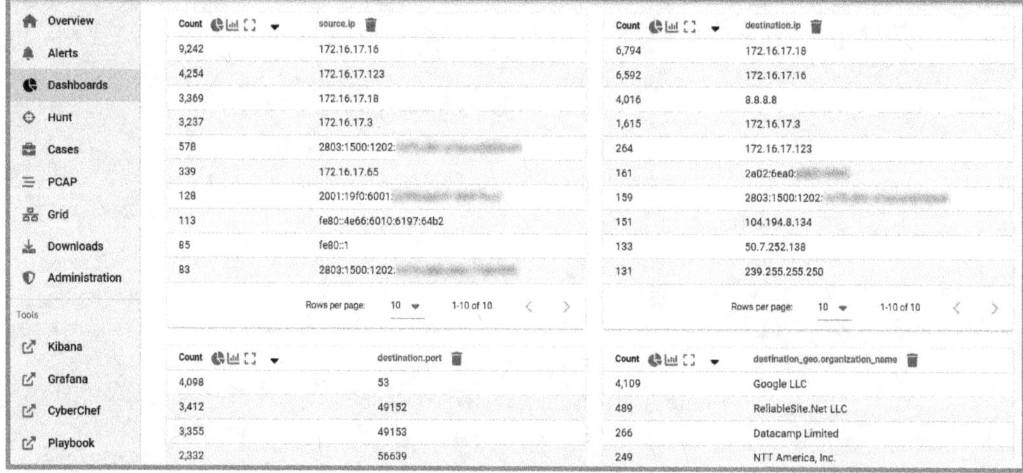

Figure 10.62 – Dashboards

8. To collect security intelligence from hosts on the network, you can download and install **Wazuh Agents** from the **Downloads** page as shown:

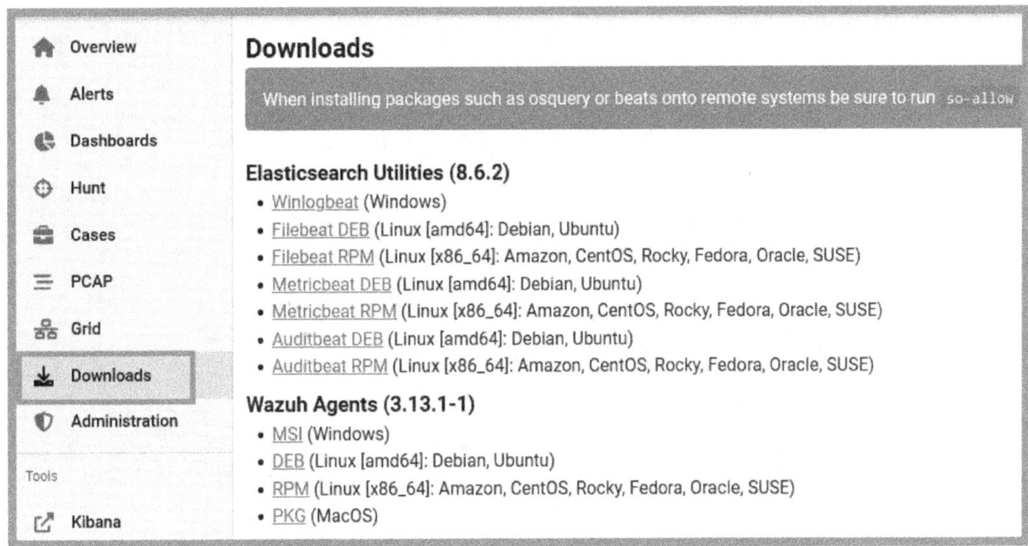

Figure 10.63 – Wazuh Agents

9. For data visualization on all alerts and threats detected by Security Onion, click on **Tools** | **Kibana** and log in using your account. The following snippet shows the default Kibana dashboard:

Figure 10.64 – Kibana dashboard

As shown in the preceding snippet, the Kibana dashboard provides cybersecurity professionals with an overall view of events and incidents, enabling a security analyst to drill down further into the activity during analysis.

10. To view all network connections and traffic flows, click on **Security Onion | Dataset | conn** as shown:

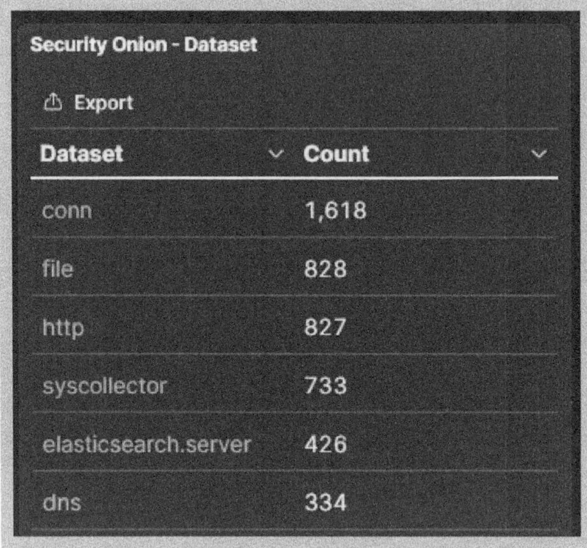

Figure 10.65 – Datasets

The connections dashboard will appear, enabling you to click on IP addresses and port numbers to gain additional information as shown:

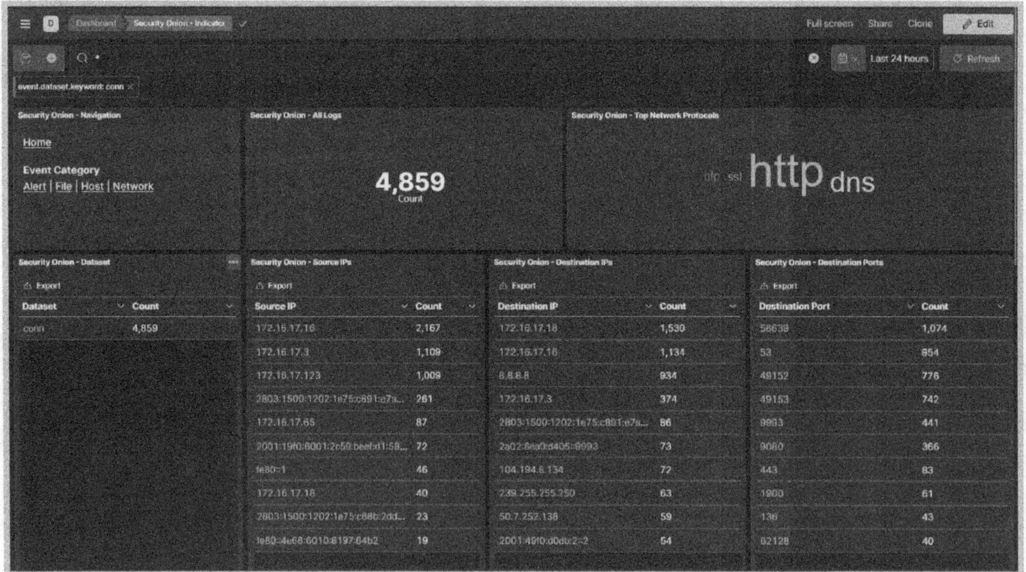

Figure 10.66 – Connections dashboard

11. Expand the left-hidden menu and click on **Analytics** to access more data visualization tools within Kibana:

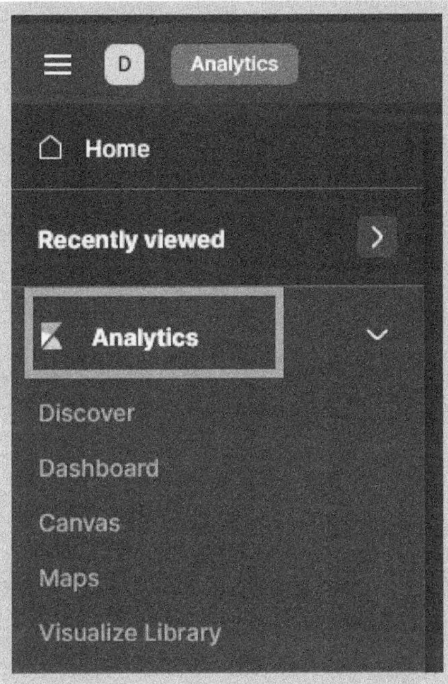

Figure 10.67 – Analytics menu

The **Discover** dashboard helps you determine when an event has been detected and how often it is occurring, as shown:

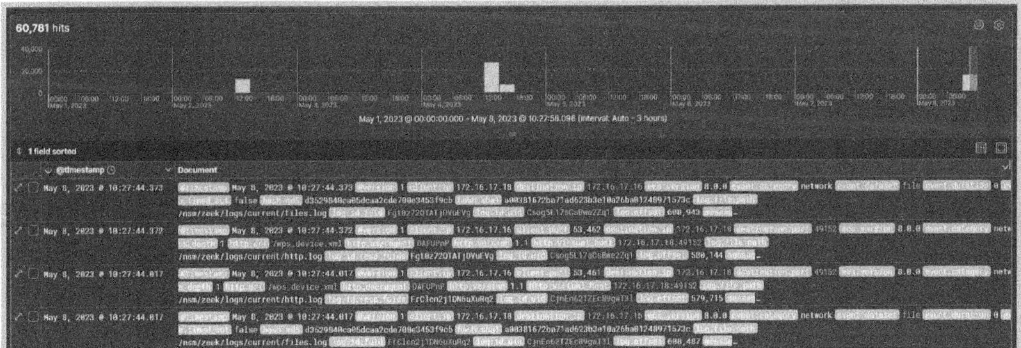

Figure 10.68 – Discovery

Lastly, **Visualize Library** provides the option to create custom dashboards or use the existing dashboards to view specific details as shown:

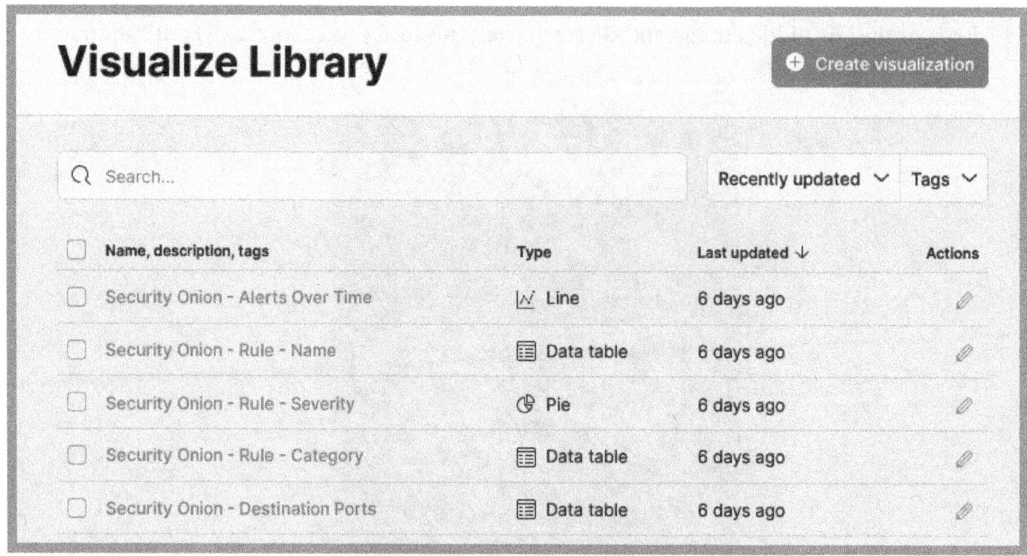

Figure 10.69 – Visualize Library

> Be sure to explore all the features and dashboards within Security Onion, as it uses many open source security tools to help cybersecurity professionals to monitor their network infrastructure to identify potential threats on their systems.

Having completed this section, you have learned how to set up and use Security Onion and its tools to monitor a network and identify potential threats.

Summary

Over the course of this chapter, you gained the skills and hands-on experience in using Wireshark to perform packet analysis to identify potentially malicious traffic, and learned how to set up and use Security Onion to monitor and detect potential threats in a network.

Lastly, I know the journey of becoming a cybersecurity professional, such as an ethical hacker and penetration tester, has many challenges; there are so many new things to learn and keeping up to date with the ever-changing cyber-threat landscape can be tough. These challenges will help you expand your potential and become an industry expert in your field of study and carry you along the path to success, as long as there's a balance of prioritization, focus, dedication, and time management.

I would personally like to thank you very much for your support in purchasing a copy of my book and congratulations on making it to the end and acquiring all these amazing new skills in performing reconnaissance as an ethical hacker. I do hope everything you have learned throughout this book has been informative for you and helpful in your journey to becoming super-awesome in the cybersecurity industry and beyond.

Further reading

- Wireshark documentation – `https://www.wireshark.org/docs/`
- Security Onion documentation – `https://docs.securityonion.net/en/2.3/index.html`
- Setting up Zeek – `https://docs.zeek.org/en/master/`
- Setting up Suricata – `https://suricata.io/documentation/`
- Setting up Wazuh – `https://documentation.wazuh.com/current/index.html`

Index

Symbols

A

B

C

D

E

P

Q

R

S

T

U

V

W

Y

Z

www.packtpub.com

Subscribe to our online digital library for full access to over 7,000 books and videos, as well as industry leading tools to help you plan your personal development and advance your career. For more information, please visit our website.

Why subscribe?

- Spend less time learning and more time coding with practical eBooks and Videos from over 4,000 industry professionals
- Improve your learning with Skill Plans built especially for you
- Get a free eBook or video every month
- Fully searchable for easy access to vital information
- Copy and paste, print, and bookmark content

Did you know that Packt offers eBook versions of every book published, with PDF and ePub files available? You can upgrade to the eBook version at packtpub.com and as a print book customer, you are entitled to a discount on the eBook copy. Get in touch with us at customercare@packtpub.com for more details.

At www.packtpub.com, you can also read a collection of free technical articles, sign up for a range of free newsletters, and receive exclusive discounts and offers on Packt books and eBooks.

Other Books You May Enjoy

If you enjoyed this book, you may be interested in these other books by Packt:

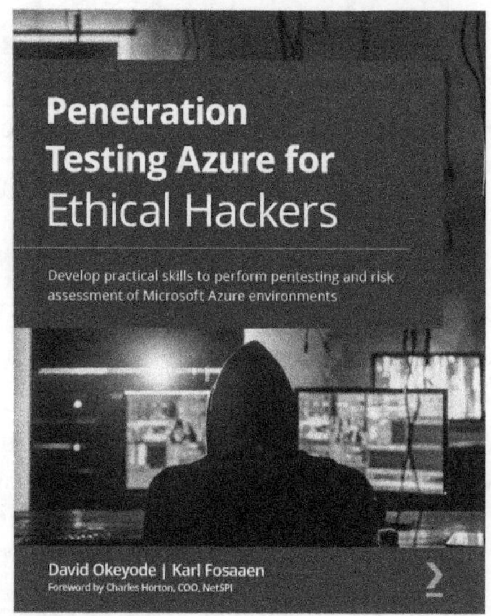

Penetration Testing Azure for Ethical Hackers

David Okeyode | Karl Fosaaen

ISBN: 978-1-83921-293-2

- Identify how administrators misconfigure Azure services, leaving them open to exploitation
- Understand how to detect cloud infrastructure, service, and application misconfigurations
- Explore processes and techniques for exploiting common Azure security issues
- Use on-premises networks to pivot and escalate access within Azure
- Diagnose gaps and weaknesses in Azure security implementations
- Understand how attackers can escalate privileges in Azure AD

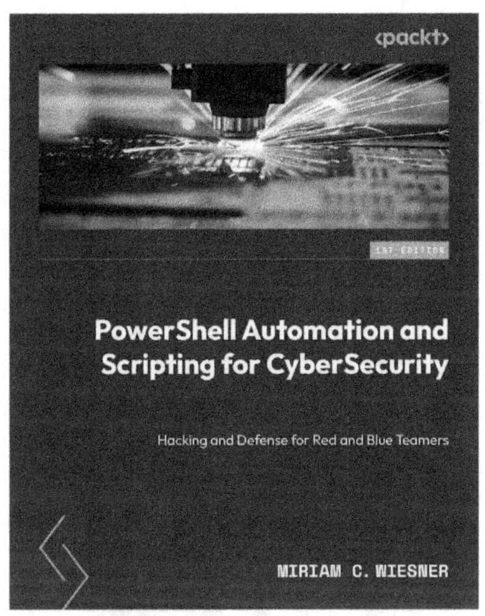

PowerShell Automation and Scripting for CyberSecurity

Miriam C. Wiesner

ISBN: 978-1-80056-637-8

- Leverage PowerShell and PowerShell related mitigations to detect attacks
- Fortify your environment and systems against threats using PowerShell
- Get unique insights into Event Logs and IDs in relation to PowerShell
- Configure PSRemoting and learn about risks, bypasses & best practices
- Leverage PowerShell for system access, exploitation and hijacking
- Red and blue team introduction to Active Directory and Azure AD Security
- Leverage PowerShell for attacks that go deeper than simple PowerShell commands
- Explore JEA: a lesser-known but powerful PowerShell management tool

Packt is searching for authors like you

If you're interested in becoming an author for Packt, please visit authors.packtpub.com and apply today. We have worked with thousands of developers and tech professionals, just like you, to help them share their insight with the global tech community. You can make a general application, apply for a specific hot topic that we are recruiting an author for, or submit your own idea.

Share your thoughts

Now you've finished *Reconnaissance For Ethical Hackers*, we'd love to hear your thoughts! Scan the QR code below to go straight to the Amazon review page for this book and share your feedback or leave a review on the site that you purchased it from.

https://packt.link/r/1837630631

Your review is important to us and the tech community and will help us make sure we're delivering excellent quality content.

Download a free PDF copy of this book

Thanks for purchasing this book!

Do you like to read on the go but are unable to carry your print books everywhere?

Is your eBook purchase not compatible with the device of your choice?

Don't worry, now with every Packt book you get a DRM-free PDF version of that book at no cost.

Read anywhere, any place, on any device. Search, copy, and paste code from your favorite technical books directly into your application.

The perks don't stop there, you can get exclusive access to discounts, newsletters, and great free content in your inbox daily

Follow these simple steps to get the benefits:

1. Scan the QR code or visit the link below

`https://packt.link/free-ebook/9781837630639`

2. Submit your proof of purchase
3. That's it! We'll send your free PDF and other benefits to your email directly

www.ingramcontent.com/pod-product-compliance
Lightning Source LLC
LaVergne TN
LVHW062301060326
832902LV00013B/1993

* 9 7 8 1 8 3 7 6 3 0 6 3 9 *